Innovation *in* India

Combining Economic Growth with Inclusive Development

Edited by
Shyama V. Ramani

CAMBRIDGE
UNIVERSITY PRESS

CAMBRIDGE
UNIVERSITY PRESS

Cambridge House, 4381/4 Ansari Road, Daryaganj, Delhi 110002, India

Cambridge University Press is part of the University of Cambridge.

It furthers the University's mission by disseminating knowledge in the pursuit of education, learning and research at the highest international levels of excellence.

www.cambridge.org
Information on this title: www.cambridge.org/9781107037564

First published 2014

Printed in India by Shree Maitrey Printech Pvt. Ltd., Noida

A catalogue record for this publication is available from the British Library

Library of Congress Cataloging-in-Publication Data

Innovation in India : combining economic growth with inclusive development / edited by Shyama V. Ramani.
pages cm
Includes bibliographical references and index.
Summary: "Examines the evolution of sectoral system of innovation in industries that are important to India's economic development"-- Provided by publisher.
ISBN 978-1-107-03756-4 (hardback)
1. Technological innovations--Economic aspects--India. 2. Economic development--India. I. Ramani, Shyama V.
HC440.T4I563 2014
338.954--dc23
2013038168

ISBN 978-1-107-03756-4 Hardback

I would like to dedicate this volume to

All the contributors of this book and their families,

And in particular to my own family and parents -
Kalyani and M.S. Venkataramani,

And to my cheer-leading team: Raji, Raja, Divya, and Vivek.

Contents

List of Figures

List of Tables

Prologue

This collective volume is a testimony to how an informal network bound by a common motivation can generate an innovation – namely, this book! It all started with a visit to New York in 2005 at the invitation of Professor Claude Henry of the Ecole Polytechnique in Paris. He told me that Professor Richard Nelson was organising a meeting with Professor Jeffrey Sachs to get economists together to discuss how technology and innovation could contribute to the attainment of the Millennium Development Goals and that I would certainly enjoy the brainstorming. Dick, as Professor Nelson is fondly called, is one of the fathers of evolutionary economics and the national system of innovation approach – subjects I knew only by name at that time. The meeting was fascinating. I had never before gone to a meeting with so many economists to simply brainstorm and debate about what would be good to study. After listening to Dick, I have become his devoted disciple. I totally loved his style of teaching and his generosity in imparting knowledge and holding stimulating discussions. I began to work with him and Franco Malerba, one of his many loyal students, on a book dealing with sectoral systems of innovation in a set of developing countries[1]. Now, Dick did not just solicit authors to submit chapters. He organised meetings and there was a lot of debate on the contents and it was a collective process of knowledge generation, coordinated by him and Franco. I learnt a lot from working with Dick. And it just occurred to me that we should be doing something similar on India.

Why did I want a team? Well, India is a big country. It has a lot of industries and a lot of poor people. The dynamics of innovation generation is different in different sectors, and when targeting different communities. This multidimensional subject cannot be deeply analysed by only one scholar. It requires a team and a collective process of knowledge generation.

[1] Franco Malerba and Richard Nelson (editors), *Economic Development As a Learning Process: Variation Across Sectoral Systems* (Edward Elgar, 2012).

Right from 2006, I started talking to those who I thought might be interested in working on this book. They were academics I met in different conferences, especially GLOBELICS[2], which is the main international meeting place for scholars interested in sharing ideas on how technology and innovation can spur development and growth. But I did not have money to organise meetings. I had no research contracts. I just had strong intentions. So the process of putting the book together has been somewhat slow. There were some wonderful takers and a few leavers. Indeed, every individual who has contributed a chapter to this book is a scholar who was personally motivated to do it too – who believed that it is a good idea to explain the dynamics of innovation generation in India to the world! So this book is the fruit of cooperation between a set of academicians who were spread over three continents but managed to work as a team off and on for four years, exchanging ideas over emails, and intermittent meetings, to bring out the present volume with its detailed examination of the Indian innovation system. I would like to thank all the contributors for their wonderful work, and particularly Gita and Smita for egging me on when I felt like giving up!

I would also like to extend my sincere thanks to Cambridge University Press team and the editors Sana Bhanot and Suvadip Bhattacharjee for their kind support throughout!

Last but not the least – thanks Dick for your enthusiasm!

[2] http://www.globelics.org/

CHAPTER 1

Innovation in India

The Challenge of Combining Economic Growth with Inclusive Development

Shyama V. Ramani and Adam Szirmai

When India attained Independence in 1947, the first concern of its policymakers was to invest and create capacity in heavy industries, such as power, iron, steel, machinery, and chemicals. The post-Independence development strategy focused on the creation of a public sector capital goods industry that would be the motor of its industrialisation. The private sector was left to cater to the demand for consumer durables and non-durables. At the same time, the founding fathers of the nation were convinced that a country could not develop industrial capabilities without first acquiring scientific and technological capabilities. Thus, the government invested in the creation of a network of public universities and institutes for advanced research to provide qualified labour to burgeoning industries. After nearly nine centuries of policies focusing on the extraction of economic surplus for the benefit of domestic and colonial elites (Maddison 1974; Lal 1988), this marked a new beginning for the acquisition of scientific, technological, and innovation capabilities as a national prerogative. Now, it is a little more than 60 years since the foundation of India's national system of innovation was laid, and it is time to look back and examine what form it has taken. What are the achievements of the Indian system of innovation? How has it performed in terms of building industrial capabilities and promoting development? What are its shortcomings? What does the future hold? These are the questions that we seek to answer in this book through a study of several sectors from different perspectives. In this introduction, we outline the elements of a conceptual framework that brings the different chapters together.

The broad conceptual framework of the book: The national system of innovation

In mainstream economics, development is seen as a derivative of economic growth, whereby growth sustains and fuels development. The determinants of economic growth are spelt out in different theories, among which the neoclassical models on the relationships between inputs like land, labour, capital, and technology and outputs such as national income are the most widely accepted and taught in economics departments all over the world. This book adopts an alternative approach to examine the processes of capability accumulation in India, termed as the national system of innovation (NSI) approach, spearheaded by the seminal works of Lundvall (1992), Nelson (1993), and Freeman (1995). This approach has also inspired the notion of a sectoral system of innovation (SSI), incorporating sectoral specificities in an innovation system, including the impact of economic actors within and outside of the NSI (Lee and Lim 2001; Malerba 2002; Malerba and Nelson 2012).

The NSI approach itself emerged from an older stream of literature of the evolutionary school of economics on the industrial 'catching-up' of the presently developed countries in the form of a set of rich and well-documented historical case studies (Rosenberg and Birdzell 1986). This approach was then applied to explain the rise of the 'newly industrialising countries' of Asia in the 1980s, and is now also applied to understand the emerging economies of today (Fagerberg and Godinho 2005; Lundvall et al. 2009).

Interestingly, the 'catch-up' and 'economic growth' models share a common assumption that if knowledge is codified and freely available, late-comer countries can acquire existing technologies at a low cost. However, thereafter, their reasoning and forecasts of the consequences diverge totally. Post-war neo-classical models of growth assume that if knowledge is codified and freely available, latecomer countries can converge to the same steady-state equilibrium growth rate determined by the rate of exogenous technological change. As capital moves to low-income countries, where it is scarcer and returns are higher, the low-income countries start growing more rapidly than the high-income countries. Thereby, the gap between the two is reduced. However, this 'convergence hypothesis' has been invalidated by decades of uneven economic growth and persistent gaps in income per capita between the low-income and high-income countries (Landes 1998).[1] Endogenous growth theory, a later

[1] According to Landes, over the past 250 years, the difference in income per capita between the richest and the poorest country in the world has increased from 5:1 to 400:1. Based on PPP dollars from the World Bank's World Development Indicators, the ratio between

version of the neo-classical growth theory, advocates endogenised technological change as a result of purposeful human investments, and predicts divergence between the rich and poor countries based on increasing returns to scale in human capital and knowledge production (Aghion and Howitt 1993). However, endogenous growth theory, in turn, cannot account for spectacular cases of catch-up.

In contrast to the deductive approach of macro-economic modelling, the catch-up literature tries to generate inductive theory via historical case studies of economic development and the accumulation of capabilities. In catch-up theory, knowledge may be freely available, but its absorption and integration depend on a range of institutional characteristics, and social and technological capabilities. The basic difference between the catch-up theory and the standard neo-classical growth theory is that the former does not assume general convergence. Instead, it supposes that specific countries with special characteristics can profit from the advantages of backwardness and achieve accelerated catch-up. The main message of the catch-up literature is that technological catching-up cannot be taken for granted because a variety of necessary and complementary capabilities may be needed for effective absorption of existing technological knowledge, even if it is freely available. For example, they may include financial-institution capabilities to bear the costs of risky investment (Gerschenkron 1962), an educated workforce with social capabilities (Abramovitz 1986), public labs and firms with technological capabilities (Lall 1992), etc. Furthermore, building a platform of favourable capabilities may require sweeping institutional and organisational changes, in the absence of which 'catching-up' may be stalled. Thus, rather than being a homogeneous or linear process, catching-up in terms of scientific, technological, and industrial capabilities is likely to be costly, difficult, nation-specific, and non-systematic with sectoral and cluster idiosyncrasies. Acquiring, adapting, and implementing technologies are creative acts of innovation, and countries and firms have to invest heavily in building capabilities. But once the conditions for catch-up have been realized, late-comer countries normally grow much more rapidly than the lead economies because they can absorb state-of-the-art technology (when freely or quasi-freely available), without bearing the costs and risks of its development. This is what Gerschenkron referred to as the 'advantages of technological backwardness.'

Like catch-up theory, evolutionary economics also allows for both processes of *catch-up* and *falling behind*. An important notion here is that of the size of

---

the richest and the poorest country in 2008 was 200 to 1, i.e., between Norway, the richest country and the Democratic Republic of Congo, the poorest country.

the technological gap. If the technological gap is too large, it is very difficult to creatively absorb technology, as the conditions in the countries of origin and the countries of destination are too different. However, if the gap is not too extreme, advantages of backwardness will tend to prevail, and catch-up will be possible (e.g. Verspagen 1993). Finally, in contrast to neo-classical growth theory, evolutionary economics emphasises the heterogeneity of economic actors, who can respond in different ways to the incentives provided by their environments.

The catch-up literature demonstrates beyond doubt that national environments influence the processes of accumulation of knowledge and technological capabilities and that such trajectories are path dependent – even when countries are well-connected to international markets. After all, the institutions and public policies that generate the incentives for knowledge creation and accumulation are highly country-specific. Thus, the NSI assumes that the commercialization of innovations in any country in a new science-based sector is a collective process embedded within a system specific to the country. In other words, the creation, development, adoption, and diffusion of innovations evolve as a function of the existence and functioning of networks between the state and a variety of organisations, such as firms, consumers, public laboratories, universities, financial institutions, and civic associations. The catch-up process is then traced as the outcome of the strategies implemented by the actors in the innovation system, taking into account the interdependence between their actions.

The evolution of the national system of innovation as a game

In the last three decades, the systems approach has emerged as a useful framework to organise historical evidence on the accumulation of scientific, technical, innovation, and industrial capabilities of 'late-comer' countries in 'catching-up'. At the same time, it remains a conceptual framework rather than a theory, open to many forms of interpretation and investigation, as regional, national, and sectoral path-dependent trajectories can be studied in many ways (Edquist 2001; Lundvall 1998).

In the present book, for instance, the evolution of capacity building in any sector is considered as a game played between a set of players, whose strategies may be inter-dependent and whose choices jointly determine final outcomes. In other words, outcomes such as innovation generation are not viewed as being due to the efforts of just one actor, say a firm, but as the result of the profile of actions chosen by the State, other firms, laboratories, intermediaries, and so forth, in the NSI and SSI. The choices of the actors are influenced by the institutional and regulatory frameworks, which provide incentives for the actors. Typical actors in the NSI and SSI are presented in Fig. 1.1.

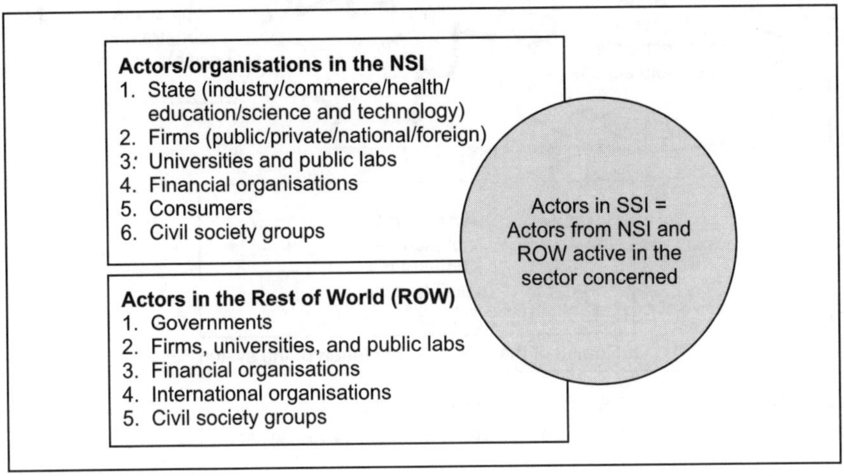

Fig. 1.1: Actors considered in the national system of innovation (NSI) and sectoral system of innovation (SSI)

In such games, the rules are set by the national and international institutions, policies, and regulation. For each sector, only some of the regulations or some of the rules of the game may be pertinent. As regulations change, the rules of the game change, and the outcomes may also change. Each actor has a set of objectives that it tries to attain. It is also endowed with a set of resources, constraints of all kinds, beliefs, cognitive structures, and a knowledge and information base. The constraints might take the form of behavioural norms, limited resources and skill, and incomplete or imperfect information base too. Each actor chooses its strategy so as to move closer to its objectives, given its constraints. The final outcome in terms of capabilities of all actors – and hence economic growth as well as inclusive development – will depend on the joint play or actions mobilised by all actors in the game, as given in Fig. 1.2. Outcomes of the game also include innovation performance, changes in the system of innovation, and ultimately, rates of economic growth of a more or less inclusive nature.

A game corresponds to a set of rules, actors, objectives, and constraints. Whenever any of these changes, a new game is set in motion. Thus, it is to be kept in mind that no notion of 'equilibrium' can be evoked in this framework.

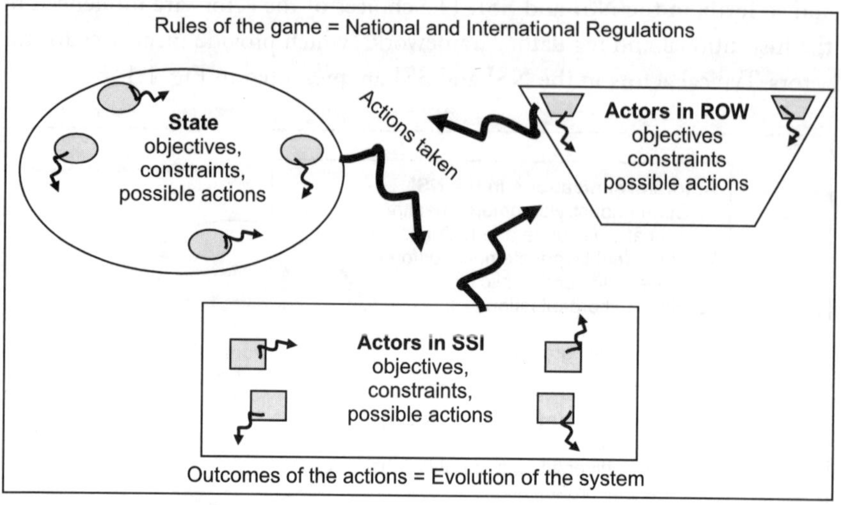

Fig. 1.2: Games within games = Capacity building
Note: ROW – rest of World; SSI – sectoral system of innovation

Rather, instead of optimising, agents continuously adapt to a continuously shifting environment while pursuing their goals. Consequently, with such continuous evolution, the discourse cannot be in terms of static equilibrium but in terms of outcomes over time, which may or may not converge.[2] Finally, these outcomes need not be socially optimal or even economically efficient at either a niche or sector level.

We illustrate this process in Fig. 1.3. Suppose, at a point of time, we start with a particular state of capabilities (in absolute terms and growth rates) in a sector, corresponding to some game (i.e., rules, actors, objectives, and constraints). The State sets the rules of the game so as to encourage 'catch-up' vis-à-vis some region of reference. As the game proceeds, the existing trajectory may continue undisturbed or there may be a new stimulus in the sectoral or national system of innovation that triggers a new game. The response of some actors provokes other changes in the system, all of which finally results in a new state of capabilities. Then the change in the size of the gap between the region of reference and the country in question reflects how successful the country has been at 'catching-up.' Ultimately, catching-up in terms of capabilities is reflected in catching-up in terms of gross domestic product (GDP) per capita.

We illustrate these notions with two examples. Many more will be elaborated in the book.

[2] See Surie (2011) for more detailed illustrative examples.

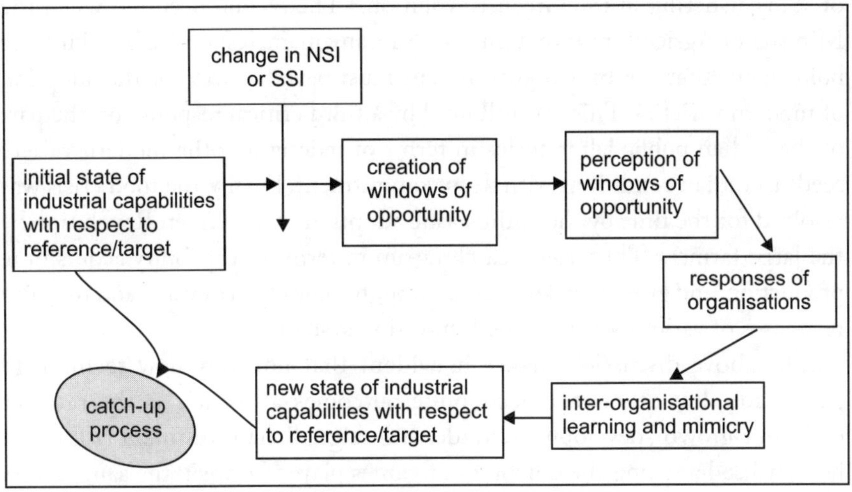

Fig. 1.3: Dynamics of catch-up at sector level
Note: NSI – national system of innovation; SSI – sectoral system of innovation

Till 1972, the main rule for innovation in the Indian pharmaceutical sector was that no Indian firm could re-engineer any branded or patented drug. However, there was a health crisis due to the lack of availability of essential drugs, and in order to come closer to its objective of ensuring access to basic drugs to its citizens, the Indian government changed the rules of the game. It was decreed that the Indian firms could produce patented drugs if they could produce them in ways different from those of the original innovator. Now, the same rule in Latin America had not had any impact. But, somehow in India, the private firms responded by investing in developing innovation capabilities. This triggered domino effects in the entire sector, resulting in a robust indigenous pharmaceutical industry by the mid-1980s. There was definitely catching-up.

From the beginning of the 1960s, when India's population rose to about 480 million, severe food shortage was experienced and India started importing about 10 per cent of its indigenous food grains production from the USA under the PL480 (Public Law 480) programme. The strategy of the State was to invest in the public agriculture research centres and universities, but this had little impact. However, a series of four unforeseen and/or uncontrollable events radically changed this situation. The first stimulus came from outside the country. The creative research of Norman Borlaug, an American professor of agriculture science, led to the creation of a new dwarf variety of wheat with 'short legs' that could support a greater amount of wheat grains on any stalk. This gave rise to a set of new 'high-yielding varieties' or 'modern varieties' of seeds, ushering in the Green Revolution.[3] The second occurred when the Minister of Agriculture at that time, C. Subramaniam, responded by taking the bold stance that the Indian government must pave the way for the adoption of modern varieties. This was followed by a third critical response on the part of the Indian public laboratories in terms of redesigning the modern variety seeds to Indian conditions with deep commitment. Finally, the food crisis was resolved for the time by the enthusiastic adoption of the Green Revolution by the large farmers. Thus, again catching-up in terms of capability acquisition, production, and growth had occurred through some chance events and the joint responses of various actors in the innovation system.

The above discussion makes it evident that not only new technology generation, but other macroeconomic outcomes also such as the rates of economic growth, development, trade, or foreign direct investment (FDI) can be visualised as being the outcomes of games played between the same actors as those mentioned in the NSI and SSI. However, in these cases, it is far more difficult to pinpoint and trace the role of each actor in the final outcomes. Thus, instead of looking into such games, we take a bird's eye snapshot of the evolution of the rules of the games and the macroeconomic outcomes in the following two sections.

Rules of the game that have impacted all sectors: Going from socialist-licence-Raj to market liberalisation in an era of globalisation

The set of rules and the strategy of the State guiding the building of industrial capabilities can be considered to fall into two distinct phases in India. At the same time, within each phase, there have been a number of changes, of which we can mention only a few in this chapter. We briefly outline the evolution here, noting that the impact of this radical transition between phase 1 and phase 2 persistently rears its head in many of the studies presented in the book.

Game 1: Building indigenous capabilities with an import substitution policy

During the 1950s, the perceived success of the economic growth model of Soviet Russia had a strong ideological impact and set the tone for the rules of the game and the strategy of the Indian State. Inward-looking trade and

[3] Professor Borlaug was awarded the Noble Peace Prize in 1970 for his role in the creation and diffusion of this life-saving innovation throughout the world.

investment policies were adopted (as by many other developing countries) to minimise dependence on imports and develop a publicly owned industrial base to serve the needs of its citizens while curbing monopolistic and oligopolistic tendencies of the private sector. The public sector was viewed as the motor of economic growth from the 1950s to the 1990s, and the private sector was mistrusted as being made up of entities whose profiteering and growth had to be controlled through rules and regulations (Ahluwalia 1991; Bardhan 1984).

In order to monitor and control the process of industrialisation, the Indian government presided over what was in many respects a 'closed command economy' as distinct from an 'open market economy.' The ensuing 'import substitution' policy was marked by five major industrial policies. First, ceilings were set on the overall profits of the companies in many sectors. Second, the Foreign Exchange Regulation Act of 1973 (FERA) restricted the foreign equity holdings. Third, the Monopolistic and Restrictive Trade Practices Act of 1969 (MRTP) was implemented to protect against undue concentration of market power. Fourth, a '*license Raj*' (or rule of the license regime) stipulated that licences had to be obtained from the concerned ministries for any expansion in the manufacturing base, imports, and exports. Fifth, final market prices were controlled in a number of non-luxury goods sectors, such as pharmaceuticals, in order to facilitate accessibility.

In the above context, the response of the industrialists to these rules was to get deeply involved in getting permits, licenses, and quotas and clamouring for fiscal and customs duty concessions for themselves rather than formulating strategies for innovation or growth. The largest monopolistic enterprises with access to the government paradoxically turned out to be best at playing the license game, even though one of the explicit aims of public policy was to control the large private firms. Neither State nor industry was inclined to invest in the development of innovation capabilities in the private sector. Public investment was channelled into building basic infrastructure and manufacturing industries, leaving technological learning to take its own course through 'learning by doing or learning by growing.' The business vision of both Indian firms and multinationals in India was oriented towards maximising very short-run profits with minimal R&D investment.

A policy change during the 1970s, however, changed the game rules in some sectors, giving them a first impetus for the development of innovation capabilities. Thus far, India's intellectual property rights (IPR) system had been defined by the 'Indian Patents and Designs Act of 1911' of colonial times, which was based on the British 'Patent Act of 1852', permitting only product patents. This was changed by the 'Patent Act of 1970' to an IPR regime, which recognized only process patents for food, medicine, and chemical processes. Such a policy experiment was initiated to promote the accumulation of technological capabilities in the public sector and induce private sector investment in these key industries. That the Government of India made its move a quarter of a century after the country attained its freedom testified to its inadequate awareness and appreciation of the reality that even in 1972, countries like Sweden, Switzerland, Spain, Italy, Japan, China, Brazil, and the erstwhile USSR either did not have an IPR or allowed only process patents in key sectors in order to catch-up.

Transition towards Game 2 of economic liberalisation

Throughout the 1980s, there were changes in regulation and State policy that took the rules of the game more and more away from its original format of import substitution with strict monitoring and control of investment (Bradford DeLong 2003). This culminated in 1990 in a series of policy jolts, with the impetus coming again from outside events. Just as during the 1950s, India had been inspired by the economic growth models of the former USSR and China; during the 1980s, it could not remain inert as these regions embraced 'market systems' and the Chinese high-command introduced pragmatic 'State capitalism', following the classic dictum of Deng Xiaoping, 'it doesn't matter if a cat is black or white as long as it catches mice.' Following these worldwide trends, Prime Minister Rajiv Gandhi pushed for economic reforms during the late 1980s, and this was fully realized with liberalisation and de-licensing in 1991 under the leadership of Prime Minister Narasimha Rao (Kotwal, Ramaswami, and Wadhwa 2011). Liberalisation of national and international financial transactions followed in 1995. Thereafter, government regulation via manufacturing and marketing licenses only served to monitor the quality and safety of the final products arriving in the market. Price control on commodities, including drugs, was eased. Procedures to obtain foreign technology agreement, imports, and exports were greatly streamlined and 100 per cent foreign ownership was permitted in most sectors. Excise duty was slashed on imports, while a value-added tax was added on domestic product. Lastly, in order to maximise the gains from globalisation and promote its exports, India signed the Uruguay round of GATT, which concluded in 1994, to become a member of the World Trade Organization (WTO) in 1995. India was thereby obliged to meet all provisions of the Trade Related Aspects of Intellectual Property Rights (TRIPs) by 2005, including a return to a uniform product patent regime in all manufacturing sectors.

Policy efforts to meld growth with inclusive development

'Inclusive development' refers to a growth process that benefits all sections of society without excluding any specific group and renders 'economic opportunities,' generated by the growth process, accessible to all. One of its main challenges is to ensure that the poor and marginalised populations that are often hidden in the informal economy, i.e., in productive activities not formally registered under local laws or as vulnerable informal workers, are also able to improve their conditions of life in the growth process. In other words, development is inclusive when the different sections of society, including the disadvantaged sections such as rural populations, women, youth, elderly, ethnic/religious groups, or immigrants, benefit from economic growth and have opportunities to participate in the production process.

Promotion of inclusive development has been one of the stated aims of the Indian policy since national independence. In order to contribute to the attainment of this objective, the strategy of the main player, the State, has been to invest in the national programmes under the 'Five-year Plans', starting from 1951. However, India's performance in terms of expanding inclusion is poor as compared to other emerging countries, especially China (for more details see the next section). It is beyond the scope of the book to explore why State policy has performed less well in India than in other emerging countries. But in this introduction, we will highlight the types of strategies and rules implemented in the past and at present for inclusive development, without delving into the details (which can be found in many other text books).

During the pre-liberalisation period, from the 1950s to the 1990s, the main focus of the inclusive policy was on rural India. Four types of strategies were deployed. First, to lower inequity stemming from skewed land ownership, land reforms were carried out to abolish intermediary institutions and feudal systems of land holdings, and the land was transferred from feudal landlords to indigent populations in a series of bold steps. However, while land reform was successful in some places, it was very unsuccessful in others due to landowner interests at local and regional levels (Frankel 1978). Second, to promote balanced development, public sector manufacturing units were set up in peri-urban areas and incentives were provided for private industry to be set up in the less-developed states. Small-scale industry received government support and protection, particularly in spinning and textiles. A variety of credit agencies were created to help small farmers buy seed, fertiliser, and pesticides. Third, access to essential goods like food grains, fuel, and medicines was established through a public distribution system and a public health care system.

The public distribution system functioned via 'ration cards' that provided essential goods according to whether the household was below or above the poverty line. Fourth, there was a policy of positive discrimination in education and public sector employment for marginalised social groups.

Following liberalisation, many of these measures to promote balanced regional development and access to essentials were continued. In addition, some of the diffused programmes were transformed into mission mode projects with specific targets in terms of not only provision of commodities but also their usage and maintenance. Unlike the strategies in the pre-liberalisation phase, these programmes aimed to facilitate the 'participation' of the disadvantaged in the growth process, and thereby, increase their 'income generation capacity' through ensuring the satisfaction of their basic needs. Initially, the focus remained on rural India. For instance, the 'Indira Awas Yojana – Rural Housing Mission' was launched in 1996, followed by the 'National Drinking Water Mission' and the 'Total Sanitation Campaign' in 1999 to ensure that the rural population had housing, access to drinking water, and a functioning toilet. Another major programme, the 'National Rural Health Mission', was initiated in 2005 to ensure healthcare for all. Similarly, in 2007, the 'National Food Security Mission' was initiated to lower hunger.

New and committed government programmes are being initiated to increase the scope of inclusive development. These represent real radical breaks from the past and are emblematic of efforts being taken to redistribute the gains of economic growth more equitably. For instance, the 'Mahatma Gandhi National Rural Employment Guarantee Act' of 2005 ensures adult members of every rural household 100 days of employment in every financial year as unskilled manual labour in public works at minimum wages fixed by the government. Though there is a debate on the efficiency of its implementation and the labour shortages it is leading to in rural areas, it has definitely had a positive impact on women's empowerment (Vij 2011) In many families, women are being allowed to break socio-cultural barriers and work outside of the confines of their homes for the first time in order to augment the earnings of the family. Also, for the first time, urban slums and urban development are being explicitly taken into account in State plans in the 'Jawaharlal Nehru National Urban Renewal Mission', also initiated in 2005. The objective of this programme is to develop the infrastructure of cities in a planned and integrated fashion, build capabilities in municipalities, renew inner city areas, and ensure universal access to basic services to city dwellers, including those in slums.

These measures constituted the action-investments of the State in the game. However, what contributed the most to the melding of 'catch-up' with 'inclusive development' was a new game-rule in the form of a system of positive discrimination. In educational institutions and public sector firms and agencies, seats were reserved by quota for those belonging to a 'Scheduled Tribe,' a 'Scheduled Caste,' a 'backward class' or a 'religious or linguistic minority.' For example, according to the website of the Ministry of Social Justice and Empowerment:

> Backward Classes means such backward classes of citizens other than the Scheduled Castes and Scheduled Tribes as may be specified by the Central Government in the lists prepared by the Government of India from time to time for purposes of making provision for the reservation of appointments or posts in favour of backward classes of citizens which, in the opinion of that Government, are not adequately represented in the services under the Government of India and any local or other authority within the territory of India or under the control of the Government of India.[4]

By inducing a large-scale participation of previously marginalised groups, the quota system enormously strengthened the social capabilities[5] that form the base for the accumulation of scientific and technical capabilities. Though little known outside of India, this system of positive discrimination, introduced in the 1950s and gradually expanded, is surely among the most complex, extensive, and inclusive in the world. The system varies between states, sectors, and academic disciplines. It has yielded from good to excellent returns in terms of increasing the caste and religious diversity of staff and students in academic institutions, public administration, and public agencies, as well as empowered the socially disadvantaged sections of society.

Such bold moves to develop social capabilities also have their down side. First, given the reservation quotas in the public sector, many of the Scheduled Caste graduates prefer to get job in the public sector, while the other graduates aim for jobs in the private sector, where social identity is surely important too, but does not determine recruitment. Thereafter, since the private sector pays higher wages than the public sector, economic inequalities are not reduced. Second, the system is used by economically well-off members of the lower castes

---

[4] http://socialjustice.nic.in/aboutdivision4.php

[5] Abramovitz (1986) describes 'social capabilities' as the quantity and quality of the educated labour force.

and religious minorities, who are more aware of the advantages provided by the system, but who have no real need of economic support for either access to education or employment, thereby blocking seats for the poor from the same communities. Third, given that access to higher degrees and professional careers in universities and public laboratories is also subject to the official reservation policy, many talented students and researchers seek to leave India for other countries where their social identity is less of a burden for their professional evolution. Fourth, it doubly marginalises the poor of the upper castes or other communities not designated as being 'backward,' who have neither the funds to go to private academic institutions, nor the requisite social identities to gain access to publicly funded institutions.

Most of the chapters in this book do not focus on the impact of innovation on inclusive development per se, but all evoke the consequences of innovation generation for the poor, either directly or indirectly. They show how in agriculture, pharmaceuticals, and energy, the low-income communities have benefited from new products, improvement in existing products, and better access to products through lowered prices. These chapters are complemented by a few that directly address pro-poor innovation-related issues such as sanitation, traditional medicine, demand drivers, and inclusive development.

Game outcomes: Economic growth, inclusive development, and innovation[6]

Growth trends: Satisfactory but not extraordinary

Innovation systems, capability building, and innovative performance are not the goals in themselves. Ultimately they should contribute to better economic performance or the realization of inclusive societal goals. The proof of the pudding is in the eating. In this section, we therefore provide a brief overview of the long-run performance trends in the Indian economy, which serve as a backdrop to the sectoral chapters in the book.[7] We put the Indian performance in perspective, comparing it with the performance of developing countries in general and with the performance of China, the other Asian giant, in particular.

[6] Most of the data in this section is derived from a database for the BRICS economies constructed by Alejandro Lavopa (see Naude, Lavopa, and Szirmai 2012). We thank Alejandro Lavopa for making the data available for this paper.

[7] The sectoral systems of innovation literature tends to neglect quantitative macro-economic trends, focusing on the evolution of the sectoral innovation systems themselves.

In Fig. 1.4, we highlight the long-run trends in GDP per capita since 1950. In this figure, one can see the dramatic acceleration of growth in GDP per capita, since the reforms of 1991. Between 1950 and 1991, the per capita growth of GDP was 1.8 per cent. This increased to an average growth rate of 5.2 per cent between 1991 and 2011. There is an interesting debate on the role of the 1991 reforms in triggering the acceleration of growth. Bradford DeLong (2003) and Rodrik and Subramanian (2005) argue that growth already started to accelerate well before 1991, after the smaller wave of reforms in the early 1980s. Indeed, we see in Fig. 1.4 that the growth in the 1980s increased from

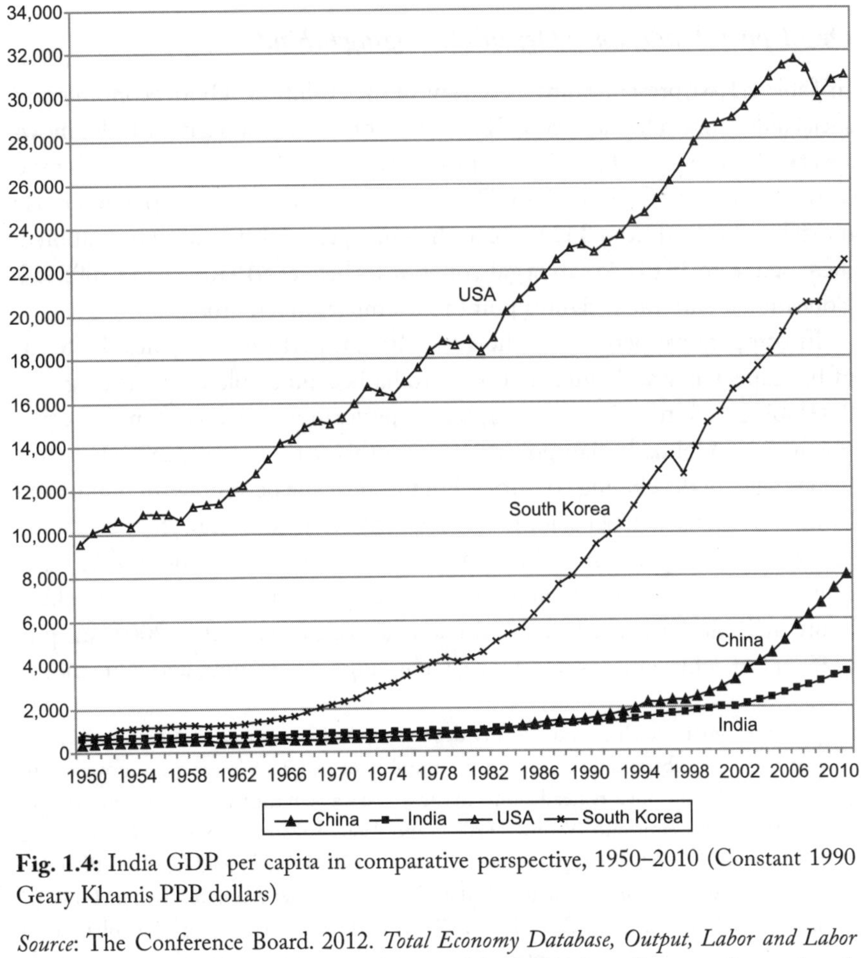

Fig. 1.4: India GDP per capita in comparative perspective, 1950–2010 (Constant 1990 Geary Khamis PPP dollars)

Source: The Conference Board. 2012. *Total Economy Database, Output, Labor and Labor Productivity Country Details, 1950–2011*. Accessed: June 2012; http://www.conference-board.org/ data/economydatabase/

a mere 1.4 per cent per annum between 1950 and 1980 to around 3 per cent per annum between 1980 and 1981.

But it is also clear that growth further accelerated after 1991, pinpointing the importance of the 1991 reforms (see also Mani 2011). Especially, after 2000, growth was very rapid, a clear case of catch-up. However, since 1978, China has been forging ahead of India from very similar levels of per capita income. In 2011, China's per capita GDP was 2.3 times as high as that of India (at 1990 constant Gheary Khamis Purchasing Power Parities). The figure also includes South Korea as an example of really spectacular catch-up.

Development indicators: Definitely lagging behind

In Table 1.1, we present a bird's-eye view of the evolution of both economic and social indicators in India since 1950. In terms of its social indicators, India shows spectacular progress. Thus, infant mortality dropped from 165 per thousand births in the period 1950–55 to 52.9 per thousand births in 2006. Under-five mortality dropped from 332 to 72 in the same period. Life expectancy at birth increased from 38 to 64 years, a gain of 26 years between 1950–55 and 2005–10. Educational enrolment figures point to a similar success story.

In comparative perspective, however, Indian performance pales. In terms of life expectancy and child mortality, India is comparable to the average for developing countries in 2005–10, but is performing far less than the other Asian giant, China. Indian poverty rates in this period are much higher than the average for developing countries. For instance, the percentage of population living on less than a dollar a day is 41.6 per cent, versus an average for developing countries of 25.2 per cent.[8] China has realized a poverty headcount of 15.9 per cent, well below the developing country average. Perhaps the most shocking figure in Table 1.1 is the Indian illiteracy rate of 37 per cent in 2005. In spite of its grand achievements in science, technology, innovation, and education –

[8] Poverty estimates for India vary greatly, depending on the source used. All estimates agree that there are substantial declines in the number of people below a given poverty line over time. But the poverty headcounts in the first decade of the twenty-first century vary. Kotwal, Ramaswami, and Wadhwa (2011) and Aggarwal and Kumar (2012) provide an estimate of 27.5 per cent of the total population in 2004/05, derived from the Planning Commission of the Government of India. The much higher estimates used in Table 1.1 are derived from Chen and Ravallion (2008), based on a poverty line of 1.25 dollars a day at 2005 PPP dollars. The advantage of this dataset is that it uses standard international poverty lines. Its estimate for China is also much higher than the estimates using national Chinese sources.

Table 1.1: Economic and social indicators of India in comparative perspective, 1950–2010

	1950–60			*1980/81*			*2005–10*		
	India	*China*	*Developing countries*	*India*	*China*	*Developing countries*	*India*	*China*	*Developing countries*
GDP per capita (1990 PPP$)	619.0	346.8	766.6	976.7	859.3	1748.4	3366.0	7395.8	4883.2
Manufactures as % of commodity production	15.0	19.7	22.9	28.0	49.3	33.5	34.8	55.7	35.4
Manufactured exports as % of commodity exports	46.0		5.0	51.0	50.0	25.0	54.4	92.8	37.0
Life expectancy at birth	37.9	44.6	42.3	56.2	67.7	59.5	64.2	72.7	66.0
Child mortality by age 1	165.0	122.0	180.0	100.1	39.8	84.0	52.9	22.0	50.2
Child mortality by age 5	332.0	266.0	281.0	135.0	57.0	117.0	72.4	26.4	72.4
Gross enrolment rate, primary education	41.0		75.8	81.7	111.6		116.0	111.0	107.8
Gross enrolment rate, secondary education	23.0		15.7	29.0	46.0		63.0	81.0	64.0
Gross enrolment rate, tertiary education	2.0		2.1	5.0	1.0		18.0	26.0	21.4
Net enrolment rate, primary education							92.1	96.4	86.9
Illiteracy rate (15+)	81.7		55.5	59.0	32.9	35.3	37.2	5.7	19.8
Percentage of population, with less than 1 dollar a day				59.8	84.0	51.8	41.6	15.9	25.2
Number of persons with less than 1 dollar a day (million)				420.5	835.1	1896.2	455.8	207.7	1376.7
Percentage of population, with less than 2 dollars a day				86.6	97.8	69.2	75.6	36.3	47.0
Number of persons with less than 2 dollars a day (million)				608.9	972.1	2535.1	827.7	473.7	2561.5

Sources: Szirmai (2012): www.dynamicsofdevelopment.com; GDP see source note Fig. 1.1; poverty rates from Chen and Ravallion (2008); Life expectancy and child mortality from UN (2010); Enrolment statistics and illiteracy from UNESCO (2012); Illiteracy India 1951 from India Stat; Manufacturing export data from UNCTAD (2012) and World Bank (2012a); Manufactures as % of commodity production from De Vries et al. (2012)

many of them documented in the chapters of this book – close to 40 per cent of the Indian population cannot read or write. Of course, all such figures can be contested, and the rankings may change according to different methods and data sources. But what is overwhelmingly clear is that Indian development has not been very inclusive, in spite of the stated aims of public policy.

Structural change: Whatever happened to manufacturing?

Returning to economic indicators, Tables 1.2a and 1.2b summarize the process of structural change in the Indian economy since 1950 (at current and constant prices, respectively). Many of these changes are rather familiar (see Kotwal, Ramaswami, and Wadhwa 2011; Aggarwal and Kumar 2012). We see a dramatic decline of the share of agriculture in GDP and corresponding increases in the shares of industry and services. What is less familiar is the rather modest increase in the share of manufacturing. In current prices, this increased from 10 per cent in 1950 to 16 per cent in 2009. Its share peaked in 1995, after which India even experienced relative de-industrialisation.

The sector that has expanded the most is the services sector. By 2009, this accounted for 55 per cent of GDP. Thus, India has become a service economy, seemingly skipping the traditional development stage whereby manufacturing dominates the economy before services take over. Optimistically, one could interpret this as a process of leapfrogging to a new development path. But a more pessimistic interpretation is that India represents a case of premature de-industrialisation (Tregenna 2013), where the share of manufacturing starts declining at low or intermediate levels of per capita income. If one accepts the notion that manufacturing is one of the key sectors driving technological change, growth, and catch up in emerging economies (Cornwall 1977; Naudé and Szirmai 2012; Szirmai 2013), then premature de-industrialisation is a serious challenge. In Fig. 1.4, we have documented how India has been falling behind China. Panagariya (2004) has explicitly argued that the reason why Indian growth has been so much slower than that of China is the weak performance of the Indian manufacturing sector. A similar argument has recently been put forward by Naude, Lavopa, and Szirmai (2012). In order to realize India's full growth potential, manufacturing needs to play a more important role than it is playing at present.

Structural change also takes place within the manufacturing sector (see Tables 1.3a and 1.3b). At constant prices, there are declines in the shares of primary-based sectors such food, textiles, and wood products. There are very

Table 1.2a: Structural change in India, 1950–2009

Sector	*Sectoral shares in gross value added (in current prices) (%)*												
	1950	*1955*	*1960*	*1965*	*1970*	*1975*	*1980*	*1985*	*1990*	*1995*	*2000*	*2005*	*2009*
Agriculture	55.3	46.5	42.8	41.3	43.5	39.0	36.8	31.9	30.0	26.8	23.2	18.7	17.1
Industry	13.9	16.7	19.6	20.8	20.3	22.0	24.2	26.5	27.6	28.0	26.4	28.3	28.2
• mining													
• manufacturing	10.4	12.0	14.1	14.7	13.8	15.3	16.3	16.5	17.2	18.0	15.8	15.5	15.9
• utilities													
• construction					4.3	4.1	4.4	4.9	5.3	4.8	5.7	8.0	8.3
Services	30.8	36.9	37.6	37.9	36.3	39.0	39.0	41.6	42.4	45.2	50.4	53.0	54.6
Total	100	100	100	100	100	100	100	100	100	100	100	100	100

Sources:
1950 and 1955: UN national accounts, 1957, 1967
1960–1970: World development indicators online
1970–2010: BRICS database constructed by Alejandro Lavopa

Table 1.2b: Structural change in India, 1980–2008

Sector	*Sectoral shares in Indian gross value added (at constant 1995 prices) (%)*						
	1980	*1985*	*1990*	*1995*	*2000*	*2005*	*2008*
Agriculture	37.4	34.3	30.6	26.5	23.0	18.6	16.3
Industry	24.4	25.0	26.7	27.8	26.7	26.9	26.6
• mining	1.9	2.1	2.5	2.3	2.1	1.9	1.8
• manufacturing	14.9	15.7	16.5	17.9	16.9	16.6	16.4
• utilities	1.8	2.1	2.4	2.7	2.6	2.4	2.2
• construction	5.8	5.2	5.4	4.9	5.0	6.0	6.3
Services	38.2	40.7	42.7	45.7	50.3	54.5	57.1
• wholesale, retail trade, restaurants, and hotels	11.9	12.4	12.4	13.8	14.6	15.8	16.1
• transport, storage, and communications	5.9	6.1	6.1	6.9	8.1	11.6	13.7
• financing, insurance, real estate, and business services	7.7	9.3	10.9	12.3	13.4	14.1	15.5
Other services	12.7	12.8	13.2	12.7	14.2	13.0	11.8
Total	100.0	100.0	100.0	100.0	100.0	100.0	100.0

Source: De Vries et al. (2012)

Table 1.3a: Structural change within Indian manufacturing, 1980–2008

	Value added at constant 1995 prices (%)						
	1980	*1985*	*1990*	*1995*	*2000*	*2005*	*2008*
Food, beverages, and tobacco	12.33	12.98	12.34	11.85	12.16	11.25	10.46
Textiles and textile products	21.96	19.16	17.31	16.00	17.16	15.33	15.15
Leather and footwear	–	–	–	–	–	–	–
Wood and products of wood and cork	8.53	5.81	3.55	2.87	2.04	0.93	1.38
Pulp, paper, paper products, and publishing	3.94	3.96	4.48	3.82	2.89	2.93	2.75
Coke, refined petroleum products, nuclear fuel	1.23	2.00	3.08	3.22	2.78	3.76	3.27
Chemicals and chemical products	8.33	9.97	11.61	13.93	15.11	16.18	15.92
Rubber and plastic products	2.27	2.56	3.42	2.96	3.13	1.88	1.76
Other non-metallic mineral products	4.56	5.38	5.77	5.36	6.18	5.60	5.55
Basic metals and metal products	15.82	14.42	14.78	14.96	14.24	15.17	16.34
Machinery nec	7.15	7.70	6.43	6.40	5.79	4.97	4.86
Electrical and optical equipment	6.37	7.27	9.85	8.70	9.35	11.57	11.57
Transport equipment	4.83	4.86	4.93	6.48	5.10	6.27	6.45
Furniture, manufacturing nec and recycling	2.67	3.94	2.43	3.46	4.06	4.16	4.54
Total manufacturing	100.0	100.0	100.0	100.0	100.0	100.0	100.0

Source: Groningen Growth and Development Centre, Ten Sector Database. (http://www.rug.nl/research/ggdc/data/10-sector-database)

Table 1.3b: Structural change within Indian manufacturing, 1970–2007

	Value added at current prices 1970–2007 (%)								
	1970	*1975*	*1980*	*1985*	*1990*	*1995*	*2000*	*2005*	*2007*
Food, beverages, and tobacco	12.8	10.8	9.1	11.6	11.7	10.1	13.2	9.3	8.8
Textiles and textile products	21.0	19.5	21.3	15.0	15.2	11.2	11.8	8.3	7.1
Leather and footwear	–	–	–	–	–	–	0.8	0.6	0.6
Wood and products of wood and cork	0.6	0.6	0.6	0.5	0.4	0.3	0.2	0.2	0.2
Pulp, paper, paper products, and publishing	5.2	4.9	4.2	3.3	3.6	4.0	4.3	3.0	2.7
Coke, refined petroleum products, nuclear fuel	2.0	2.6	2.7	3.2	4.9	5.1	5.0	13.4	13.3
Chemicals and chemical products	13.6	15.1	14.1	15.1	13.9	20.2	21.1	16.2	13.6
Rubber and plastic products	3.2	2.5	2.5	3.4	3.4	2.7	3.4	2.5	2.6
Other non-metallic mineral products	4.1	3.9	3.9	5.8	5.1	4.9	5.6	4.5	6.8
Basic metals and metal products	15.7	16.3	15.2	15.0	15.2	14.6	13.4	17.2	22.2
Machinery nec	6.8	8.1	8.6	9.7	8.0	7.9	6.3	5.6	5.9
Electrical and optical equipment	6.9	8.2	8.8	8.5	8.7	8.0	7.1	7.2	7.1
Transport equipment	7.1	6.9	8.3	7.9	9.5	10.2	6.6	10.6	7.7
Furniture, manufacturing nec and recycling	0.8	0.5	0.6	1.0	0.4	0.9	1.3	1.3	1.3
	100.0	100.0	100.0	100.0	100.0	100.0	100.0	100.0	100.0

Source: UNIDO, Indstat Database, Geneva, Various Issues

substantial increases in the shares of chemical products and electrical and optical equipment.[9] The latter is typically a high-tech sector.

In current prices, there are similar declines in food, beverages, tobacco, textiles, and wood and pulp products and similar increases in refined petroleum products. The main difference with the constant price series is that there is no increase in the share of electrical and optical equipment. This is not surprising. Over time, the prices of computers, office equipment, and other electronic equipment are increasing less relative to the prices of other sectors (or even decreasing in absolute terms), so that the current price shares in later years tend to be lower.

Trade: More exports but even more imports

Between 1970 and 2009, there has been a spectacular increase in the openness of the Indian economy, as documented in Table 1.4. The share of exports in GDP increased from 3.8 per cent in 1970 to 20 per cent in 2009. The share of manufactured exports increased from 2.2 per cent in 1980 to 11.7 per cent in 2009. Trade openness (the sum of exports and imports as percentage of GDP) increased from 7.8 per cent in 1970 to 45.8 per cent in 2009. It should be noted that in spite of increasing export success, imports continued to exceed exports, indicating a persistent negative balance on the current account of the balance of payments.

Tables 1.5a and 1.5b provide information on the changes in the structure of exports. The manufacturing sector turns out to be much more important in terms of export share than in terms of GDP. For instance, in spite of all the discussion about the Indian IT sector and service-led growth, manufactured exports in 2009 account for no less than 58 per cent of the total gross value of exports, up from 36 per cent in 1980. This is somewhat at odds with the pessimistic conclusions about premature de-industrialisation derived from Table 1.4.

Within the manufacturing sector, the long-run trends in export shares are rather similar to those for GDP. We see a decline in the importance of traditional low-tech sectors such as food, textiles, and leather and an increase in the shares of refined petroleum products and electrical and optical instruments.

---

[9] In the international standard industrial classification (ISIC), revision 3, the economy is broken down into 17 one digit groups, of which manufacturing is one. At two-digit level, 99 divisions are distinguished of which 22 are within manufacturing (starting with division 15 manufacture of food products and beverages and ending with division 37). In Tables 1.3a and 1.3b, 22 divisions have been collapsed into 14 categories.

Table 1.4: Indian export performance, 1970–2009

	1970	*1980*	*1990*	*2000*	*2009*
Exports of goods and services (million current dollars)	2,362	11,557	22,594	59,064	257,957
GDP (value added) (million current dollars)	6,1470	184,761	326,796	467,788	1,287,292
Exports as percentage of GDP in current prices	3.8	6.3	6.9	12.6	20.0
As percentage of world exports	NA	0.5	0.5	0.7	1.6
Exports of manufactured goods (million current dollars)	NA	4,154	14,447	37,687	150,517
Manufactured exports as percentage of GDP	NA	2.2	4.4	8.1	11.7
Manufactured exports as percentage of world exports	NA	0.3	0.5	0.7	1.4
Total exports (million current dollars)	2,362	11,488	23,225	61,886	265,022
Total imports (million current dollars)	2,430	17,352	27,919	66,202	325,072
Sum of exports and imports in million current dollars	4,792	28,839	51,144	128,088	590,094
Sum of exports and imports as percentage of GDP	7.8	15.6	15.7	27.4	45.8

Source: UNCTAD (2012)

Table 1.5a: Structure of exports in India, 1980–2009 (%)

Sector	*1980*	*1990*	*2000*	*2009*
Agriculture	27.0	11.6	6.5	2.9
Industry	47.3	67.9	66.1	61.4
• mining	11.3	3.9	1.5	2.7
• manufacturing	36.0	64.0	63.8	58.3
• utilities	0.0	0.0	0.0	0.0
• construction	0.0	0.0	0.8	0.3
Services	25.7	20.5	27.4	35.7
• wholesale, retail trade, restaurants, and hotels	0.0	0.0	0.0	0.0
• transport, storage, and communications	17.3	11.1	10.2	9.1
• financing, insurance, real estate, and business services	7.5	9.3	16.1	26.3
Other services				
Total	100.0	100.0	100.0	100.0

Source: UNCTAD (2012)

Table 1.5b: Structure of manufactured exports in India, 1980–2009 (%)

Manufacturing sector	*1980*	*1990*	*2000*	*2009*
Food, beverages, and tobacco	9.6	6.9	6.4	5.2
Textiles and textile products	26.6	15.0	14.4	5.5
Leather and footwear	12.1	22.8	17.2	8.0
Wood and products of wood and cork	0.8	0.1	0.1	0.1
Pulp, paper, paper products, and publishing	0.2	0.3	0.6	0.5
Coke, refined petroleum products, nuclear fuel	2.1	3.9	4.3	16.2
Chemicals and chemical products	10.6	10.9	12.3	12.6
Rubber and plastic products	2.2	1.3	1.5	1.2
Other non-metallic mineral products	1.2	0.7	1.7	1.2
Basic metals and metal products	7.6	5.7	8.5	9.9
Machinery nec	6.1	4.7	3.5	4.8
Electrical and optical equipment	3.5	2.6	2.9	6.6
Transport equipment	14.4	3.0	2.7	6.5
Furniture, manufacturing nec and recycling	3.1	22.3	24.0	21.6
	100.0	100.0	100.0	100.0

Sources: UNCTAD (2012); World Bank (2012b)

Foreign direct investment

One of the important aspects of economic reform in India is the opening up of the economy to FDI. Before the economic reforms of 1991, India was rather hostile to foreign investment, but its policy stance has since changed and has become more FDI friendly. Table 1.6 provides evidence of the success of India in attracting increasing flows of FDI. By 2010, foreign investment accounted for 4.5 per cent of total gross fixed capital formation (GFCF). India now attracts 4.3 per cent of all foreign investment flowing into developing countries and emerging economies. A very similar transformation occurred in China, but starting from a completely closed economy, China has now realized a far larger share in total FDI flows than India. In 2010, it attracted more than 15 per cent of total FDI. As percentage of GFCF, FDI in China is lower than that in India. But this is misleading, as China has far higher rates of savings and investment than India (over 40 per cent of GDP). It is interesting to note the high rates of Greenfield investment[10] in both countries, which indicates that foreign investment contributes to the direct creation of new capital and facilities, rather than taking the form of acquisition of existing firms and assets.

Of course, the role of FDI in the economic development of India is debated. Many studies argue that the contribution of FDI to technological change and upgrading is limited because there are few technology and knowledge spillovers from foreign to domestic firms (also shown in some chapters). But on the other hand, other works suggest that there are other kinds of spillovers, which may be just as important in the form of organisational capabilities, marketing capabilities, advertising capabilities, and logistics.

Table 1.6: FDI trends in India and China, 1970–2010

	1970	*1980*	*1990*	*2000*	*2010*
India					
• FDI (current US$, million)	45	79	237	3,588	24,640
• Greenfield investment as % of FDI			97.9	70.3	77.5
• FDI as % of GFCF	0.5	0.2	0.3	3.3	4.5
• FDI as % of GDP	0.1	0.0	0.1	0.8	1.4
• FDI as % of developing countries	1.2	1.1	0.7	1.4	4.3
China					
• FDI (current US$ million)	–	57	3,487	40,715	05,735
• Greenfield investment as % of FDI			100.0	83.5	94.4
• FDI as % of GFCF	–	0.1	3.5	10.0	3.9
• FDI as % of GDP		0.0	0.9	3.4	1.8

Contd.

[10] A form of foreign direct investment where a parent company starts a new venture in a foreign country by constructing new operational facilities from the ground up. In addition to building new facilities, most parent companies also create new long-term jobs in the foreign country by hiring new employees. http://www.investopedia.com/terms/g/greenfield.asp

Contd.

	1970	*1980*	*1990*	*2000*	*2010*
• FDI as % of developing countries	-	0.8	10.0	15.8	18.4
Developing Countries					
• FDI (current US$, million)	3,854	7,479	34,853	257,625	73,568
• FDI as % of world FDI	28.9	13.8	16.8	18.4	46.1
Advanced economies					
• FDI (current US$, million)	9,491	6,599	72,602	1,145,055	70,103
• FDI as % of world FDI	71.1	86.2	83.2	81.6	53.9

Note: FDI – foreign direct investment; GFCF – gross fixed capital formation; GDP – gross domestic product

Source: UNCTADstat

Investment in R&D: Getting better but still insufficient

The above facts and figures reflect the progress made in terms of industrial capabilities catch-up. But what has been the investment in R&D and the impact of new technology generation?

Both India and China are conscious of the importance of technological upgrading. Though formal investment in R&D is only one of the ways by which innovation capabilities can be developed (others being through 'learning by doing,' 'licensing' or 'firm or asset acquisitions'), it remains a good indicator of the potential for new technology generation. Table 1.7 shows how successful India and China have been in this respect. In India, R&D investment as percentage of GDP increased from 0.3 per cent in 1980 to close to 1 per cent in 2008. In the same period, however, China started from a lower level of 0.1 per cent and succeeded in raising its R&D expenditure to 1.4 per cent of GDP. Both countries still have some way to go before they achieve advanced country levels of R&D expenditure, which typically vary between 2 and 3 per cent of GDP.

Table 1.7: Total R&D investment as percentage of GDP

	1980	*1985*	*1990*	*1995*	*2000*	*2005*	*2008*
India	0.3	0.4	0.6	0.7	0.8	0.8	0.9
China	0.1	0.2	0.5	0.6	0.9	1.3	1.4
South Korea		1.5	2.0	2.6	2.7	3.1	3.8
Taiwan		1.0	1.7	1.8	2.0	2.5	2.7
Average 16 advanced economies	1.7	2.0	2.1	2.1	2.5	2.6	2.7

Sources: Castellaci and Natera (2011); South Korea, Taiwan, and advanced economies from Szirmai, www.dynamicsofdevelopment.com

Innovation: Lots of patents but no acceleration in productivity increases

Table 1.8 illustrates that India and China are achieving spectacular success in increasing the number of patents granted in the USA. From almost nothing in 1980, the number of patents increased to 403 in 2005 in India and to 565 in China. After 2005, the increase accelerated in both countries, but far more rapidly in China than in India. In 2010, the number of patents in China was 3303, and that in India, 1137. The issue of innovation performance in India will be elaborated in more detail in the subsequent chapters.

Table 1.8: Patent performance, 1980–2010

	(Number of patents granted in the USA)							
	1980	*1985*	*1990*	*1995*	*2000*	*2005*	*2008*	*2010*
India	8	16	23	38	131	403	672	1137
China	4	76	48	63	162	565	1874	3303
South Korea			290	1,240	3,472	4,591	8,730	12,508
Taiwan			861	2,087	5,806	5,993	7,779	9,635
USA			52,977	64,510	97,011	82,586	92,001	121,179

Source: Castellaci and Natera (2011); South Korea, Taiwan, and the USA from Szirmai, www.dynamicsofdevelopment.com; Original source: USPTO, TAF database

Ultimately, what we are interested in are the effects of innovation performance on economic productivity. Fig. 1.5 presents data on the long-run evolution of labour productivity in the manufacturing sector, measured as GDP per worker. Labour productivity can be taken as a very rough proxy of the level of technological capabilities of a country. It indicates the extent to which machinery, education, efficiency, organisation, and technology augment the productivity of raw unskilled labour.[11] Fig. 1.4 represents the labour productivity of four selected countries as a percentage of labour productivity in the USA, the world productivity leader. This means that a graph, which is running horizontally, does not indicate a lack of productivity growth in an absolute sense. Rather, it implies a rate of productivity growth, which is equal to that of the USA, in other words, it demonstrates the absence of catch up.

[11] Here we can only present comparative figures for labour productivity in manufacturing, rather than for labour productivity in the total economy. As manufacturing is one of the R&D intensive sectors of the economy, it nevertheless provides relevant information about the economic outcomes of innovative performance.

Applying this logic to Fig. 1.5, we can note that the productivity performance of Indian manufacturing is very weak. It is only barely keeping up with the USA, but there are no indications of catch-up whatsoever. Contrast this with China, which started at much lower levels of relative productivity than India, but overtook India in the year 2000 and is now experiencing a spectacular catch-up spurt, comparable to that of Korea and Taiwan, and earlier Japan.

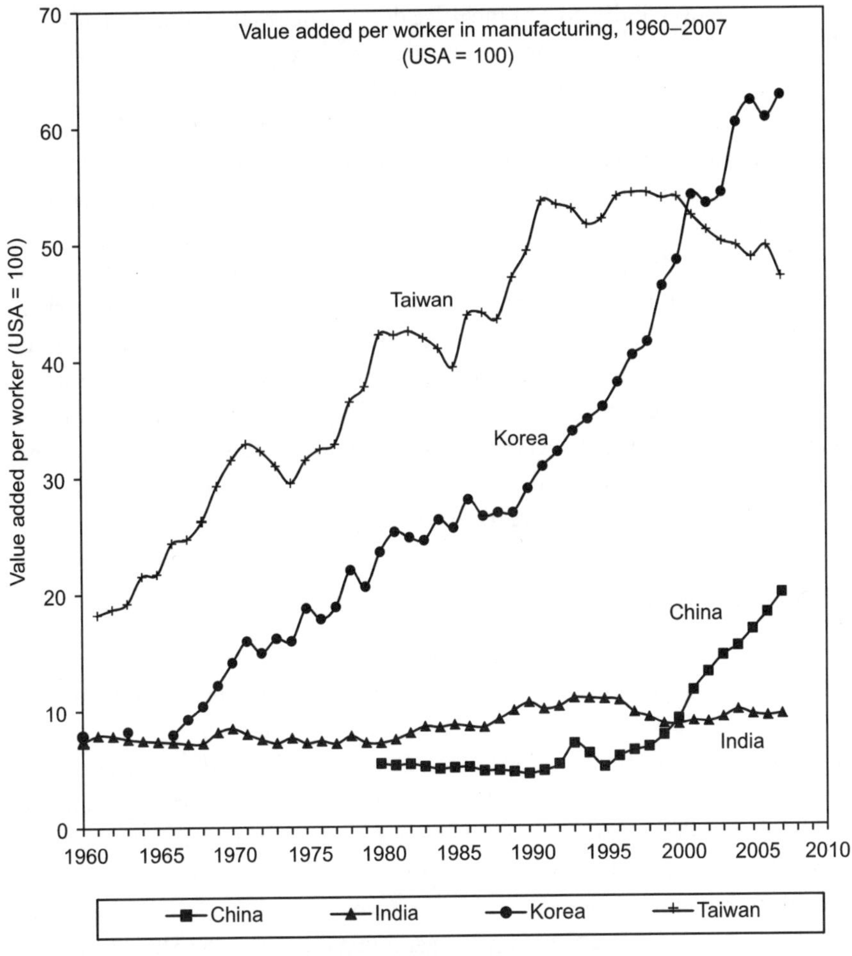

Fig. 1.5: Comparative labour productivity trends, 1960–2007

Source: Own calculations from Groningen Growth and Development Centre data, using country working papers, 10-sector data-base (http://www.rug.nl/research/ggdc/data/10-sector-database); EUKLEMS database, supplement with data from WDI, ILO, Laborsta and UNIDO industrial statistics yearbooks

Concluding remarks

How good are the final outcomes of the game?

In absolute terms, the economic and social indicators reviewed in the preceding section point to a considerable dynamism in the Indian economy. But a comparative perspective shows that in terms of growth, India is rapidly falling behind China and is not doing much better than the average of the developing world. Moreover, indicators of inclusive development reveal that India is even lagging behind other developing countries. With respect to structural change, given its level of per capita income, India provides an example of premature de-industrialisation. There is an interesting debate about service-led growth and whether services could replace manufacturing as an engine of growth. It is clear that the service sector has played and is playing an important and dynamic role in Indian economic development.

What is really worrisome, however, is that the manufacturing sector is stagnant both in terms of productivity performance and in terms of its share in GDP, though its share in exports remains important. Sustained economic growth will also require a more vigorous growth in manufacturing and industry. For this innovation in terms of new technology design, new product design, new technology, new products and processes, new services, new organisational routines – all will crucially determine whether or not there is acceleration in productivity increases in the years to come.

What does this book cover on innovation in India?

The book starts with two chapters that set the background. Though it is an over simplification, for every product or service, including innovations, sold in a market, there is a supply side and a demand side. On the supply side, increasingly innovations are not being generated ad-hoc during the process of manufacturing, but by calculation and rational design prior to production. At a country level, the rate of innovation generation by rational design is strongly correlated with its scientific capabilities. Thus, Chapter 2 by Gita Surie examines how the innovation system has evolved with respect to the formal academic establishment. She starts by examining the role of the State in creating scientific capabilities, then she outlines the major achievements of universities and public laboratories, and finally, she illustrates how universities can be a catalyst of technological innovation, entrepreneurship, and new market creation through a detailed case study. On the demand side, the challenge for India is to improve the conditions of its poor and support them so that their needs can be satisfied via markets and public distributions systems. In Chapter 3, Smita Srinivas peers into the nature of demand and its implications for innovation generation. She highlights the dilemmas faced and the solutions possible, when innovations concern essential commodities that must be made available to all people, irrespective of whether or not they can pay for it.

Part II continues its focus on innovation by zooming into five different sectors. Chapter 4 by Carl E. Pray and Latha Nagarajan examines innovations in the Indian seed industry and the impacts of the biotechnology revolution. Chapter 5 by Nagesh Kumar traces the take-off of the Indian software industry. This is followed in Chapter 6, by a study of the Indian pharmaceutical industry by Shyama V. Ramani and Samira Guennif. The potential of innovations in the emerging nanotechnology sector is explored in detail by Manish Anand in Chapter 7. Finally, in Chapter 8, Ambuj D. Sagar and Ankur Chaudhary reveal the complexity of the energy sector, analysing some of its sub-systems in which innovations are generated. These five chapters essentially consider supply side dynamics. They either trace the evolution of the sector as a whole or the main technologies used in the industry, as a function of State policies and regulations, exogenous events, and the responses of the relevant economic actors concerned. They also assess the impact of the innovations on the ecosystem in which they are generated and commercialized.

After this excursion through the supply side innovation dynamics of sectoral evolution, in Part III, the reader is taken back to the impact of innovations on inclusive development in three chapters. Arijita Dutta starts Chapter 9 with an examination of innovation in the traditional medicine healthcare system. Then, in Chapter 10, Shyama V. Ramani and Shuan Sadre Ghazi look into innovations in low-cost toilets and the reasons for the sanitation crisis in India. Finally, in Chapter 11, Dinesh Kumar Abrol rounds off this discussion, taking a panoramic view of the lessons from India on innovation for the poor.

To conclude, in Chapter 12, Shyama V. Ramani wraps up with a short note on how the game of innovation creation and diffusion has changed in India, using examples of findings of the different chapters of the book.

On the rules of the game for future innovations

In the new millennium, one of the key assumptions of the macroeconomic theories of growth and that of the catch-up theories no longer holds – namely that of quasi-free access to knowledge. It has been noted repeatedly in the catch-up literature that loosely designed IPR regimes gave rise to 'quasi-free scientific and technical knowledge' that rendered international and sectoral diffusion of technology possible, and these were crucial for the catching-up of follower countries (Fagerberg and Godinho 2005). At the end of WWII, most countries followed the Paris Convention of 1883, the oldest international IPR convention of the time. The Paris Convention was quite open and gave freedom to the signatories to set up their own IPR systems, according to their nation's individual needs. Countries could choose to patent either a product or a process, define their own criteria for patentability, establish the length of duration for the patent, and set up individual patent granting and patent publication services. Most catching-up countries either chose to have a loose IPR regime or switched to one without product patents paving the way for exploitation of 'existing superior foreign technologies.'

However, these stories may be relics of the past, as in the post-TRIPS era, product patents reign in all sectors, restricting the possibilities for catching-up through re-engineering of original innovations. The traditional route of catching-up of the now-developed countries and the post-war emerging economies has become more difficult. Today, developing countries like India are engaged in technology races with developed countries while being constrained in five important ways. First, the access to state-of-the-art technology has become more restricted due to the universal property rights regimes. Second, firms and public agencies are strapped for financial resources for R&D and innovation. Third, for the State, given the pressing problems of poverty and deprivation that have to be addressed, the marginal opportunity cost of every unit of money invested in an uncertain bid to win technology races is much higher than that for higher-income countries. Fourth, despite teeming populations, there are 'skills gaps' (both soft and hard) and 'knowledge gaps' even in qualified personnel. Finally, catch-up is needed also in terms of institutional capabilities, especially for intra-institutional and inter-institutional cooperation and coordination.

To sum up, the international technology races of today are like sports tournaments, where the developed countries represent the corps of well-trained athletes with the best of shoes, food, drugs, and coaches; and the developing countries, such as India, form the barefoot runners carrying heavy sacks of poverty burdens on their backs. The barefoot runners know that they cannot be the first in most of the races. Therefore, they must identify in which of the events they can make a mark for themselves and focus their attention on these. But, since races are held regularly, the runners can learn over time from racing again and again and develop new strategies to do better. Similarly, TRIPS may not spell gloom but mark the dawning of an era where new forms of catching-up may emerge. To its credit, the Indian system of innovation is stronger than ever before; it is vibrant, dense, and complex, bearing absolutely no resemblance to the sparse system it started out as at the time of Indian Independence. What an interesting race and what an interesting player for economists to study! Peering into the future by studying the past and the present, our book thus aims to offer insight on possible innovation trajectories in the future not only for India but for other developing countries as well.

References

Abramovitz, M. 1986. 'Catching-up, Forging and Falling Behind.' *The Journal of Economic History* 46 (2): 385–406.

Aggarwal, A. and N. Kumar. 2012. 'Structural Change, Industrialization and Poverty Reduction: The Case of India.' Paper prepared for the UNU-MERIT/UNIDO workshop: *The Untold Story: Structural Change, Industrialization and Poverty Reduction. The Case of the BRICS.* 16–17 August, Vienna.

Aghion, P. and P. Howitt. 1993. 'A Model of Growth through Creative Destruction.' *Econometrica* 60 (2): 323–51.

Ahluwalia, I. J. 1991. *Productivity and Growth in Indian Manufacturing.* New Delhi: Oxford University Press.

Bardhan, P. 1984. *The Political Economy of Development in India.* New Delhi: Oxford University Press.

Bradford DeLong, J. 2003. 'India since Independence: An Analytic Growth Narrative.' In *In Search of Prosperity. Analytic Narratives on Economic Growth*, edited by D. Rodrik, 184–204. Princeton: Princeton University Press.

Castellaci, F. and J. M. Natera. 2011. 'A New Panel Dataset for Cross-Country Analyses of National Systems, Growth and Development (CANA).' NUPI Working Paper 783. Oslo: Norwegian Institute of International Affairs. Accessed: June 2012; http://www.nupi.no/content/download/83264/278278/version/1/file/WP-783-Castellacci-Natera+.pdf.

Chen, S. and M. Ravallion. 2008. 'The Developing World is Poorer Than We Thought, but No Less Successful in the Fight against Poverty.' *Quarterly Journal of Economics* 125 (4): 1577–625.

Cornwall, J. 1977. *Modern Capitalism: Its Growth and Transformation.* New York: St. Martin's Press.

De Vries, G. J., A. A. Erumban, M. P. Timmer, I. Voskoboynikov, and H.X. Wu. 2012. 'Deconstructing the Brics: Structural Transformation and Aggregate Productivity Growth.' *Journal of Comparative Economics* 40 (2): 211–27.

Edquist, C. 2001. 'The Systems of Innovation Approach and Innovation Policy: An Account of the State of the Art.' DRUID conference. Aalborg. Accessed: June 2012; http://www.druid.dk/conferences/nw/paper1/edquist.pdf

Fagerberg, J. and M. M. Godinho. 2005. 'Innovation and Catching Up.' In *The Oxford Handbook of Innovation,* edited by J. Fagerberg, D. C. Mowery, and R. R. Nelson, 514–43. Oxford: Oxford University Press.

Frankel, F. R. 1978. *India's Political Economy 1947–1977.* Princeton: Princeton University Press.

Freeman, C. 1995. 'The National System of Innovation in Historical Perspective.' *Cambridge Journal of Economics* 19 (1): 5–24.

Gerschenkron, A. 1962. *Economic Backwardness in Historical Perspective.* Cambridge: Harvard University Press.

Kotwal, A., B. Ramaswami, and W. Wadhwa. 2011. 'Economic Liberalization and Indian Economic Growth: What's the Evidence?' *Journal of Economic Literature* 49 (4): 1152–199.

Lal, D. 1988. *The Hindu Equilibrium, Vol I: Cultural Stability and Economic Stagnation: India, c. 1500 BC – AD, 1980.* Oxford: Oxford University Press.

Lall, S. 1992. 'Technological Capabilities and Industrialization.' *World Development* 20 (2): 165–86.

Landes, D. 1998. *The Wealth and Poverty of Nations.* London: Abacus.

Lee, K. and C. Lim. 2001. 'Technological Regimes, Catching-up and Leapfrogging: Findings from the Korean Industries.' *Research Policy* 30 (3): 459–83.

Lundvall, B. A. 1992. *National Innovation Systems: Towards a Theory of Innovation and Interactive Learning.* London: Pinter.

———. 1998. 'Why Study National Systems and National Styles of Innovation?' *Technology Analysis and Strategic Management* 10 (4): 407–21.

Lundvall, B. A., K. J. Joseph, C. Chaminade, and J. Vang. 2009. *Handbook of Innovation Systems and Developing Countries.* Cheltenham: Edward Elgar.

Maddison, A. 1974. *Class Structure and Economic Growth in India and Pakistan.* London: Allen and Unwin.

Malerba, F. 2002. 'Sectoral Systems of Innovation and Production.' *Research Policy* 31 (2): 247–64.

Malerba, F. and R. R. Nelson (eds.). 2012. *Economic Development as a Learning Process.* Cheltenham: Edward Elgar.

Mani, S. 2011. 'Promoting Knowledge-Intensive Entrepreneurship in India.' In *Entrepreneurship, Innovation and Development,* edited by A. Szirmai, W. Naudé, and M. Goedhuys, 194–227. Oxford: Oxford University Press.

Naudé, W., A. Lavopa, and A. Szirmai. 2012. 'Industrialization and Technological Change in the BRICS: The Contribution of Multinational Enterprises and Domestic

Investment.' Paper prepared for the UNU-MERIT/UNIDO workshop on *The Untold Story: Structural Change, Industrialization and Poverty Reduction. The Case of the BRICS.* 16–17 August, Vienna.

Naudé, W. and A. Szirmai. 2012. 'The Importance of Manufacturing in Economic Development: Past, Present and Future Perspectives.' UNU-MERIT, Working Paper Series 2012–041. Maastricht, the Netherlands. Accessed: December 2012; http://www.merit.unu.edu/publications/wppdf/2012/wp2012-041.pdf.

Nelson, R. 1993. *National Innovation Systems: A Comparative Analysis.* Oxford: Oxford University Press.

Panagariya, A. 2004. 'Growth and Reforms during 1980s and 1990s.' *Economic and Political Weekly* 39 (2): 2581–94.

Rodrik D. and A. Subramanian. 2005. 'From "Hindu Growth" to Productivity Surge: The Mystery of the Indian Growth Transition.' IMF Staff Papers, Palgrave Macmillan 52 (2): 193–228.

Rosenberg N. and L. E. Birdzell jr. 1986. *How the West Grew Rich. The Economic Transformation of the Industrial World.* New York: Basic Books.

Surie, G. 2011. 'Emergence of New Markets, Distributed Entrepreneurship and the University: Fostering Development in India.' *International Journal of Technoentrepreneurship* 2 (3/4): 362–81.

Szirmai, A. 2013. 'Manufacturing and Economic Development.' In *Pathways to Industrialisation in the 21st Century, New Challenges and Emerging Paradigms*, edited by A. Szirmai, W. Naudé, and L. Alcorta, 53–75. Oxford: Oxford University Press.

———. 2012. *Statistics on Socio-economic Development.* Cambridge: Cambridge University Press. Accessed: June 2012; www.dynamicsofdevelopment.com <http://www.dynamicsofdevelopment.com.

The Conference Board. 2012. *Total Economy Database, Output, Labor and Labor Productivity Country Details, 1950–2011.* Accessed: June 2012; http://www.conference-board.org/data/economydatabase/.

Tregenna, F. 2013. 'Deindustrialization and Reindustrialization.' In *Pathways to Industrialization in the 21st Century*, edited by A. Szirmai, W. Naude´, and L. Alcorta, 76–101. Oxford: Oxford University Press.

UNCTAD. 2012. *UNCTADSTATS.* Accessed: June 2012; http://unctadstat.unctad.org/ReportFolders/reportFolders.aspx?sCS_referer=&sCS_ChosenLang=en).

UNESCO. 2012. *UNESCO Institute of Statistics, Data Centre.* Accessed: June 2012; *http://* stats.uis.unesco.org/unesco/tableviewer/document.aspx.

UN (United Nations). 2010. 'World Population Prospects: The 2010 Revision.' CD-ROM edition. Department of Economic and Social Affairs, Population Division. Accessed: June 2012; http://esa.un.org/unpd/wpp/Excel-Data/mortality.htm.

Verspagen, B. 1993. *Uneven Growth between Interdependent Economies. An Evolutionary View on Technology Gaps, Trade and Growth.* Aldershot: Avebury.

Vij, N. 2011. 'Building Capacities for Social Empowerment: The Missing Link between Social Protection and Social Justice.' International Conference on *Social Protection for Social Justice, Institute of Development Studies*, UK. Accessed: June 2012; http://www.ids.ac.uk/files/dmfile/Vij2011SocialauditsinMGNREGAinIndiaCSPconferencedraft.pdf.

World Bank. 2012a *World Development Indicators Online*. Accessed: June 2012; http://data.worldbank.org/data-catalog/world-development-indicators.

World Bank. 2012b. *World Integrated Trade Solution (WITS)*. Accessed: June 2012; http://data.worldbank.org/data-catalog/wits;.

Three major trends mark the current scenario vis-à-vis the evolution of the role of university in national innovation systems of emerging economies such as India. First, with increasing participation in global markets, firms recognize that in addition to technical capabilities of personnel and the quality of local knowledge, an entrepreneurial, collaborative, and leadership mindset is crucial to enhance competitiveness. In other words, universities need to generate 'social capabilities' in addition to 'scientific' and 'technical' ones. Second, firms in emerging and developed economies are collaborating more with academic institutions in terms of R&D and knowledge exchange due to heightened competition, abbreviated product life cycles, and rising costs together with increased connectivity. New forms of technical, social, and entrepreneurial skills are being developed to exploit such opportunities for learning and innovation. Third, even within countries, although knowledge production was hitherto largely concentrated within the firm, market leaders are increasingly acquiring knowledge from external sources via joint ventures, alliances, and hybrid relationships with other organisations, including the university, a key resource institution.[1] Given the above trends, the imperative to accelerate gains from technological innovation has also led to policy changes promoting university–industry partnerships in India, especially in knowledge-intensive sectors.

There is extensive organisational research exploring the impact of university–industry partnerships on the diffusion of knowledge and commercialization of technologies in industrialised countries (Mowery and Sampat 2005; Mowery et al. 2001; Owen-Smith et al. 2002; Woolgar 2007). However, not enough is known about their evolution or their impact or how such partnerships can foster the emergence of new industries and entrepreneurship to address societal challenges in emerging economies like India [an exception is Datta and Saad's (2011) research on the Indian innovation system and hurdles that constrain universities].

Therefore, using insights from the national systems literature (Nelson 1993; Lundvall 2007, 1992) and a systems approach (Carroll and Burton 2000; Holland 1998; Kauffman 1993; Simon 1965), this chapter examines the evolution of innovation and entrepreneurship capabilities in India, that stem from the science and technology (S&T) training and research institutes. It focuses on three issues. First, the evolution of the construction of scientific and technological capabilities under the traditional paradigm is traced, starting

[1] In this article, university laboratories and government R&D laboratories are treated interchangeably.

from the colonial period up to the adoption of economic reform during the early 1990s. Second, the evolution of complexity of the knowledge generation system during the last two decades is explained. Finally, I provide a case study of technology incubation and dissemination by an Indian technical institution to demonstrate how university–industry partnership can meet the dual challenge of global competition and economic development by catalysing the emergence of entrepreneurial capabilities and new markets to solve socio-economic problems. Evidence from the case suggests that university–industry partnerships contribute to matching global competition and addressing local development challenges by developing and deploying low-cost innovations and diffusing entrepreneurial capabilities to create new markets.

The chapter is also organised along the same lines. The first section outlines the historical evolution of India's S&T system and the role of the universities and government research laboratories in advancing science and technological innovation. The following section highlights the challenges faced by the S&T establishments in promoting innovation capabilities. The next section contains the case study. The final section discusses the results and concludes.

The university in the evolution of Indian science and technology

In this section, the role of public S&T establishments are examined in the historical evolution of the Indian system of innovation in three periods: the early history under colonialism; post-Independence period from 1947 to 1990, and post-liberalisation period from 1991 to the present day.

Colonial science and technology

Research on pre-colonial S&T in India suggests that Indian astronomy, mathematics, medicine, textile, and steel-making processes impressed pre-colonial travellers to India (Kumar 1995; Arnold 2004). In addition, scholars (Rahman 1982; Dharampal 1971) argue that S&T also flourished in medieval India and cite as evidence a number of source materials in Sanskrit, Arabic, and Persian in the field of astronomy, mathematics, and medicine as well as a wide range of scientific and technological subjects (Kumar 1995). Tributes paid to Indian textiles, metal-works, construction, and ship-building techniques by the early European travellers suggest that India had attained the distinct status of a technologically advanced country by the beginning of the eighteenth century (Sangwan 1991). Likewise, in agriculture, ecological farming was

and industrial development. After independence, among the first national laboratories and institutes to be established were the following: National Physical Laboratory, National Chemical Laboratory, National Metallurgical Laboratory, Fuel Research Institute, Central Food Technological Research, the Central Drug Research Institute, Central Glass and Ceramics Research Institute, Central Road Research Institute, Central Building Research Institute, Central Leather Research Institute, and the Central Electro-chemical Research Institute (First Five-year Plan, 1951[2]). The Atomic Energy Commission, established in 1948, was assisted by the Tata Institute of Fundamental Research (founded in 1945) in its aim to develop electric power from nuclear energy, apply nuclear science in agriculture, industry, medicine, and health, and build scientific capabilities in different fields related to atomic energy (Second Five-year Plan, 1956[3]). The Scientific Policy Resolution (1958) along with the Third Five-year Plan (1961) recognized the need to foster the development of scientific capabilities by increasing the number of scientists and promoting scientific research in line with the objective of developing the economy, reducing the burden of imports through import substitution, raising exports, establishing new enterprises, and improving technological know-how. The Fourth (1970) and Fifth (1976) Five-year plans continued this trend of broadening the scope of existing institutions and introducing new ones such as the National Institute of Oceanography, a Regional Research Laboratory at Bhubaneshwar, the National Geophysical Research Institute, Structural Engineering Research Centre, and the Industrial Toxicology Research Centre (GOI). During the Sixth Five-year Plan (1981), the government issued a technology policy statement in 1983, covering a host of issues from indigenous development, assessment, forecasting and import, to absorption, adaptation, and further development of technology.

At the ministerial level, from 1947 to 1990, the number of departments conducting R&D under the aegis of the Ministry of Science and Technology grew. Table 2.1 lists selected S&T departments of various ministries and the year they were established. However, in many instances, these activities were undertaken under other names prior to the department being formally established. Besides these, several other ministries support R&D in agriculture, information technology, health, rural development, environment and forests, and water resources. Also, two public enterprises, the National Research

[2] Government of India. 1951. Accessed: 23 January 2012; http://planningcommission.nic.in/plans/planrel/fiveyr/default. html.

[3] Government of India. 1956. Accessed: 23 January 2012; http://planningcommission.nic.in/plans/planrel/fiveyr/default. html.

Development Corporation (1953) and Central Electronics Limited (1974), were established to promote technology commercialization.

Table 2.1: Selected departments supporting R&D activity under various ministries

Ministry	*Department*	*Year established*
Ministry of Science and Technology	Department of Science and Technology	1971
	Council of Scientific and Industrial Research	1942
	Department of Space	1972
	Department of Atomic Energy	1954
	Department of Scientific and Industrial Research	1985
	Department of Biotechnology	1986
Ministry of Defence	Defence Research and Development Organization	1958
Ministry of Communication and Information Technology	Department of Electronics and Information Technology	1970
Ministry of Earth Sciences	Department of Ocean Development	1981

Universities were set up to produce trained personnel, and a large network of institutions for advanced research was also established outside the university system. From the 1950s to the 1980s, as detailed in the other chapters, the Indian university system and network of engineering schools were crucial in providing qualified labour for industrial growth in medium, heavy, and high-technology industries. The emphasis on building S&T infrastructure resulted in an increase in the number of universities from 27 in 1950–51 to 82 in 1970–71 (Patel 1996) and 128 in 1980–81 (GOI 2009). After the 1980s, the financially constrained Government of India found it difficult to set up new universities and colleges. Lack of funds also led public universities and colleges to focus on setting up self-financing courses and distance education programmes. In contrast, non-subsidised education became more viable due to the growth of a middle class that could afford higher fees (Agarwal 2007). Thus, private institutions proliferated over the years with nearly 30 per cent of enrolment in private unaided institutions (Agarwal 2007). However, despite the large number of institutions, only a few are of high quality.

In the field of engineering, a network of 'Indian Institutes of Technology' (IITs) was established in the 1960s in collaboration with institutes in the USA, Germany, UK, and USSR to provide high-quality S&T training, conduct research, and ensure adequate manpower for strategic technology-based industries (Agarwal 2007). Table 2.2 provides the number of S&T personnel from universities, by scientific discipline and qualifications. The Department of

- soil and water management, human and animal nutrition, health, education, industry, energy, communication and transportation, besides agriculture.
- high-tech general purpose technologies such as information technology and biotechnology.
- policy implementation that also included commercializing traditional knowledge and intellectual property management.

The Tenth Five-year Plan (2002–07) also emphasised on indigenous technologies and frontier technologies from other countries, strengthening application-oriented R&D for technology generation, and integrating S&T developments with all spheres of activity to aid economic development. Similarly, the Eleventh Five-year Plan (2007–12) aimed to encourage university–industry partnerships to foster innovation, and disseminate technology and entrepreneurial capabilities; the strategy also called for new emphasis on education, health, and other socially relevant issues to ensure technology-led accelerated and inclusive growth. The allocation of funds, by sector for the various national laboratories and approved outlays are outlined in Fig. 2.1 and Table 2.3.

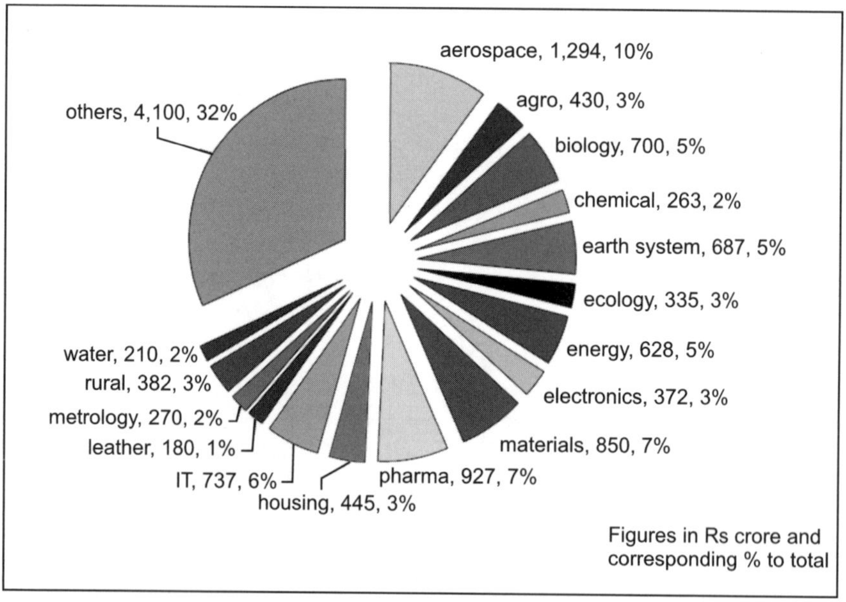

Fig. 2.1: Distribution of proposed allocation, by sector, for national laboratories under the Eleventh Five-year Plan

Source: CSIR (2006)

Table 2.3: Distribution of approved outlay in the Tenth and Eleventh Five-year Plans for National Laboratories

Heads	*Tenth Plan*		*Eleventh Plan*	
	Approved allocation	*% of total allocations*	*Projected*	*% of total allocations*
(i) Sectoral distribution				
1. Aerospace science and engineering	161	8.47	1294	10.10
2. Agro, food processing, and nutrition	62	3.26	430	3.36
3. Biology and biotechnology	192	10.11	700	5.46
4. Chemical science and technology	83	4.37	263	2.05
5. Earth system science	188	9.89	687	5.36
6. Ecology and environment	45	2.37	335	2.62
7. Energy resources and technology	52	2.74	628	4.90
8. Electronics and instrumentation	96	5.05	372	2.90
9. Engineering materials, minerals, and manufacturing	104	5.47	850	6.64
10. Pharma, healthcare, and drugs	341	17.95	927	7.24
11. Housing, road, and construction	63	3.32	445	3.47
12. Information technology	79	4.16	737	5.75
13. Leather	24	1.26	180	1.41
14. Metrology	21	1.11	270	2.11
15. Rural development	0	0.00	382	2.98
16. Water resources and technology	0	0.00	210	1.64
Total	1,511	79.53	8,710	67.99
Others				
(ii) Ninth/Tenth Plan commitments	39	2.05	600	4.68
(iii) Creating, nurturing, and sustaining the core knowledge frontier	250	13.16	1,000	7.81
(iv) House	0	0.00	500	3.90
(v) Open source drug discovery programme for infectious disease	0	0.00	500	3.90
(vi) Laboratory modernisation for eco-friendly sustainable growth	0	0.00	500	3.90
(vii) Civil infrastructure and refurbishment	100	5.26	1,000	7.81
Total	389	20.47	4100	32.01
Grand Total	1,900	100.00	12,810	100.00

Source: CSIR (2006)

Recent estimates from the Department of Science and Technology indicate that R&D expenditure increased from Rs 18,088.16 crore[4] in 2002–03 to

[4] A crore is equal to 10, 000,000.

Challenges faced by science and technology establishments in promoting innovation capabilities

The findings of the previous section make it evident that by the start of the new millennium, the S&T infrastructure had been firmly established. It had made a real contribution to the growth of several sectors and it had grown in complexity. Next, we briefly assess the current strengths and weaknesses of the universities and S&T institutions.

Strengths of the existing science and technology infrastructure

Numbers count

The major strength of India's scientific and technological capabilities lies in its sheer numbers of qualified labour and the institutions teaching S&T or applying it to improve production processes. Higher education was imparted by over 300 universities, 13 institutions of national importance, and nearly 17,000 colleges in 2003–04, producing more than 60 per cent of doctorates in science; approximately 300,000 people were employed in R&D establishments, including in-house R&D units in public and private sector firms (EXIM Bank 2006, 38). There are about 3,000 R&D institutions in India comprising national laboratories, R&D institutes in the central, state, and non-government sectors, and R&D units in industry (EXIM Bank 2006). As noted earlier, private educational institutions are growing at a faster rate than public universities (Agarwal 2007).

Increasing accumulation of intellectual property

While slow to patent inventions internationally, the USA has granted 2,150 patents to Indian organisations (USPTO).[5] Out of more than 300 Indian universities, the number of academic institutions that filed for patents was in the range of 22–29 per year (a total of 62 over a four-year period, 2001–04), disproportionately small when compared with the high number of universities engaged in R&D activities. The performance of the CSIR is exceptional (Nezu 2005; see Table 2.5) and the number of patents filed and granted doubled or tripled after 2001 as compared to previous years. Likewise, patenting has increased in several other Indian institutions (see Table 2.6 for patents granted to selected institutions).

[5] Data collected from USPTO, November 2008 (http://www.uspto.gov/).

Table 2.5: CSIR patenting activity (1997–2006)

Year	*India*		*Abroad*	
	Filed	*Granted*	*Filed*	*Granted*
1997–98	264	155	94	28
1998–99	310	134	114	48
1999–2000	377	112	208	47
2000–01	408	117	490	62
2001–02	413	341	569	89
2002–03	421	166	771	189
2003–04	406	276	566	217
2004–05	415	175	529	275
2005–06	407	276	645	179
Total	3,421	1,752	3,986	1,134
Total filed: 7,407			*Total granted:* 2,886	

Source: CSIR (2006)

Among the teaching institutes, most patent applications originate from the IITs, the Indian Institute of Science (IISc), University of Delhi, and All India Institute of Medical Sciences (AIIMS), New Delhi. More than 70 per cent of filings in India [either as conventional applications or through the Patent Cooperation Treaty (PCT)] are from foreign inventors or their assignees, while the contribution from Indian inventors and assignees has been less than 30 per cent for 2006–07 (GOI 2009).

Support from the State

Ministerial infrastructure for innovation promotion is in place. The DSIR (one of the three departments of the Ministry of Science and Technology, formed in 1985) endeavoured to encourage industry to increase its share in the country's R&D expenditure and catalyse faster commercialization of lab-scale R&D from 2002 onwards by providing financial support for selected R&D activities. These include technology development, absorption and up-gradation of imported technology, patent filing, consortium projects for industry groups, conferences, and technology missions in selected areas such as healthcare, telecom, and environmentally sound and socially relevant technologies for rural areas.

In addition to supporting the accumulation of innovation capabilities at the firm level, the DSIR also promotes techno-entrepreneurship at an individual level via a collaborative effort with Technology Information, Forecasting and Assessment Council (TIFAC) of the Department of Science and Technology;

Table 2.6: Patenting by selected institutions in India

Institution	*2000*	*2001*	*2002*	*2003*	*2004*	*2005*	*2006*	*2007*	*2008*	*2009*	*Total (2000–09)*
Indian Institute of Science			1	2	1	1	4	6	2		17
Indian Institute of Technology		3	3	7	7	4	3	1	6		34
CSIR	3	2	7	8	4	7	5	11	4	1	52
Jawaharlal Nehru University								1			1
University of Delhi	2		1	6	4	2	1	2			18
Jadavpur University					2						2
University of Madras	2										2
Bose Institute	2				2	1	1				6
Total											132

Sources: WIPO: World Intellectual Property Organization; Controller General of Patents Designs and Trademarks of India

the Technopreneurship Programme (TePP) is the only Government of India programme that funds individual innovators. Examples of technologies funded include a motorised auto-distractor – a technology for regenerating bone and soft tissue, devised overseas and brought to India by Prasad Narayan Kulkarni of Sangli and developed with partial support from TePP (DSIR 2006). Other projects supported by TePP include an environmentally friendly solid biomass furnace, which demonstrated savings in fuel consumption and was installed in various schools; a vertical cylindrical solar water heater designed by a teacher innovator; and a small three-wheeled 10 HP tractor designed by an innovative farmer (DSIR 2006).

During the Eleventh Plan (2007–12) and beyond, the DSIR aimed to strengthen the Indian system of innovation by establishing an ecosystem to support creativity and boost the rate of innovation. Some measures in this line include the following:

- Amendment of the Indian Patent Act to comply with the TRIPS agreement; intellectual property rights (IPR) protection was further strengthened through policies such as the Designs Act (2000), the Trade Marks Act (1999), and the Geographical Indications of Goods Act (1999), harmonising the intellectual property regime with the global system.
- Provision of fiscal incentives to promote industrial R&D and encourage the utilisation of indigenously developed R&D (EXIM Bank 2006).
- Enhancement of information capabilities and creation and maintenance of digital databases. In addition, DSIR promoted documentation of traditional knowledge, began electronic publishing of selected S&T material, and established an open archive initiative providing research resources in selected areas such as molecular informatics (see Chapter 9 on traditional medicine).
- Establishment of an early-stage venture fund (Fund for Accelerating Start-ups in Technology; DSIR 2006) to provide seed capital to start-ups, particularly in the information and communication technology (ICT) sector for innovative products and services and leveraging linkages with Indians working in high-technology firms in regions like Silicon Valley in the USA.
- Promotion of industrial consultancy services in India and overseas to accelerate technology-intensive exports. Technology-intensive exports are supported by creating awareness about the technological capabilities of industries and R&D establishments among potential foreign buyers and establishing mechanisms to catalyse international technology transfer and trade.

Additionally, the IITs and other leading universities and national laboratories began providing training in intellectual property management and encouraging their scientists to patent (CSIR 2006).

Increasing support for public–private partnerships

The CSIR's 'New Millenium Indian Technology Leadership Initiative' (NMITLI) aims to achieve global leadership in technology through public–private partnerships by funding time-bound and result-oriented projects with short cycle times; as of 2005, these included 14 core projects involving over 110 R&D institutions and about 45 industry partners. This initiative also served as the largest funding route for the biotechnology sector and was worth Rs 1500 million (Ganguli 2005). Some examples of public–private partnerships for technology development are presented in Table 2.7.

Table 2.7: Some public–private technology partnerships

	Partnerships	*Purpose*
1.	CSIR (public) Tata Consultancy Services (private)	bioinformatics educational software development; development of bio-suite for applications such as genome sequence analysis, genome analysis, 3-D modelling, simulations, manipulations, and structural changes for drug design, pathway modelling and comparative genomics
2.	CSIR (public) Lupin Laboratories (private) International Centre for Genetic Engineering and Biotechnology (international organisation) Bose Institute (public) Central Drug Research Institute (public) Indian Institute of Science (public-private partnership) Center for DNA Fingerprinting and Diagnostics (public) Indian Institute of Chemical Technology (public) Institute of Genomics and Integrative Biology (public) National Chemical Laboratory (public) National Institute of Immunology (public) Regional Research Laboratory, Jammu (public) University of Hyderabad (public)	new anti-tuberculosis drug

Contd.

Contd.

	Partnerships	*Purpose*
3.	Dabur India (private) Centre for Molecular Biology (public) Indian Institute of Science (public-private partnership) CSIR (public; provided entire funding of Rs 150 million for the project)	collaboration to develop vaccine constructs, and gene therapy candidates for cancer of the head, neck, and gall bladder.

Source: Ganguli (2005)

Support for business incubators and technology transfer

Technical institutions such as the various IITs have set up business incubators to develop and commercialize technologies in collaboration with industry. In turn, industry has sought university collaborations for manpower development to ensure that relevant skills are imparted to newly trained scientists, enabling them to become productive in the shortest possible time. Firms also forge collaborative agreements for drug discovery and to strengthen the faculty's research activities (e.g., collaboration between Ranbaxy and Anna University was initiated for the former and with Jamia Hamdard to support a PhD programme in pharmaceutical medicine). Several companies collaborated with academics by initiating support programmes with scholarships and research awards.

Success stories: Growth of knowledge–intensive industries, information and communication technologies, and biotechnology

By taking risks, incumbent firms and new start-ups emphasising innovation are largely responsible for the emergence and growth of knowledge-intensive industries such as ICT and biotechnology during the 1990s and into the new millennium. Firms in these sectors have generally followed an alliance strategy to acquire knowledge and complementary capabilities from universities, both in India and elsewhere. Also, biotechnology firms have been accumulated capabilities by undertaking contract research for global firms seeking to reduce R&D costs (Surie 2008).

By 2005, India had emerged as a key player in the ICT sector. It was the first developing country to join the top 10 exporters of ICT-enabled services with exports worth US$41 billion and a market share of 3.8 per cent and an annual growth of 37 per cent (2000–05), followed by China ($26 billion in 2005, market share of 2.4 per cent and annual growth of 22 per cent from 2000 to 2005). International sourcing of business activities and processes has become an integral element in trade and investment in ICT-enabled services, especially between developed and developing countries. India accounted for 19 per cent of all IT service projects in 2002–03 (OECD 2006), 65 per cent of the global market in ICT offshoring, and 46 per cent of global business process outsourcing (NASSCOM-McKinsey Report 2005).

What is the role of Indian academic establishments in the ICT success story? First and foremost, the general trend of increasing participation by developing countries in outsourcing and a broadening of the geographical scope of foreign direct investment (FDI) by industrialised countries, is due to the availability of low-cost, technically skilled labour and presence of world-class universities like the IITs, in addition to competitive pressures, improving host country environments, and the promise of new markets (Surie 2008; Nezu 2005). Although the Indian software industry grew rapidly from the late 1980s onwards, and software services rose, particularly in the post-liberalisation period of the 1990s, most of the services were based on labour arbitrage. Over time, however, companies such as Infosys and Tata Consultancy Services developed good capabilities in doing design work and achieved a strong understanding of software development processes. Although lagging, innovation and product development are currently being emphasised as critical for development and competitiveness in global markets.

Similar success was replicated in the biotechnology industry. The Indian biotechnology sector was mainly a government-sponsored effort in the mid-1980s, and it took more than a decade to induce private sector participation [see Ghose and Bisaria (2000) and Ramani and Visalakshi (2001) for an overview of the Indian biotechnology industry]. However, by 2011, close to 400 firms employing about 50,000 scientists generated nearly US$4 billion through domestic sales and exports (Reid and Ramani 2012).

During the 1990s, the biotechnology industry focused on plant tissue culture, diagnostic kits based on poly- and mono-clonal antibodies for some diseases, and research biologicals such as restriction enzymes and reagents (Visalakshi and Sandhya 2000). By the end of the 1990s, Indian firms were producing biogenerics or re-engineered therapeutic products based on genetically engineered or recombinant technologies already in the international market. Interestingly, in the biogenerics market, dominated by imports of established global brands, the first Indian players were in fact new companies created specifically to exploit the opportunity offered by recombinant therapeutics. With the entry of local competitors with a critical cost advantage, the prices of some life-saving drugs came down drastically in India. For example, Shantha Biotech (now part of Sanofi-Aventis) first introduced its locally developed hepatitis B vaccine in 1997 through an innovative expression system that made it possible to lower the cost by 30-fold, from US$15 to US$0.5 (Ramani and Maria 2005).

Visalakshi and Sandhya (2000) note that although the Indian biotech industry introduced only a few new biotechnology- and fermentation-based products, the efforts of Indian R&D institutions have been recognized worldwide in the areas of immunology, genetic engineering, molecular biology, and birth control measures. Besides providing qualified labour, universities and public laboratories have contributed 195 out of 208 of USPTO patents for India (Reid and Ramani 2011). Finally, a handful of biotechnology firms such as the niche player Strand Life Sciences incubated by the Indian Institute of Science, have been created as spin-offs from academic institutions. Armed with the required scientific expertise, but lacking knowledge of the scaling-up process and downstream competencies like marketing, they start as contract research service providers, usually for multinational firms, either in biotechnology research, bioinformatics, or clinical trials (Ramani and Maria 2005).

Also, despite low commercialization and reduction in FDI since the financial crisis, non-equity modes of investment (NEM) such as contract research that often lead to technology and skills acquisition by NEM companies in the bio-pharmaceuticals sector continued and have led to their transformation into transnational firms and technology leaders in their own right (UNCTAD 2011).

Weaknesses in the science and technology system within the national system of innovation

Unprecedented expansion of universities since Independence in 1947 resulted in the neglect of quality, research, and management aspects of the Indian university system (Gupta 2003; Saxena 1990). Though in terms of size, India's output in terms of the number of qualified personnel is impressive (see Table 2.8), most universities are teaching institutions. The few institutions that do conduct research face a shortage of faculty and funds, while the rest lack the requisite infrastructure and culture for conducting scientific research. Expansion in engineering enrollment by 8.2 per cent per annum from 1995 to 2000 and 21.9 per cent from 2000 to 2004 (Shukla 2005) due to increased demand for engineers was perceived to have led to a decline in the quality of students taking admission, the quality of teaching, and the quality of graduating engineers (Sayeda, Rajendran, and Lokachari 2010). The total number of researchers in India was 154,840, compared with 1,390, 649 in USA and 1,214,912 in China (GOI 2009). Despite the large absolute number of researchers, India fares poorly in international comparisons, with only 140 researchers per million versus 7,681

Table 2.8: Number of S&T graduates from universities by field of science and level of qualification

	Graduate			*Postgraduate*			*Doctorates*			*Total*		
	1979	1995	2003	1979	1995	2003	1979	1995	2003	1979	1995	2003
Science	99,749	139,257	334,865	17,638	23,807	75,459	2,262	3,155	4,740	119,649 (70.6)	166,219 (70.5)	415,064 (68.1)
Engineering and technology	18,364	32,250	127,499	3,155	3,667	12,249	506	546	779	22,025 (13.0)	36,463 (15.4)	140,527 (23.1)
Medicine	15,090	19,613	31,697	3,485[a]	4,634[a]	7,071[a]	–	–	–	18,575 (11.0)	24,247 (10.3)	38,768 (6.4)
Agriculture and veterinary science	6,280	5,752	9,109	2,384	2,284	4,616	480	827	1,195	9,144 (5.4)	8,863 (3.8)	14,720 (2.4)
Total	139,483 (82.3)	196,872 (83.5)	503,170 (82.6)	26,662 (15.7)	34,392 (14.6)	99,195 (16.3)	3,248 (1.9)	4,528 (1.9)	6,714 (1.1)	169,393 (100)	235,792 (100)	609,079 (100)

Note:
[a]MD/MS include doctorates in medical science
Numbers in brackets represent percentages
Source: Department of Science and Technology (2009)

in Finland, 6,139 in Sweden, 5,713 in Singapore, 5,546 in Japan, and 5,277 in Denmark (see Table 2.9). Realization of the fact that this poses a serious problem for India's future competitiveness, has led the government to consider massively increasing S&T education going forward as even a 10-fold increase by 2030 would barely match China's current output (GOI 2010). Additionally, the number of institutions across regions is insufficient to fulfil demand in India, while wide disparity exists across states in terms of access, gender equity, and gross enrolment ratios (MHRD 2012; Agarwal 2007).

Table 2.9: Number of researchers (per million people) in selected countries

Sl. No.	*Country*	*No. of researchers per million people*	*Population in million*	*Total researchers*
1.	Argentina	895	39	34,905
2.	Australia	4,053	21	85,113
3.	Austria	3,657	8	29,256
4.	Brazil	461	189	87,129
5.	Canada	3,922	33	129,426
6.	China	926	1312	1,214,912
7.	Czech Rep.	2,578	10	25,780
8.	Denmark	5,277	5	26,385
9.	Finland	7,681	5	38,405
10.	France	3,353	61	204,533
11.	Germany	3,386	82	277,652
12.	Hungary	1,745	10	17,450
13.	India	140	1106	154,840
14.	Italy	1,407	59	83,013
15.	Japan	5,546	128	709,888
16.	Korea, Rep.	4,162	48	199,776
17.	Mexico	464	104	48,256
18.	Netherlands, The	2,524	16	40,384
19.	Norway	4,668	5	23,340
20.	Pakistan	80	159	12,720
21.	Russian Fed.	3,255	143	465,465
22.	Singapore	5,713	4	22,852
23.	Spain	2,639	44	116,116
24.	Sri Lanka	141	20	2,820
25.	Sweden	6,139	9	55,251
26.	United Kingdom	3,033	61	185,013
27.	United States	4,651	299	1390,649
28.	Venezuela	86	27	2,322

Source: Researchers per million – UIS, UNESCO (website accessed on 15 October 2008) Population figures – World Development Indicators 2008, The World Bank

Moreover, incentives are lacking for S&T undergraduates to pursue research as a career. Indeed, Chatterjea and Moulik (2012, 2006) note that as of 2010, no Indian university or institute has ranked among the world's top 300 universities in the 'Academic Ranking of World Universities', produced since 2003 by China's Shanghai Jiao Tong University. IISc Bangalore ranked between 301 and 400, the IIT Kharagpur ranked between 401 and 500, and only two Indian universities were in the top 500 list. Among the diverse reasons for the low ranking, starting with the omnipresent lack of funds, three central problems are identified: (i) politicisation of the academic establishment; (ii) lack of recognition of either meritorious or dismal performance with suitable reward or penalty, for the lowest in the ranks to the highest and (iii) lost opportunities in catching-up through providing incentives for non-resident Indian academics to return to India and contribute to improving its scientific capabilities.

Several shortcomings were revealed by the heads of public laboratories in research conducted at the start of the millennium (Ramani and Visalakshi 2001). These include lack of incentives for excelling in science, hierarchy-driven research institutions, university departments plagued by nepotism and cronyism, political appointment of heads of public academic institutions such that institutional strategy fluctuates according to the predilections of the state and central government ministers, the existence of a quota system for admission to many universities and for positions (50–85 per cent) in public research centres for socially disadvantaged communities, low accountability for poor performance, and lack of recognition for good performance. Ramani and Visalakshi (2001) suggest that currently the *incentive* system in public research is so distorted that the standards of Indian science are maintained by dedicated individuals in isolated pockets in different laboratories of the country, despite the system and not because it.

However, recent evidence from public R&D laboratories in India provides a more nuanced picture (Choudhury and Khanna 2009). Policy changes have been introduced to improve incentives for scientific research and the Department of Science and Technology has attempted to promote science education and careers in science[6]. Choudhury and Khanna (2009) found that 42 Indian state-owned laboratories over 1993–2006 significantly improved performance, as they started with a base of negligible USA patents, but were granted more patents than all domestic private firms combined. Moreover,

[6] As expressed by the Head of the Department of Science and Technology to the author in 2009.

patent licensing revenue increased from 3 per cent to 15 per cent as a fraction of government budget, without significantly affecting publication quality and quantity following intensive policy change and leadership change at labs. There was a sharp increase in Indian patents filed by CSIR starting 1996, a trend that flattened out around year 2000. In contrast, foreign patents filed and granted continued to increase consistently and the CSIR had a disproportionate share compared with USA patents filed by Indian entities and competed with Indian private firms for the highest share of patents filed in India. Thus, these labs were able to leverage their knowledge assets to effect a turnaround. Choudhury and Khanna (2009, 15) argue that the finding that national labs are able to increase revenue from licensing patents to multinationals is contrary to the view that emerging market of R&D entities ('the South') is capable of costless imitation alone and not innovation.

Nevertheless, the challenge remains of inculcating a culture conducive to research more generally in the vast number of Indian universities that are isolated from the national laboratories. Moreover, 'pure basic research' output has generally been regarded in academia as being more intellectual and yielding higher dividends to both researcher and the institution concerned. Mohan and Gupta (2003) argue that Indian researchers have tended to focus on addressing problems that are important internationally, but divorced from the socio-economic needs of a developing country. Since research was de-linked from the social context and production processes, the development of indigenous technologies on a large scale was inhibited. Consequently, Indian science has made little contribution to solving the development problems of the country. Even in the area of teaching, Arun Kumar (2008) notes that the present engineering system in India is a distorted legacy of the British colonial top-down approach, wherein instructional material and teaching methods are centrally decided and implemented. Thus, teaching methods are theoretical with little practical relevance, the course content is obsolete; and assignments, project work, and industrial training programmes are mainly for obtaining credentials rather than skills and knowledge.

Also, until recently, Indian universities were not aware of the importance of intellectual property and lacked the resources to manage them (Ganguli 2005).[7] Funds for filing patents from TIFAC and sponsored projects were not

[7] This point was earlier noted in 2000 by Dr Mashelkar, former head of the CSIR, at a conference at XLRI, Jamshedpur.

sufficient for filing patent applications abroad (Ganguli 2005). IPR policy in Indian academic institutions is still evolving, and knowledge about patenting is currently being disseminated. A law similar to the Bayh-Dole Act of 1980 in the USA, specifying the ownership of the inventions that arise from publicly funded R&D for all academic units is currently being debated in the Indian Parliament.

While collaboration with industry and efforts to commercialize new technologies are being encouraged, particularly in the IITs, such efforts need to be institutionalised across research laboratories. Support for research from industry, mostly in the form of consultancies, does not usually involve large-scale projects (Ganguli 2005), although there are some exceptions in biotechnology and other industries.

As far as skills and capabilities are concerned, an employer survey to determine the employability and skill gaps of Indian engineers, conducted by the World Bank (Blom and Saeki 2011), suggests that there is substantial dissatisfaction with the quality of graduates. Besides, improvement is needed in core employability skills and communication skills of graduates (the so-called 'soft skills' including team working skills and willingness to learn). Graduates also seem to lack higher order thinking skills (analysing, evaluating, creating) and the ability to attempt complex tasks such as applying appropriate tools to solve a problem, analyse and interpret. This raises the question whether the Indian engineering education system overly trains students to memorise science and engineering knowledge without adequately emphasising the applicability, analysis, and out-of-the-box thinking that employers look for (Blom and Saeki 2011). Similarly, Reid and Ramani (2012) found in their interviews with biotech firms that while students are good at the theoretical level, they are not skilled either in terms of practical knowledge or in terms of social capabilities to work well in teams. Therefore, it takes a year or two to train newly hired graduates before they can be fully integrated.

In summary, despite progress, challenges of coordinating across research institutions and leveraging and disseminating knowledge to achieve global technology leadership remain. Therefore, improving the Indian system of innovation in terms of transforming scientific knowledge into technology with commercial potential requires that institutions become more meritocratic and collaborative and less political, and foster higher order thinking skills and the social capabilities of graduates.

Torch bearers of the future? A case study of innovation for rural markets

Although academic–industry collaborations were slow to begin, and initially yielded only few innovations, currently there is greater emphasis on accelerating innovation via such collaborations. While the consensus is that India needs to boost indigenous production of knowledge and technology for low-income communities, in the early stages, most innovations targeted mainstream markets. However, the realization that new products and services can be paired with innovations in distribution and new business models that address both urban and rural markets – as suggested by scholars like Prahalad (2004) – has led to a focus on developing technological solutions for low-income communities. To illustrate this premise, we now turn to a case study of technology dissemination via academia targeting this segment.

This case study highlights technology dissemination by the Telecommunications and Networking (TeNet) group at IIT-Madras, a leading Indian technical institute. Industry-funded projects, faculty, and BTech and PhD students were involved together with full-time project staff. While the TeNet group aimed initially to support entrepreneurship by incubating alumni start-ups, by the mid-1990s, some ventures such as Tejas Networks were started by non-alumni. Scientists applied their expertise in the telecommunications and networking areas to develop technologies and products for mass use in urban and rural settings. Subsequently, as telecommunications infrastructure improved and the subscriber base increased even in villages, by 2005, motivation shifted to developing applications to solve socio-economic problems in surrounding rural areas. These solutions were also disseminated to similar regions elsewhere in India. By 2007, the group had associated 16 faculty members and incubated 15 companies with about 1500 engineers. The case study is drawn from interviews conducted by the author during 2008 and 2009.

The expertise and focus of the TeNet group were aligned with the government's mission of accelerating innovation in telecommunications and deploying this for rural development. The group's mission of scaling telephone connectivity from 7 to 100 million by developing a world-class but affordable access network (AN) was reflected in the selection of start-ups for incubation. Several R&D companies were incubated, largely but not solely, by IIT-Madras alumni, including Midas Communications, Banyan Networks, Vembu Systems (AdventNet Systems), Nilgiri Networks, and N-Logue Communications. The group partnered with these companies to develop a wide range of access technologies. The products designed by the group include Fibre Access Network (optiMA), Wireless in Local Loop System [corDECT (Digital Enhanced Cordless Telecommunications)], Direct Internet Access System (DIAS), CygNet Network Management System, and a host of protocol stacks. All were aimed at significant reduction in access cost and simultaneous large-scale Internet use. Products were designed to allow subscriber access via multiple media and separate the telecom and Internet traffic at the access unit, and often spanned multiple technologies and projects.

The first company incubated in 1994, Midas Communications, had about 700 employees and US$50 million in annual sales by 2006. The dream of 100 million telephones was achieved by 2005 and sales of its flagship product, corDECT, the locally developed wireless local loop systems, reached about Rs 1000 crore (US$235 million) in India and about 20 other countries. The group also developed RFICs to be deployed with the global Digital Enhanced Cordless Telecommunications (DECT) standard for short-range cordless communications in Midas' flagship product. Using standards increased the likelihood of gaining economies of scale and wider acceptance of the product as well as use in other applications.

Until the mid-1980s, a local loop or an AN in India consisted of a pair of copper wires connecting the subscribers at home or office to the nearest exchange. The local loop in urban areas was expensive, and difficult and time-consuming to deploy. It was typically 6–8 km and used 0.5–0.6 mm copper gauge that was difficult to maintain and the source of almost 85 per cent of faults. With this technology, the per-line local loop cost rose to over 85 per cent of the total cost of installing the telecom network. However, new technologies introduced from the mid-1980s changed the AN and lowered the cost by reducing the length of the loop to 3–4 km and that of the copper gauge to 0.4 mm, but the high cost of copper continued to make this solution expensive (see Fig. 2.3).

In mid-1990s, international standardisation of access signalling protocols enabled the AN to be independently deployed using innovative technologies like fibre, wireless, DSL on copper, hybrid fibre-coaxial cable or even power lines. Since access dominated cost and was most fault-prone and time-consuming to deploy, availability of new access solutions became the key to expanding the telecom network, especially in developing countries. Moreover, telecommunication networks needed to focus on providing both telephone and Internet services.

The first access product developed by the TeNet Group, corDECT WLL (Fig. 2.4a), was a subscriber unit called Wallset with Internet Port (WS-IP), located at subscribers' premises, with a standard telephone interface and a serial port to connect a PC without a telephone modem.

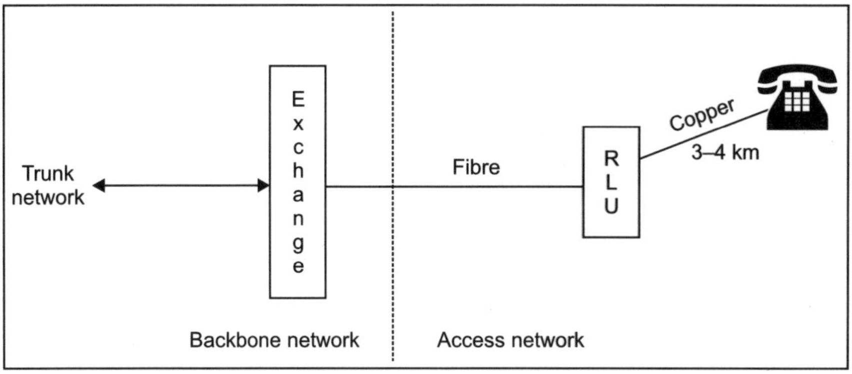

Fig. 2.3: Fundamentals of telecommunication networks
Source: Ashok Jhunjhunwala, Devendra Jalihal, and K.Giridhar, TeNet Group, IIT-Madras

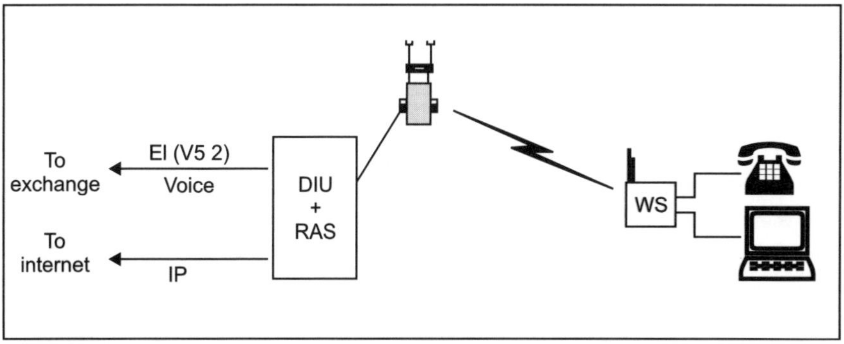

Fig. 2.4a: corDECT Wireless Local Loop
Source: Jhunjhunwala (2000)

The system provides Internet and simultaneous telephone conversation. The WS-IP is connected to a base station (CBS) on wireless and the base station is connected to the DECT Interface Unit (DIU) and a Remote Access Switch (RAS). A typical Access Unit serves 200 to 1000 subscribers. Figs 2.4b and 2.4c show the pictures WS-IP, CBS, and the DIU. While the Wallset-base station wireless link distance could be as large as 10 km, it was generally 1–2 km in urban areas. For sparsely connected rural areas, a Relay Base Station (RBS) could be installed on a tower up to 25 km away from a central base station and in turn could serve subscribers in a 10 km radius. The cost of an installed corDECT line in India varied from Rs 14,000 (US$325) to Rs 18,000 (US$425), depending upon the deployment scenario. A multiple subscriber unit, referred to as Multi-Wallset (MWS), was designed to serve four independent subscribers in the same building, reducing the per-line cost of corDECT by half.

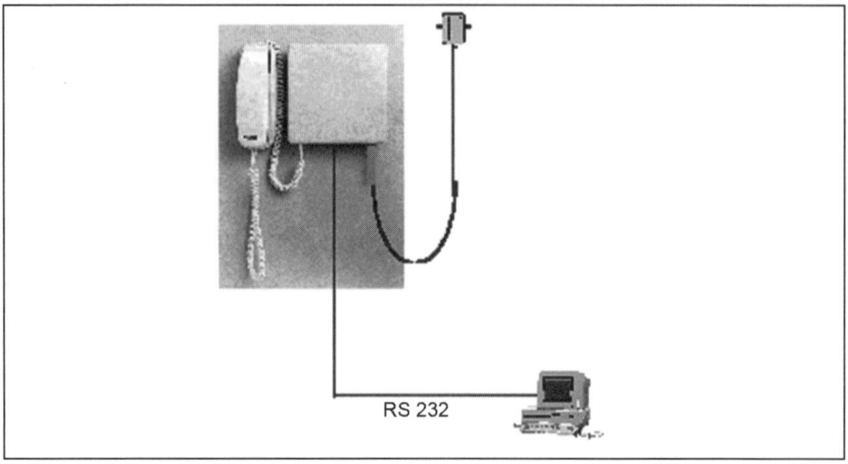

Fig. 2.4b: WS-IP: Wallset with Internet Port
Source: Jhunjhunwala (2000)

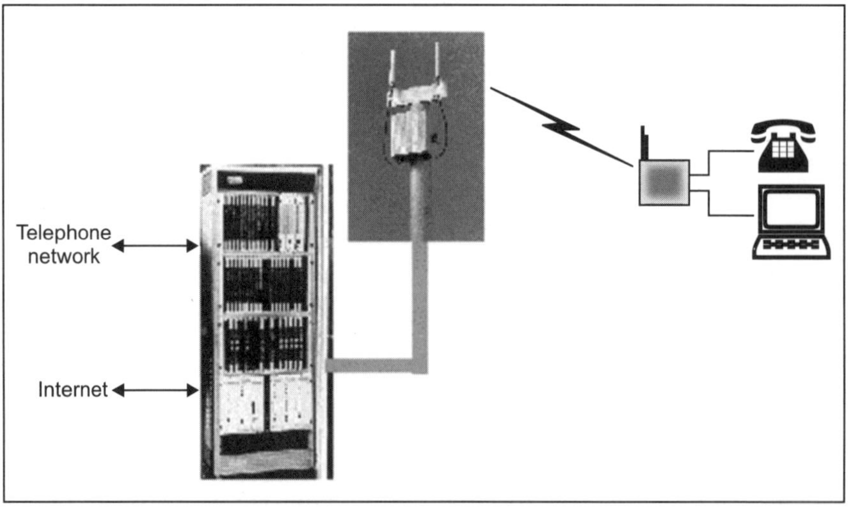

Fig. 2.4c: DIU, Base Station, and WS-IP
Source: Jhunjhunwala (2000)

Innovation for rural development

Next, the TeNet group focused on incubating other ventures and developing technology-based products and services to provide rural solutions. For example, N-Logue Communication was launched to provide Internet and voice services in every village using the corDECT wireless access system as only one-third of India's 600,000 villages had a telephone connection. By establishing links between the university and rural areas through companies like N-Logue Communications, the TeNet group facilitated the adoption of new digital technologies and helped transform these areas. The communication infrastructure provided by N-Logue enabled local entrepreneurs to run Internet kiosks in the villages and share revenues with the telecommunications infrastructure provider.

Additionally, N-Logue's IT infrastructure facilitated livelihood generation. The village Internet kiosk helped aggregate demand for education, healthcare, and finance services. Further, kiosk operators could act as facilitators for skill development and enable urban buyers to source services from remote villages. Thus, kiosk operators could earn a living even in small villages. The kiosk was available at a cost of about Rs 53,000 (US$1200) and included a multimedia PC with local language software, back-up power supply, training, and six months of Internet connection. Financial support was available from N-Logue and the State Bank of India. Typically, the entrepreneur would obtain a bank loan to set up operations. Availability of local language software allowed users to create local content. Moreover, to prevent software and content piracy, a web-based system ensured that only legitimate account holders could access these services. Thus, the technologies spawned by the TeNet group via N-Logue and other firms helped form linkages with the rural population. Lastly, IIT-Madras formalised the rural venture incubation process, and under the aegis of its Rural Technology Business Incubator (RTBI), various stages of innovation and venture creation were clearly identified and demarcated.

Innovation via technological recombination

Innovations for rural markets were created by recombining elements of frontier technologies to address local needs. For example, although new wireless network access technologies were developed to enable connectivity via mobile phones, existing cellular networks did not extend to most rural areas. To address this problem, TeNet scientists developed a micro GSM[8] base station to provide low-cost cellular coverage in villages. The base station uses the wireless system in the village kiosk to connect to the national telephone network. It provides coverage for GSM cell phones in a 2-km radius and allows villagers to make

[8] Global System for Mobile Communications (originally from *Groupe Spécial Mobile*).

local telephone calls using a standard GSM handset at one-fourth the cost per minute; villagers could also make international calls. Using a micro GSM base station brought the cost per user down from Rs 350 per month to about Rs 100 per month. Moreover, the availability and low cost of second-hand GSM handsets made cell phones more affordable.

Thus, technologies available in industrialised countries were reconfigured for India. Cellular base stations are very expensive due to the high cost of the tower; since the electronics are cheaper than the cost of the tower, there is little incentive to reduce the cost of electronics drastically. Consequently, large base stations use expensive electronics with high power consumption. In redesigning the electronics to make it more affordable, scientists made the base station out of handset electronics, thus reducing the power usage and cost substantially.

Diffusion of innovations and entrepreneurship via new ventures

Using the Internet kiosk as a platform, the operator of the kiosk could generate enough money to service the bank loan, pay rent and electricity, and also obtain additional income by providing a range of services for a small fee. Examples of services include tutorials for rural areas[9] where failure rates of school children are high as they lack access to after-school coaching in English and mathematics. Internet-based tutorials were introduced via the kiosks and almost 100 per cent of the children taking the tutorials passed; a few did not pass because kiosk operators crowded each computer with 20 children, allowing them little time to learn. By early 2007, kiosks had been established in about 30 districts in three states.

RTBI incubated several other start-ups. Samviteducation provided on-line education, while DesiCrew, a rural business process outsourcing venture, focused on skill-based training in information technology. ROPE, a home décor and lifestyle start-up, worked with villagers to produce artisanal products for the home, thereby increasing the latter's livelihood generation capacity. In healthcare, the kiosk could also provide remote medical services via the Internet through a telemedicine kit available for Rs 12,500 (US$300) developed by yet another venture, Neurosynaptic Communications Pvt Ltd in collaboration with the TeNet group. The telemedicine kit enables monitoring of temperature, blood pressure, pulse rate, and other vital signs under the control of a remotely located doctor who can see and communicate with the patient via video-conferencing. Similarly, in financial services, the Gramateller, a low-cost ATM machine, was

[9] Developed by an IIT-Madras scientist in collaboration with school teachers.

developed by Vortex Engineering Private Limited in collaboration with the TeNet Group at IIT-Madras at a fraction (1/16th) of the cost of a traditional ATM machine (Rs 50,000 versus about Rs 800,000). The low-cost machine was deployed in villages by major Indian banks like ICICI; it differed from traditional ATM machines in requiring biometric authentication for security (to dispense with secret user-identification numbers, since secrecy is not part of village culture) and in dispensing only one denomination of notes.

Similarly, TeNet/RTBI developed a low-cost Weather Monitoring Kit (available for Rs 20,000) to enable insurance companies obtain accurate weather-data. The kit helps collect temperature, humidity, pressure, wind speed and direction, and rainfall data that is transmitted via the Internet or a cell phone network. Financing was obtained to develop and implement this project in 50–100 villages. These new products and services targeted at rural populations helped disseminate both technology and entrepreneurship by attracting more villagers to become entrepreneurs and increase the scale of technology-based ventures in rural regions.

Village business process outsourcing ventures generated income by taking advantage of lower cost resources for local language content and data entry, including entry of drawings using software such as AutoCAD. The IT infrastructure also allowed distributed, outsourced craft production as design specifications could be sent as images over the Internet to villagers to ensure meeting quality standards. TeNet/RTBI anticipated that outsourcing such work would create employment opportunities for women and others unable to migrate to cities, further accelerating rural entrepreneurship. While these ventures are currently being piloted, scaling up to large volumes is still a challenge.

TeNet/RTBI also anticipated that deployment of low-cost communication technologies would help stabilise links between the university/incubator (the technology source) and the village, thus enabling the system to be adaptive and function at the 'edge of chaos' (Kauffman 1993: Carroll and Burton 2000).

From locally to globally competitive firms

As noted earlier, the TeNet group did not limit itself to innovating for the rural sector. For urban areas, the group developed similar low-cost technologies such as a cable wireless system to provide broadband Internet connection over cable, which became available on a small scale in 2006. This technology addressed the problem of providing Internet access in homes with cable TV connections (about 60 million homes) but lacking the quality required for

Internet connections. Likewise, the RFICs developed for deployment in the corDECT system were subsequently licensed to Analog Devices, indicating that technology developed by the TeNet group and commercialized by Midas Communications was globally competitive.

At the macro level, the TeNet group aspired to disseminate its technologies globally via their incubated ventures, and increase their size and scale by diffusing entrepreneurship. The TeNet/RTBI also aspired to facilitate the growth of billion dollar telecommunication product companies in India. One way to achieve this was to consolidate several of the incubated telecom software companies and focus on the larger ones with products associated with telecom access throughout the value chain – from the exchange to the subscriber. To this end, the group focused on harnessing its technical capabilities in the wireless area, and industrial expertise gleaned from incubating new ventures, to create a public-private partnership to develop the next generation of wireless standards for products targeted at the needs of the Indian population. Thus, the emergence of new firms that are initially locally competitive, and subsequently, globally competitive suggests the importance of linkages and interactions at both micro and macro levels in enabling adaptation and innovation in a complex system.

Conclusion

This chapter outlined the evolution of the Indian system of innovation in so far as the S&T institutions are concerned and then illustrated what universities could do through a case study of technological innovation and new venture incubation by IIT-Madras to promote inclusive development.

The evolution of the Indian S&T system and evidence from the case study of university–industry linkages support the conceptual framework developed in the introduction by illustrating how the S&T infrastructure in India has so greatly evolved in terms of complexity. As the five-year plans progressed, both the number of departments and the scope of the plans increased. The focus on adoption of new technology and import substitution dominated R&D efforts in the pre-liberalisation period and led to limited interactions with both foreign technology suppliers and amongst Indian organisations; however, post liberalisation in 1991, the desire to become globally competitive and to learn led to an increase in the number of interactions and collaborative exchanges with global multinational partners. Academic–industry partnerships were also initiated within India. Heightened interactions, in turn, led to an increase in the complexity of the Indian system of innovation as Indian players (both public and private) discovered more about the nature of global competition and strategic sunrise industries like IT, biotechnology, and nanotechnology and directed their attention and investments accordingly. Recognition of the need for entrepreneurship to spur development motivated various public sector organisations, including the Department of Biotechnology, to establish schemes to assist entrepreneurial efforts, again enhancing the complexity of the R&D system.

The case study presented in this chapter exemplifies how new venture incubation via academic–industry partnerships can help diffuse technological innovations and entrepreneurship and create new markets to transform the rural ecosystem and enable sustainable growth. The case study reveals that technologies developed by scientists can be embedded in a rural system linked to the university via its incubated ventures. Technological solutions can be developed for problems faced by the rural population and the rural system can be reconfigured by embedding these solutions. Implementing these solutions required innovation via technological recombination and entrepreneurship dissemination. Various goals such as providing low-cost technologies, income generation opportunities, improved education, and healthcare at the village or micro level and rapid scaling of new ventures were successfully achieved by accelerating entrepreneurship. In addition, solutions for rural markets also resulted in new, low-cost technological products that could be globally competitive, particularly in the telecommunications industry. Moreover, university–industry interactions with rural areas enabled the transformation of the latter into new markets for innovative technology-based products and helped to integrate them into the market economy. Finally, university–industry linkages were also used to scale incubated start-up ventures via consolidation to foster the emergence of large domestic firms capable of competing in global markets.

In conclusion, the case study demonstrates that universities can go beyond just producing qualified personnel to promote inclusive development. Indeed, rural development can be promoted via market creation by fostering (i) linkages between the university and industry, both sources of expertise, and the rural regions; (ii) innovation through recombination of existing technologies to address the needs of the rural community; and (iii) dissemination of entrepreneurship and technology via new ventures. Connections between the rural community and the university allow feedback from users and help to make technological solutions more appropriate. Since rapid scaling can be accomplished by disseminating entrepreneurship, it is necessary to ensure standardisation of quality and training to prevent bottlenecks arising from poor capabilities.

Future practice can also be informed by the insights presented above. This article provides a template for developing new emerging markets via activating university–industry linkages. The template can be used by universities in other developing and industrialised countries to stimulate technological innovation and entrepreneurship.

References

Agarwal, Pawan. 2007. 'Higher Education in India: Growth, Concerns and Change Agenda.' *Higher Education Quarterly* 61 (2): 197–207.

Arnold, David. 2004. *Science, Technology and Medicine in Colonial India.* Cambridge, UK: Cambridge University Press.

Arun Kumar, D. 2008. 'Impact of Globalization on Core Engineering Education in India.' *Perspectives on Global Development and Technology* 7 (1): 55–68.

Blom, Andreas and Hiroshi Saeki. 2011. 'Employability and Skill Set of Newly Graduated Engineers in India.' Policy Research Working Paper # 5640. Washington, D.C.: World Bank. Accessed: 27 April 2012; http://www.wds.worldbank.org/external/default/WDSContentServer/IW3P/IB/2011/04/25/000158349_20110425112950/Rendered/PDF/WPS5640.pdf.

Carroll, Tim and Richard M. Burton. 2000. 'Organizations and Complexity: Searching for the Edge of Chaos.' *Computational and Mathematical Organization Theory* 6 (4): 319–37.

Chatterjea A. and S. A. Moulik. 2012. 'Academic Research.' In *The Oxford Companion to Economics in India*, edited by Kaushik Basu, 1–7. New Delhi: Oxford University Press.

Chatterjea, A. and S. P. Moulik. 2006. 'Doctoral Education and Academic Research (in India).' CHERI Working Paper #87. New York: Cornell University, ILR School. Accessed: June 2012; http://digitalcommons.ilr.cornell.edu/workingpapers/65/.

Choudhury, Prithviraj and Tarun Khanna. 2009. 'State Owned Entity Reform in Absence of Privatization: Reforming Indian National Laboratories and Role of Leadership.' Working Paper 10-006. Boston: Harvard Business School. Accessed: June 2012; http://www.hbs.edu/faculty/Publication%20Files/10-006_1051fcd3-70f6-4a5c-b388-3621d56f2375.pdf.

CSIR (Council of Scientific and Industrial Research). 2006. *Report of the Working Group on CSIR: Eleventh Five-Year Plan (2007–2012).* New Delhi: Government of India.

Datta, S. and Mohammed Saad. 2011. 'University and Innovation Systems: The Case of India.' *Science and Public Policy* 38 (1): 7–17.

Department of Science and Technology. 1997. *Research and Development Statistics, 1994–1995*, p. 75. New Delhi: Government of India

———. 2009. *Research and Development Statistics, 2007–2008.* New Delhi: Government of India.

Desai, Ashok, ed. 1988. *Technology Absorption in Indian Industry.* New Delhi: Wiley Eastern.

Dharampal. 1971. *Indian Science and Technology in the Eighteenth Century: Some Contemporary European Accounts.* Delhi: Impex India. (Reprint. 1983. Hyderabad: Academy of Gandhian Studies).

DSIR (Department of Scientific and Industrial Research). 2006. *Report of the Working Group for Eleventh Five-Year Plan (2007–2012).'* Volume II. New Delhi: DSIR, Government of India.

EXIM Bank. 2006. 'Strengthening R&D Capabilities in India.' Occasional Paper No. 115. India: Quest Publications.

Ganguli, Prabuddha. 2005. 'Industry-Academic Interaction in Technology Transfer and IPR: The Indian Scene – An Overview.' Report written for World Intellectual Property Organization. Mumbai.

Ghose, T. K. and V. S. Bisaria. 2000. 'Development of Biotechnology in India.' *Advances in Biochemical Engineering/Biotechnology* 69 : 87–124.

GOI (Government of India). 2010. *India as a Global Leader in Science.* New Delhi: Department of Science and Technology.

———. 2009. *Research and Development Statistics 2007–08.* New Delhi: Department of Science and Technology.

———. 2007. *Nano Mission.* New Delhi: Department of Science and Technology. Accessed: June 2012; http://nanomission.gov.in/.

———. 1956. *Second Five-Year Plan.* New Delhi: Planning Commission. Accessed: 23 January 2012; http://planningcommission.nic.in/plans/planrel/fiveyr/default.html.

———. 1951. *First Five-Year Plan.* New Delhi: Planning Commission. Accessed: 23 January 2012; http://planningcommission.nic.in/plans/planrel/fiveyr/default.html.

Gupta, S. 2003. 'Introduction: Scientific Research in Indian Universities: Status, Challenges and Prospects.' In *Scientific Research in Indian Universities*, edited by Sanjay Gupta, selections from University News-12, vii–xiv. New Delhi: Association of Indian Universities.

Holland, John H. 1998. *Emergence: From Chaos to Order.* New York: Basic Books.

Jhunjhunwala, Ashok. 2000. 'Unleashing Telecom and Internet in India.' Presented at India Telecom Conference at the Asia/Pacific Research Center, Stanford University, November. TeNet Group, IIT Madras. Accessed: 17 February 2012; http://www.tenet.res.in/Papers/unleash.html.

Kauffman, Stuart A. 1993. *The Origins of Order: Self Organization and Selection in Evolution.* New York: Oxford University Press.

Kumar, Deepak. 1995. *Science and the Raj.* New Delhi: Oxford University Press.

Lall, Sanjay. 1987. *Learning to Industrialize: The Acquisition of Technological Capability by India.* Basingstoke, UK: MacMillan Press.

Lundvall, B. Å. 2007. 'National Innovation Systems—Analytical Concept and Development Tool.' *Industry and Innovation* 14 (1): 95–119.

———. ed. 1992. *National Systems of Innovation: Towards a Theory of Innovation and Interactive Learning.* London: Pinter.

MHRD (Ministry of Human Resource Development). 2012. *Report to the People on Education 2010-2011.* New Delhi: MHRD, Government of India. Accessed: June 2012; http://mhrd.gov.in/sites/upload_files/mhrd/files/RPE-2010-11.pdf.

Mohan, S. and B. M. Gupta. 2003. 'Relevance and Quality of Indian Science.' In *Scientific Research in Indian Universities*, edited by Sanjay Gupta, selections from University News-12, 13–20. New Delhi: Association of Indian Universities.

Mowery, David C. and Bhaven N. Sampat. 2005. 'The Bayh-Dole Act of 1980 and University-Industry Technology Transfer: A Model for Other OECD Governments?' *Journal of Technology Transfer* 30 (1/2): 115–27.

Mowery, David C., Richard R. Nelson, Bhaven N. Sampat, and Arvids A. Ziedonis. 2001. 'The Growth of Patenting and Licensing by U.S. Universities: An Assessment of the Effects of the Bayh–Dole Act of 1980.' *Research Policy* 30 (1): 99–119.

NASSCOM-McKinsey Report. 2005. *Extending India's Leadership of the Global IT and BPO Industries.* New Delhi: NASSCOM.

Nelson R., ed. 1993. *National Innovation Systems: A Comparative Analysis.* New York: Oxford University Press.

Nezu, Risaburo. 2005. 'Technology Transfer, Intellectual Property and Effective University-Industry Partnerships: The Experience of China, India, Japan, Philippines, The Republic of Korea, Singapore and Thailand.' *World Intellectual Property Organization.* Accessed: June 2012; http://www.wipo.int/export/sites/www/freepublications/en/intproperty/928/wipo_pub_928.pdf.

OECD. 2006. *Information Technology Outlook 2006.* Paris: OECD.

Owen-Smith, Jason, Masimo Riccaboni, Fabio Pammolli, and Walter W. Powell. 2002. 'A Comparison of U.S. and European University-Industry Relations in the Life Sciences.' *Management Science* 48 (1): 24–43.

Patel, Ila. 1996. 'India, International Review of Education/Internationale Zeitschrift für Erziehungswissenschaft.' *Revue Internationale de l'Education* 42 (1/3): 75–96.

Prahalad, C. K. 2004. *Fortune at the Bottom of the Pyramid: Eradicating Poverty through Profits.* Philadelphia, PA: Wharton School Publishing.

Rahman, A., ed. 1982. *Science and Technology in Medieval India: A Bibliography of Source Materials in Sanskrit, Arabic and Persian.* New Delhi: Indian National Science Academy.

Ramani, S.V. and A. Maria. 2005. 'TRIPS and its Possible Impact on the Biotech Based Segment of the Indian Pharmaceutical Industry.' *Economic and Political Weekly of India* 40 (7): 675–83.

Ramani, S.V. and Visalakshi. 2001. 'The Role of Resources and Incentives in the Integration of Biotechnology in Developing Countries: An Indian Case Study.' *International Journal of Biotechnology* 2 (4): 297–312.

Reid, S. and S. V. Ramani. 2012. 'The Harnessing of Biotechnology in India: Which Roads to Travel?' *Technological Forecasting and Social Change* 79 (4): 648–64.

Sangwan, S. 1991. *Science, Technology and Colonisation: An Indian Experience 1757–1857*. Delhi: Anamika Prakashan.

Sarkar, Benoy Kumar. 1937. *India in Exact Science Old and New: Presidential Address at the Positive Sciences Section of the Second Indian Cultural Conference, Calcutta* (Held under the Auspices of the Indian Research Institute, Calcutta). Calcutta: Indian Research Institute.

Saxena, R. P. 1990. 'Governance of Indian Universities: From Decay to Dynamism?' *Higher Education* 20 (1): 91–111.

Sayeda, B., C. Rajendran, and P. S. Lokachari. 2010. 'An Empirical Study of Total Quality Management in Engineering Educational Institutions of India.' *Benchmarking* 17 (5): 728–67.

Shukla, R. K. 2005. *India Science Report*. New Delhi: National Council of Applied Economic Research.

Simon, H. A. 1965. 'The Architecture of Complexity.' *General Systems* 10: 63–73.

Surie, G. 2008. *Knowledge, Organizational Evolution and Market Creation: The Globalization of Indian Firms from Steel to Software*. Northampton, MA: Edward Elgar.

Surie, Gita. 1996. 'The Creation of Organizational Capabilities through International Transfers of Technology.' PhD dissertation. Pennsylvania: The Wharton School, University of Pennsylvania.

UNCTAD. 2011. *World Investment Report: Non-Equity Modes of International Production and Development*. Accessed: 14 February 2012; http://www.unctad-docs.org/files/UNCTAD-WIR2011-Full-en.pdf.

Visalakshi, S. and G. D. Sandhya. 2000. 'R&D in Pharma Industry in Context of Biotech Commercialization.' *Economic and Political Weekly* 35 (4): 4223–233.

Woolgar, Lee. 2007. 'New Institutional Policies for University-Industry Links in Japan.' *Research Policy* 36 (8): 1261–274.

CHAPTER 3

Demand and Innovation

Paths to Inclusive Development

*Smita Srinivas**

Introduction

The world is so full of technological innovations, and countries such as India have so many of them that we might be tempted to forget the genesis and effect of such innovations on 'development.' What kinds of inclusive societies can we plan for through such innovations? While most of the chapters in this book are concerned with knowledge and technology capability accumulation in different industrial sectors in India, this chapter starts at the other end, with questions: Whom is it for? How does it permeate our societies? Will it be enough? What analytical frames may help us? This chapter analyses the different ways in which demand and needs are connected: demand for innovation is often assumed to exist, and innovative societies are assumed to satisfy their internal needs, both of which may be overly strong assumptions to make. It shows how examining the process of development from the point of view of 'demand' has been a somewhat neglected pursuit in innovation studies and the economics of technical change. Examples drawn from cases in India, with a brief reference to some international ones, have been used to illustrate some categories of demand. This chapter is not exclusively on India but is intended to provide fodder for reflection on how the gamut of policies promoting knowledge and new technology generation in a country like India can better address the needs and demands of its citizens.

* I thank Shyama Ramani for the invitation to submit a chapter and for her very constructive editorial commentary, which has helped this chapter take shape. Several of the themes here on markets and the political economy of demand and supply are more completely addressed in the book: Srinivas, Smita. *Market Menagerie: Health and Development in Late Industrial States* (Palo Alto, CA: Stanford University Press, 2012). I thank Oyebanke Oyeyinka for research assistance on electrification.

Though it is self-evident that effective demand is only one among several possible factors that shape innovation, it is important to debate it more fully (Edquist and Hommen 1999). It is indeed true that several technological innovations have arisen from alternate routes, sometimes more supply driven and sometimes because of more resources having been explicitly earmarked for such efforts: for instance, the atom bomb, the automobile, and (one could debate this) even the steam engine. Nevertheless, to varying degrees, the imperative to create demand for the innovation is a pivotal element of innovation itself: we see this in the case of automobiles, electrical systems, certain vaccines, and consumer electronics.

The national system of innovation (NSI) and other Schumpeterian and neo-Schumpeterian innovation frameworks have strongly contributed to our understanding of how supply-side institutions have created sectoral advances and technological investments, and how to see the entrepreneurial function within innovation. Nevertheless, they fall short when explaining the role of demand in innovation generation and diffusion 'for' development, the latter has been primarily treated as an undifferentiated 'pull' for innovation supply. Studies such as those of Lotz (1993) on medical instruments remind us of the conditions under which actors identify and create effective demand for hearing aids. The advances of science too often play an important role in determining the timing of supply and lateral innovations budding off a primary innovation. It is not that demand is absent in NSI scholarship, however. Even in those early strands of such scholarship, we see the importance of social cohesion, welfare states, and demand (e.g., Edquist and Lundvall 1993). Richard Nelson in his earlier work, *The Moon and the Ghetto* (Nelson 1977), asks how it is that we can have the capacities to establish missions to the moon but are unable to deal with our social ghettos. Therefore, at least to an extent, the burden for the study of innovation is to establish the social issues that we can attend to with increasing technological sophistication. These concerns certainly move beyond the realm of innovation study to those on citizenship rights and other forms of justice, but innovation study itself has a language for understanding demand that may get us nearer to these goals.

Demand, both as a concept and as a quantity, has an organisational and institutional – more systemic – aspect. Much of the present research on innovation tends to focus on the effects of demand on individual firms (the quantity, size of the market), neglecting the examination of its larger role, even though considerable evidence exists that demand also shapes the broader, shared environment of industrial sectors, regions, and collective clusters of firms (e.g., research on development blocs, French *filieres,* and studies on procurement or clustered innovations). From the perspective of development linkages in an industry of firms with some product and process differentiation, there are *collective* responses when demand rewrites the market environment in significant ways. And as we will see, in such cases demand may or may not be exclusively expressed through markets.

It is thus particularly important to understand the nature of demand in industrialising economies, where markets are especially various, complex, and uncertain while technological capabilities on the supply side are being rapidly built. While these economies may well be compared internationally on the supply side, their demand sides are diverse and sectorally distinct (Srinivas 2012). In the seminal 'catch-up' literature (which the introduction to this book discussed), the emphasis has been on building technological capabilities through satisfaction of demand, without too much probing of the choice of 'demand' to satisfy. But there have been traditions in Latin America, for example, with the Economic Commission for Latin America and the Caribbean (ECLAC) structuralists, such as Raul Prebisch, deeply concerned that industrial policies were tilted toward promoting export growth and the satisfaction of domestic demand from elites, and not the lower-income groups. This can lead, as the Latin Americans (and leaders such as Gandhi too) worried about, to deep inequalities built into production priorities, skewed toward emulating consumers of industrialised economies.

Thus, demand is homogeneous neither between nor within countries. Possibly the most important debate to have is on the types of innovations states should support, and the wider role that policy plays whenever a commodity or innovation is a necessity or quasi-necessity. At the heart of this debate is how scarcities or gaps are identified and how the resources for different technological categories are planned and shaped to begin with (Srinivas and Sutz 2008).[1] Because of cognitive and structural reasons, some firms innovate in environments where several taken-for-granted resources are scarce and are somehow able to identify resources and demand where others cannot see them. The importance of deepening our comparative social context for innovation thus cannot be over-emphasised. Although the 'catch-up' literature has its own limitations, it has successfully emphasised the importance of detailed qualitative

[1] As should be clear from the language of this chapter, the frameworks derive from those strands of 'old' institutionalism, evolutionary economics, and political economy that are focused on technical change, knowledge, and institutions. Therefore I use the terms institutionalism and political economy as synonymous for this body of scholarship although there are several schools of thought within both.

case studies around innovative products and processes. What it has made visible are the methodological limitations of more traditional economic techniques in studying innovation that do not capture its full institutional and evolutionary process. In this spirit, demand too has been done an injustice because more orthodox studies depict it primarily through econometric studies and price segmentation, making strong assumptions about the homogeneity of underlying institutions and politics. Such methodological neglect has serious implications for development because if demand realities are overly dependent on markets and price signals alone, and insulated from actual evolution of institutions, politics, policies, physical plans, and built investments, then innovation supply may be misdirected and we may miss crucial inter-linkages that matter for inclusive societies.

Indeed, as Mowery and Rosenberg (1979, 105) contend, demand may have been

> ...overextended and mis-represented, with serious possible consequences for our understanding of the innovative process and of appropriate government policy alternatives to foster innovation. Both the underlying, evolving knowledge base of science and technology, as well as the structure of market demand, play central roles in innovation in an interactive fashion, and neglect of either is bound to lead to faulty conclusions and policies.

Although demand acts as one important type of inducement, or 'focusing device' (Rosenberg 1969; Ruttan 1997), we need more clarity on market mechanisms as dynamic, evolving processes (Nelson 2005; Srinivas 2012). Therefore, the present chapter focuses on the nature of demand in shaping innovation. India presents a mixed picture in this regard. These policies need not be 'innovation' policies; they may be any policy that spurs innovation. This chapter uses several brief examples to show how strategic demand-led policies and national welfare regimes (or their neglect) can influence the nature and rate of innovations generated, and transform (or leave unsatisfied) needs into demand and create new markets. The inability to bridge demand with diverse policies can generate social dilemmas or lower innovation generation potential.

The remainder of this chapter is organised as follows: The next two sections comment on the existing theories with respect to their contribution towards a better understanding of the nexus between needs, demand, innovation generation, and diffusion. The following section briefly illustrates some of the propositions discussed with both international and Indian examples. Final section is the conclusion to the chapter.

The political economy of innovation supply, needs, and effective demand

Broadly speaking, innovations are products or processes new to a firm, a community, or a region if not new to the world market (Schmookler 1966; Nelson and Rosenberg 1993). For instance, the innovation may be an iPod (completely new to the world market), a diagnostic malaria kit (which may be either new to the firm or the nation, sometimes new to the world) or electric car (perhaps new to a region of the world, but has been created in multiple places independently). Schumpeter stretches the notion of innovation further to include five elements: new products, processes, markets, input sources, and organisation (Schumpeter 1934). 'Development,' for this chapter, is a process of connecting many of the elements to questions of inequality and needs. Studying demand instead of supply may seem like the classical chicken and egg problem: which comes first? Whatever the analogy, most studies of innovation resolutely begin with the product or process rather than its demand.

Demand has two prime effects: one to signal new markets (one of Schumpeter's sources of innovations), the other to signal the need for specific new products or processes (two more of his elements). Further, innovation is industrially, institutionally, and materially embedded. Scarcities in any of these characteristics, or organisational outliers to traditional capitalism and its firms require that new analytical approaches be devised for how innovation is conceived, created, and diffused, and that new policy and planning processes emerge (Srinivas and Sutz 2008). For example, if the process of innovation is very idiosyncratic in adopting new approaches to old problems, then innovating in highly adverse conditions, finding new uses for discarded items, or cognitively re-conceiving the problem itself may transform the original attributes of innovation beyond recognition (ibid). Similarly, innovation perceived narrowly as breakthrough in devices alone or as conducted by white coats in a laboratory misses the point entirely.

How is demand generated? Many factors are responsible. First and foremost, collective trends and policy push shape demand. For instance, social and political considerations (as defence and warfare priorities did for other industries) shape demand for health innovations. Moreover, country-specific delivery platforms and sub-national policies, for example, can structure urban political entitlements and determine the generation and diffusion of crucial technological innovations in health, water, sanitation, construction, and energy. These institutions derive from different patterns of demand and lead to different outcomes of inclusion even in single countries. In several other countries, governments can be important lead-users, setting demand conditions in wide-sweeping ways.

More specifically, demand is not a single signal of desire (yes or no) but has several stages and actors in the innovation and production process. In other words, technological innovation is invariably embedded in an *industrial* process. Demand can issue either as final demand catering to consumers or as intermediate demand or movement of commodities along the vertical chain of production by intermediate firms. Intermediate demand in India, Brazil, or South Africa can boost core sectors in information and communication technologies, pharmaceuticals, water, electricity, and automobiles. As the structuralists were keenly aware, if intermediate demand is heavily skewed towards consumer goods by elites, the intermediate demand process runs the risk of forever skewing the entire production chain. As Arocena and Sutz (2000) emphasise, 'systems' of innovation are effectively *ex-ante*. In suppliers of health innovations, for instance, the *ex-ante* ability to design institutions has historically been limited by the planning and policy apparatus of governments, which may poorly embed firms with other health institutions such as hospitals, or insurance systems.

Various recent studies on inequality and innovation for developing countries (e.g., Arocena and Sutz 2000; Sutz and Arocena 2006; Albuquerque 2007; Srinivas and Sutz 2008) lean towards explaining structural and institutional aspects of how inequality skews the process of innovation. While large inequality creates complicated, segmented demand, it also generates a response from diverse supply organisations, which the current theories of innovation and regulation describe only poorly. Demand is also discriminatory between income groups, a fact still misunderstood in the economics and management literature. We must therefore distinguish between 'need' and 'effective demand.' Need, although a human desire and a statement about wants, may remain unexpressed as demand in markets. Similarly, some sectors have entirely involuntary needs and demand generated by the human condition. Healthcare is one of these, often beyond the explanatory ability of neoclassical choice-based paradigms (Hodgson 2008).

More generally across sectors, innovation, which is often considered to be generated as a response to people's needs, technically responds only to effective demand, i.e., a willingness to pay. Innovating organisations and entrepreneurs, in principle, pursue innovation because of the projected demand and not existing needs. But frequently existing needs are mistaken for effective demand. Indeed, not a day goes by without estimates being made of the size of the Indian middle class and the market for basic consumer goods or for technological innovations in wireless, computers, or automotive sectors being computed accordingly. The Bottom of Pyramid (BoP) approach that is so popular in business schools today, goes even further to assume market participation of all low-income individuals as potential consumers or is quite optimistic about their conversion (singly and institutionally) to the ranks of those willing to pay. The BoP approach and developmental policies can tend to be blind to several power and developmental contexts (e.g., Arora and Romijn 2009). Government plans and policies play a crucial role in converting needs to demand, and in identifying needs and lack of demand. In addition, markets, unions, or families may also convert needs to demand by providing what Albert Hirschman calls 'exit, voice, or loyalty' institutions through which they can express collective desires.[2] Therefore, briefly, let us separate demand into four categories: first, the more traditional economics term 'effective demand'; second, need that is not recognized as a need; third, need recognized as a need but not as a demand; and fourth, recognized but unfulfilled demand.

On needs and effective demand

Effective demand is a statement about how participation in one market constrains ability to demand in another. While latent demand is what a consumer is willing to buy at any given price, effective demand implies a consumer is willing and *able* to buy at every price, i.e., demand is effectively supported by a will to purchase the concerned commodity or service at the price offered. Effective demand derives from neoclassical assumptions of general equilibrium conditions and price balancing between different markets. Effective demand is distinct from 'regular' (latent) demand in that individuals are constrained by participation in other markets that shapes their demand in yet others. For instance, when unemployed, a worker's participation in labour markets is constrained and, therefore, her effective demand for goods and services is lower than her latent demand. Thus, any welfare regime that complements her wage earnings and which substitutes for the risks she experiences in the labour market can boost her effective demand. This is the strategy of states, which commence large public

[2] The moral dilemmas of organisational form for this redistribution are of course by no means assured:

> Whatever is available to justly distribute changes as our productive forces develop through innovation. To put it another way, claims for just distribution of resources and/or capabilities co-evolve with technical change and innovation. In this sense, the restless moral and political struggle for social justice and equality cannot be divorced from innovation and the system of social relations in which innovative activity takes place. (Papaioannou 2011, 327)

spending as a national strategy to boost demand (historically the Keynesian response) thus generating increased consumption that in turn enhances supply and reinvigorates growth. But for the purposes of a book on inclusion in development, effective demand is only one category of demand and hides several additional layers that can exclude.

Need that is not recognized as a need

One challenge is that plans and policies do not officially recognize existing needs as needs to begin with. This lack of recognition is compounded by the very same inequalities and poor political representation that exacerbate the needs. Private firms are no particular help here: they take their cues from policy priorities or lack of market and profit incentives and sideline these needs. The science and technology (S&T) policy system – which bureaucrats and researchers may defend – may be especially delinked from the needs of citizens (the classic ivory tower). Technological innovations have a role to play not just in satisfying these needs, but some – such as in information and communication technologies – can also help to make needs more visible and traceable, and thus more recognized and categorised in policy circles as needs.

Need recognized as a need but not as a demand

Sometimes need is recognized as such in policy-making, but not as market demand by firms. Some surgical instruments may be desperately needed to treat some communities, and may indeed generate R&D attention from a public research institute, but not from private firms. A classic challenge is the many technological innovations that sit on the shelves of labs, but are never commercialized. Even when some firms see it as worthwhile to invest time on exploring possible needs, they may not pursue a product or process unless a demonstrable demand can be found or created. The State too has the discretion to convert a priority-recognized need, even in the absence of a robust market demand, into a new market by creating special incentives for firms to invest in particular types of R&D or for commercializing innovations.

History-changing national priorities can dramatically convert needs into demand or create new demand and new markets such as in wartime, when infrastructure, communications, and health innovations tend to surge. Both national and regional priority for specific types of innovative products can shape the industrial and social investment strategies of metropolitan areas. In Tokyo, social investments are now tied to both welfare goals for serving an ageing population and thus also to strategic innovative industries (especially healthcare, nanotechnologies, IT, and environmental). Similarly, Finland and some Baltic states connect social welfare goals to innovation priorities. Finland boosts demand from citizens and is investigating demand instruments through public procurement, urban planning, and other instruments (Srinivas and Wallin 2012; Lember et al. 2011; European Commission 2005).One cannot too easily extrapolate from urban settings in one part of the world to another. The point, however, is that there are several policy instruments that can be used to signal attention to some types of need in their conversion to demand institutions.

Recognized, but unfulfilled demand

Even recognized demands may not, however, translate to fulfilled demands. There are innumerable examples to illustrate this case.

An Indian example from healthcare: The Jaipur Foot

Healthcare arises from the complex, often involuntary demand generated by people falling ill, but with very uneven access to services, surgical instruments, vaccines, or drugs for 'neglected' diseases. Those late industrial states that have spurred technological capabilities in the health industry play a particularly important role in matching supply with weak demand institutions. This arises when the evolution of different institutions for production, demand, and delivery is unsynchronised (Srinivas 2012). It is especially remarkable that a nation can have such advanced technological capabilities but such problems with designing the diffusion of these benefits. India is a country with a high number of citizens with disabilities. The 'Jaipur Foot' prosthetic is an example of an Indian innovation that was at first not recognized as a need by the government in any substantial way, but eventually converted by an NGO to an 'actual' need and onward to an incipient but effective demand for which a potential market now exists, which the State could help legitimise and expand.

At the outset, an immense need existed for low-cost prosthetics in India, against a backdrop of relatively scarce primary care and emergency care services. Even when recognized as a need, government health clinics and hospitals had been able to address only a fraction of it, leaving the bulk of the need effectively ignored. The Jaipur Foot is provided with no cost considerations, by a non-profit organization the *Bhagwan Mahaveer Viklang Sahayata Samiti* (BMVSS, roughly translated as the Bhagwan Mahaveer Society for Disability Assistance), which has advocated and made visible the need for low-cost prosthetics, underscoring the challenge of first making the need visible, then converting the need into demand.[3] Beyond access services for the Jaipur Foot, the BMVSS has also invested in its own technological innovations within the materials, design and production of the Foot. Beginning from mere modest settings, through the remarkable efforts of a small team, the organisation has now more formal and international ties for R&D in a range of issues that will optimise the Jaipur Foot under Indian conditions. The BMVSS now generates a global product, very well suited for international conditions similar to India's, and for amputees that war generates across the globe. The organisation is able to now charge a nominal amount for the Foot and responds substantially to the need that exists. It is clear that states, firms, and NGOs could do much more to convert more need into demand, even if some of it is subsidised. Policy's role could be more powerful precisely in those areas where several people can neither afford the innovations at a specific 'market' price nor sometimes access them even if they could afford them. This may occur due to severe distribution and delivery problems, which are common in health, education, water, and also in prosthetics.

While the Jaipur Foot's success lies in recognizing needs and slowly building demand, a vast recognized demand remains unfulfilled. The organisation even at full supply capacity is unable to fulfil the demand for low-cost prosthetics in India and overseas. Technological innovation here is limited by absent policies explicitly geared to ensuring that all Indians needing prosthetics can be met, and the absence of other non-profit organisations and private firms that can fulfil the demand. The Jaipur Foot shows us several institutional gaps in unfulfilled demand even when policy recognizes the need. The same situation exists in several areas of paediatric health and in women's reproductive health, areas where India's record is quite alarming.

Demand and innovation in economic theory

How innovation creates demand

In Schumpeter's schema, effective demand is kept in the background because he was focused on elaborating on why and how innovation drove business cycles. He saw technological knowledge accumulation as rising on the upswing of long waves and thus propelling technology forward. According to him, innovations create the investment bunching and therefore the business cycles. The Schumpeterian engine of change is the firm that not only responds to demand pressures, but through its innovative entrepreneur, also entirely reworks

[3] www.Jaipurfoot.org

the scaffolding of the economy in 'creative destruction.' However, Schumpeter's innovative milieu is not immune to demand; rather it is more sensitive to supply. Schumpeter's vital point is that it is precisely because of supply-side effects and particular innovation clusters that major innovations are at the root of long-wave cycles, and subsequently demand and investment-cycles bunch around such technology systems and create the business cycles (and recessions) of the economy (Freeman and Perez 1988; Courvisanos 2009).

Both Keynes' and Kalecki's views on demand are similar in the macroeconomic context, but differently manifested in their microeconomic implications. In Keynes' view of economy, effective demand signals the strength of the business cycle and channels for firms to respond with particular products and processes. The macro–micro link to demand is expressed through different means for Kalecki, where imperfect competition and planned industrial change have important market effects on firms (Courvisanos 2009). In contrast to this supply-driven approach, Rothbarth (1942) synthesises Schumpeter with Kalecki by discussing the latter's 'adaptation mechanism' that converts the importance of supply mechanisms to a more focused attention on effective demand [see discussion in Courvisanos (2009)].

Both process and product innovations give rise to demand. There are several reasons for this. One is increased productivity in the form of process innovation that incorporates technical progress in new capital equipment, making the previous capital stock technologically obsolete. (Ibid, 1119). In addition, entrepreneurs themselves generate a new demand, by wishing to invest and act as first movers to utilise the new innovations themselves (Kalecki 1968, 269). Also, changes in technology create changes in relative profit rates and, therefore, stimulating investment in turn. All add to profit expectations that exceed profits generated from demand in the business cycle [White 1999, 347, cited in Courvisanos (2009)]. However, none of these would occur without some involvement from governments, which creates the political context for innovation.

How demand creates innovation

While innovations generated within firms create pressures for firms to engage in a wider political context to ensure markets for their goods and services, demand creates innovation as well. In neoclassical models, individual consumers express their preference for a specific product in the marketplace. Similarly, dissatisfaction is expressed by switching to substitutes, should they exist, by consuming less of this product and possibly more of another, or by earning more and buying better/more expensive products. Firms, of course, do take into account income inequality in anticipating demand for their outputs and in planning their price segmentation strategies. Yet, the more robust the demand and market stability for suppliers, the more willing they are to submit to cost-cutting to benefit consumers, governmental price regulation, and various technical standards to improve quality. By strategically building and orienting demand, policy can influence social goals such as affordability, quality, and quantity. State-induced procurement, for example, can have a systemic effect by inducing a set of tiered regulatory quality standards and technological learning (Edquist and Hommen 1999; European Commission 2005; Srinivas 2006). Thus, the demand-side can potentially act to increase technological learning and higher quality supply. It can also potentially influence product or process innovations.

When market size determines demand-led innovations, the hypothesis is that they, in turn, induce primarily process innovations (Schmookler 1966), but when they are user-led, product innovations take place (Myers and Marquis 1969; Fontana and Guerzoni 2008; Von Hippel 2005). As Von Hippel points out, the democratising of element also lies in the degree of user involvement in customising the process of innovation. However, domestic market size is only relevant to local firms if they can realize economies of scale in R&D or production, and thus use the domestic market to generate competitive advantages over other firms, if this market presages demand elsewhere (Porter 1990). If the home market can be strategically used either by firms and/or government policies to generate economies of scale via procurement, insurance, or other means, then national regulatory means as well as global management of technologies go beyond management of products to technological learning and eventually competitive advantages. Demand alone, however, is insufficient to realize full advantages to firms, customers, or populations in industrialising countries, and much more institutional design and planning are necessary in setting procurement and technical standards. Moreover, although market size is not automatically a competitive advantage, market creation and legitimisation to grow effective demand and market size constitute the late industrialising state's critical economic and political role (Srinivas 2012).

On mismatches between demand and innovation and the role of policy

The preceding sections make it clear that the textbook approach to supply and demand hides layers of institutional transformation of need and demand that complicate the generation of and access to innovations. Thus, we have listed several grades of need and demand and how they are politically and institutionally determined in a contextual relationship. But, even where supply capabilities are up to the task and need is recognized and signalled as demand, demand may still go unfulfilled because of political and investment misalignment between the firms and the government, and political contention about who bears the risks and costs of such investment.

Coincident with the fears of businesses are the scope of innovation policies that governments will generate and the industrial control they represent: e.g., support to incumbents versus newcomers, which may be manifested in patent protections and other appropriation policies; R&D subsidies and tax rebates; public procurement and public–private contracting; union and industrial relations; stimulus packages and spatial plans; and land deals and financial concessions (see also Courvisanos 2009). The back and forth between the government and the firms for control and consolidation results in the waves of 'techno-economic' revolutions (Freeman and Perez 1988) such as the rise of large business corporations in railroads, IT, steam, electricity, chemicals, and pharmaceuticals, and later automobiles. However, in a climate of distrust and uncertainty, the governments, firms, universities, non-profits organisations, and entrepreneurs may underinvest in the supply of innovations necessary to fulfil demand.

Given the conditions discussed above, how can the State reinforce the demand for innovations through improving access? On the demand side, there are two basic ways to do this: through lowered product cost, and through higher income. Effective costs can be lowered by better distribution, information, other indirect means, or through price controls and subsidies. Therefore, the true institutional question pertains to the State, markets, and other institutions in determining (i) the cost of the innovation; (ii) disposable income of the individual; and (iii) institutional access channels. States and policy-makers often try to increase consumption possibilities through two additional means: first, increasing the effective output per worker, i.e., higher income (wages) through higher productivity rates; second, through social policies of various types that raise the real income of the individual. Certain institutions may additionally make access to innovations possible through minimising income inequalities, or through fundamental rights, making income irrelevant altogether. These propositions are now briefly illustrated in two contexts pertinent to India, urbanisation and healthcare, with examples from various countries.

Urbanisation as a source of innovation

Cities provide density of demand and network and agglomeration effects that are attractive for innovators. However, the built environment of cities can also limit demand. Urban history reveals two factors as being very important for the generation of innovations via urbanisation. First, the nature of the cooperation and coordination between different levels of government is primordial. Second, competition and cooperation between different governing bodies, business interests, and 'civil society' determine the outcomes of demand and response. Urban planners and political economists have been deeply concerned with who gains and loses from certain types of investment. The more such investments create monopolies, the higher the rents and likelihood of firms investing. Innovators are attracted by this market structure, but assuring that those holding market power generate benefits to a wider community requires analysis on a case-by-case basis.

Urban history clearly points out the tensions between multiple roles for the State in amassing effective demand but leaves open several institutional paths for actors and organisations. Municipal governments have not always neatly aligned with regional or national governments whose plans and policies did not create coherent goals for innovation. For instance, in Japan, Germany, and the USA, large businesses, especially those who wished to exercise control over utilities and other goods, directly opposed and challenged local governments in order to usurp power within urban and regional plans (Hård and Misa 2008). Likewise, Metropolitan Tokyo worked against the directives of national government (which had been the general locus of innovation and industrial policies) and instead created alternate institutions and localised spatial plans to support innovation (Fujita 2003). The Tokyo Metropolitan Government (TMG) worked substantially to tie the industrial growth strategies – especially to revitalise the export sectors – to neighbourhoods, to streamline labour supply, its mobility, and infrastructure investments with continuously growing domestic demand and new markets, and also tied a rising demand to a new social compact for investment. This coincided with the imperative of an aging population with health needs and a degraded environmental context.

In India, too, as cities continue to grow rapidly, they set into motion a complex interaction between powerful incumbent companies controlling essential services, informal firms providing essential commodities, and the government (and others). In other words, effective demand need not be institutionally represented as recognized demand in a single manner as these different entities compete. Furthermore, the State will inevitably create its own political constituencies and power relationships through discretionary regulations. Certain states in India, such as Gujarat, have recently moved towards a basic services compact of electricity, water, and sanitation for all residents of the so-called 'slums' irrespective of tenure or other formal land security as long as they pay a tax of some kind. These initiatives are still being debated. In contrast, other States and cities continue to make basic services (despite the demand) contingent on formal property tenure in these areas. Therefore, municipal bodies may fragment or institutionalise in different ways the effective demand for certain product and process innovations. In doing so, the investments necessary to spur supply of innovations becomes divorced from the demand side while satisfying (or not satisfying) the basic needs. India's current dilemma about innovation and inclusion is exacerbated by a domestic demand for luxury goods such as automobiles from higher incomes and higher access to credit (the banks' supply push) to higher-income groups (Chandrasekhar 2009). This is the same type of trend that, critics argued, was evident in import-substituting periods, creating an elitist skew to demand.

Electrification: Tales of two cities

New York city in the USA provides a good example of how effective demand from a rising urban social class could be a crucial driver of innovation in utilities like electricity, gas, and construction. Population density and rising incomes for some, especially combined with municipal support from government and urban and regional planners for these innovations, created effective demand as a signal to innovators and financiers such as Edison (Hughes 1993a). These aligned the spatial, political, and business interests in a way that made it favourable for the rise of Pearl Street junction in New York City, and the spread of electric innovations from centres such as this. Not only were innovators and financiers such as Edison closely tied to the rise of the corporation and laboratories in Menlo Park, but such innovators were also closely tied to urban policy-makers and their support for vital physical infrastructure. At first, Edison and others worked strategically to raise interest from businesses to support the diffusion of electrification for entertainment purposes (lights to attract consumers) and only eventually to a status as basic needs for households. The growth of effective demand for electrification could not have occurred without the lead taken by urban and regional authorities for Edison's vision (even if it first appeared for what seems like frivoulous reasons) into what eventually emerged as a powerful utility [New York Edison Illuminating Company, later into Consolidated Edison (ConEd)] that continues to supply electricity to this author's apartment. In electricity, therefore, Edison's genius was creating a team with two others and a series of interlocking companies that combined mechanical inventions, management with finance, and integrated electric systems bundled in such a manner that business leaders and the municipal government fostered public support for his creations (Hughes 1993 a,b). The multipronged strategy created effective demand in diverse ways: government procurement, spatial design, tax incentives, advertising, city and business supports for lighting important areas, a demonstration effect par excellence, and other wider investment supports that transformed electric lighting structures into monopolistic utilities very rapidly. This occurred primarily through business needs first and only later transferred to low-income residents in tenements or other individual households with rising incomes (Ibid). Effective demand in this case did not arise from recognizing needs of households, but of commercial interests and Edison's clever leveraging of their power.

Now let us turn to India. Though not comparable given its distinct history, it is interesting to note the evolution of electrification in the former Mysore state and the absence of innovation generation and spatial penetration. Mysore state and selected cities within it (Mysore, erstwhile Bangalore, and Kolar gold fields – the last using most of the generated electricity) experienced the first onset of electrification approximately (and only) two decades after New York City was electrified and developed it on an ambitious scale. The state of Mysore furthered industries through planned electrification, powerfully switching gold mining from steam to electric power in Kolar, and transferred key technologies and capabilities from the General Electric Company's plant near Mysore. With engineer-visionary M. Visveshvaraiah at the helm, its planned industrial development combined significant infrastructural improvements, including electrification, but this remained an unfinished regional project. Reddy (2004, 889), for instance, has argued that the urban industrial bias of Indian S&T was itself problematic: 'By and large, post-Independence, Indian science became western-oriented and Indian technology focused on the needs of urban settlements, industry, the central government, and to much smaller extent, the state governments.' Thus, the institutional form Application of Science and Technology to Rural Areas (ASTRA) at the Indian Institute of Science (IISc) was formed in Bengaluru as a response to the electrification and other energy and material needs of rural India, a core value being that 'technological development and societal demands are dialectically related, each transforming the other' (Ibid, 890) and that social learning of needs would happen in the proximity of such communities through extension centres (Ibid, 891). Yet, despite this supposed urban bias to Indian S&T, the urban management of these technologies by city leaders and municipal authorities diverged quite dramatically between India and other industrialised contexts and the form of these experimental extension centres remained conservative.

A possible explanation lies in the fact that although Indian and particularly Mysore's electrification policies prioritised industry with a view to equity considerations as well, in the process they succeeded less robustly in spreading the innovation. Just as Carlson and Gorman (1990) point out that Edison did better when businesses were his customers and not income-differentiated end-users, Indian policy-makers and scientists generally did worse at estimating the needs of 'the poor,' which resulted in mixed effectiveness in regional power systems. Given State failures, civil society groups moved to address unmet needs. For instance, in electricity, water, and sanitation, the People's Science Movement is an example of a prominent grassroots response to the Indian nation-state's misplaced and unsuccessful effort in connecting S&T to people's needs (e.g., Kannan 1990). The response to demand may come from the state, from NGOs, or businesses, for example, but in each case, it represents a particular political economy of innovation and a window into the degree to which needs and demand are tended to. There is clearly much more analysis required to fully appreciate the political economy of demand and supply of electricity over the last century.

Urban waste management: A persisting social dilemma and a potential opportunity

Urban solid waste management is another tremendous opportunity to connect S&T strategies and innovation policies to effective demand. When policy-making ignores the dynamic nature of needs and demand in this sector, it leads to social dilemmas and potentially suppresses innovation rates from the diverse actors. These actors include citizen groups, municipal governments, private firms generating waste and those cleaning it up, informal workers, small and medium firms that are struggling to build economies of scale in waste processing, and high-tech firms looking for metals and other solid waste extraction and processing applications.

Indian cities struggle with untreated solid waste, and the municipal and labour context for waste management in India is considerably different from urban services in other countries (but not much unlike that of Latin America and Africa). Self-employed or other low-income workers who process this waste develop their own technology responses or other organisational solutions in their work. They, however, fall entirely out of the traditional Indian innovation 'system.' The politics of waste processing has high stakes. Larger-scale processors of waste would prefer economies of scale, monopolies in both the price and location of services. Several wish to establish monopolies in particular stages of the value chain such as paper re-sale, electronics waste collection, or even incineration. The private sector incentives are wittingly and unwittingly assisted by a lax policy context that generates substantial externalities through several conflicting policy layers and goals (for instance, industrial and retail growth, public health, labour welfare, urban planning). Several city municipal corporations – which are responsible for this waste – have been unable to reconcile the spatial dispersion and particular nature of waste (e.g., plastics, electronics, paper) and the technologies needed to process them with the social context of labour.

The social dilemma thus created is clearly expounded by *Chintan*, an advocacy organisation, which describes the importance of considering what is 'green' about recycling when these mismatches between the potential labour supply solution (over 1 lakh such informal workers in Delhi alone almost a decade ago) and the non-existence of viable demand institutions are so extreme.

> A study currently being undertaken by Chintan shows that this informal sector itself feels that cuts, beatings, vehicle hits on roads, unsegregated waste, toxic chemicals in the waste, and breathing ailments, including acute breathlessness through fumes and allergies, are amongst the most common occupational hazards they face. This is also being borne out by clinical tests. This is further compounded by poor living conditions, which are also the cause of diseases seen in the study: worm infestations and stomach ailments - a direct result of poor sanitation. Often they are unable to access adequate care for these ailments. Owing to their status/image, they are unwelcome in government hospitals and are under-confident about accessing even the most basic facilities. Thus, recycling takes place at the cost of human health. It no longer remains a green activity (Chintan Environmental Research and Action Group 2003, 6).

The challenge in building effective demand for waste management is the fragmentation of the existing system. Waste types (plastics) are difficult to reconcile with major waste generators (businesses); spatial planning regimes do not accommodate space for processing waste, or minimising its impact on all urban residents, but especially processors themselves; the technology appropriate for the separate tasks is rarely institutionally contracted in an appropriate fashion, for example, incineration may be popular and creates advantages for companies with big investments, rather than alternate, and perhaps more environmentally friendly options suitable for Indian cities. Urban spatial planning and zoning regulations can be outdated as land-use patterns change and create serious disincentives and penalties for self-employed waste workers and larger waste consolidators to find more efficient routes and points for segregating or processing the wastes. Mobility of the workers and their transport of the waste also set upper limits to what can be processed. Cities such as Chennai, Bengaluru, Delhi, and Kolkata have all generated some responses to converting need to demand, but successful actors have often been NGOs and private firms rather than the State whose municipal bureaucracy has responded slowly.

Such evident inefficiencies have often led to various new actors who arbitrage the supply chain and cost differentials set in place by non-functioning municipalities and instead establish an alternate supply chain. Some NGOs such as *Chintan* have successfully managed to initiate new municipal contracting norms to benefit informal workers, businesses, and residential areas. A substantial part of this effort goes beyond traditional organising of such workers, to detailed data gathering about how much need for waste management is going unattended. Companies such as *Sahas* in Bengaluru and widespread service organisation experiments such as *Exnora* emerged from the experiences in Chennai and other similar instances, recognizing and responding differently to the complex political economy of the demand in this sector. The technological innovations in this field can be quite diverse: from better serviced waste collection carts, to radio frequency identification devices (RFID) used to track waste collection and processing (now being tested in Brazil and other places), to sorters and incinerators better adapted to local employment practices. In the last mile, there are complex questions of skills and employment, social respect for municipal cleaners, and substantial information and coordination gaps in separating industry and household solid wastes.[4]

Health innovations and demand

Just as urbanisation provides instances of technological innovations that have thrived or failed depending on how quickly and robustly dynamic institutions have emerged to build demand, so too does the health sector. Here is a sector to which the traditional market failure arguments do not apply for several reasons, but where need and demand are not static (Hodgson 2008; Srinivas 2012). Social contexts – from prestige, to fear, to tolerance of disease, to levels of social exclusion, to market systems for procurement – are only some among a range of influences on demand. Policy, therefore, is an important element not only for attending to the demand for healthcare and health technologies, but also for suppressing the rates of innovation in them.

[4] See the recent furore in 2012 of Bangalore: Not 'Garden City' but 'Garbage City' (Accessed: 10 April 2013; http://blogs.timesofindia.indiatimes.com/life-notes/entry/bangalore-global-garbage-city).

I define 'innovation' in health as the development and introduction of new vaccines, drugs, devices, and surgical procedures into clinical practice. There is a triad of production, distribution, and delivery that defines health technologies (Moran 2000; Srinivas 2012). The challenge in health technologies is that need is difficult to study, that not all need is translated to demand, and that market demand depends on several regulatory features. Health technologies can be 'pulled' to the marketplace through a variety of inducements, including establishment of large markets and structuring of markets in particular ways (e.g., intellectual property, terms of payment and reimbursement, or price controls). Though there is no single linear path from demand to innovation, State planning and policy priorities create opportunities to recognize need and institute demand and help to create and legitimise markets and expand demand in several ways.[5]

Government policy and spending may influence the creation of innovations through ensuring ready demand. Procurement can be a dynamic element of this strategy. For instance, instrumentation innovations may come about largely independent of demand. Elsewhere, increased healthcare spending may drive the demand for certain innovations. As an illustration of the latter, it may be mentioned that innovations in antibiotics have been spurred on by governments in the interests of public health well before a marketplace was established, and India successfully set an early global example of this (Sahu 1998).

Health policy may support demand through ensuring availability at a low price. Vaccine innovations often are available at close to zero costs; markets are those established by the State as primary buyer. However, not all vaccines follow this pattern; the R&D of private sector firms may move ahead based on anticipated demand. Their goal is to establish demand from government procurement by inclusion in the country's expanded programme of immunisation. A recent example is how the state government of Karnataka has initiated the availability of generics, certain branded drugs, and free or heavily discounted surgical implants in hospital dispensaries.[6] This is a case where a need has been identified for several decades but the strategies to convert the need to effective demand and then to ensuring the fulfilment of that demand have been uneven at best.

---

[5] These technical innovations may not necessarily arise only within the private sector, but may also come from public research laboratories, community or non-profit health organisations or those working outside the formal R&D system.

[6] Covered in most national and regional dailies. See, for example, http://www.deccanherald.com/content/271681/generic-drug-stores-sell-surgical.html, accessed: 19 August 2012.

The State may act as both a crucial buyer and an intermediate user and tester to support demand. For instance, TamiFlu for H1N1 (Swine flu) is available in government dispensaries and through government supply; direct retail is prohibited. In other instances, vaccines such as for measles, mumps, and rubella (MMR) are available at a lower price in government clinics. Others, such as for cervical cancer, may need markets gradually created and expanded over time through education of the public, selective advertising, and lobbying of governments and physicians. Some agencies of the State have been stronger influences on both supply and demand of innovations: for instance, armies have always been important intermediate users and testers of vaccines well before the public, with soldiers not merely used as sacrificial lambs but because the technologies themselves were developed to respond to specific military needs and required careful customising. There are now also new efforts to decentralise telemedicine (the Rugged and Portable Telemedicine System) based on defence research advances developed by scientists at the Defence Bioengineering & Electromedical Laboratory in Bengaluru.[7]

Finally, the State may facilitate effective demand through welfare policies. In many parts of the medical system, the doctor specifies what the patient consumes and the State or the private insurance picks up the bill. Thereafter, there are three main channels for patients to access medicines: they may buy them 'out of pocket' (OOP), through partial or full reimbursements from health insurance or other programmes, and through free programmes subsidised by the State or other organisations. Thus, the channel of payments and reimbursements is an important part of any innovation model and is historically an important determinant of anticipated and actual demand. Welfare States institutionalise a wide array of citizen entitlements, including access to medicines in order to boost consumption and mitigate the effects of under- or unemployment and downturns in the economy.

Late industrial countries (LIs) such as India are distinct from earlier industrial countries in terms of how health needs are translated into demand through individual and collective buying instruments (Srinivas 2012). India has a sophisticated supply in health technologies (as does the USA), even relative attention to delivery mechanisms but very weak demand institutions. Even when supply-side capabilities begin to look increasingly similar to those

[7] http://www.thehindubusinessline.com/news/article3630627.ece, accessed: 18 August 2012.

of the USA, UK, or Germany, the demand structure and institutional form are very different. In LIs, comprehensive welfare states do not exist; instead weak systems of social protection provide only partial social protections against markets, and markets themselves may be underdeveloped or poorly regulated (Ibid). For instance, India has a fragmented system of public and private coverage of health expenses, effects of which are large OOP payments by most citizens. The Karnataka state government's generic drugs distribution programme is a step to changing market structure for the industry, and this has been followed by an announcement by the central government of access to free drugs in all government primary health clinics and at hospitals. Similarly, Andhra Pradesh and other governments have been experimenting with the expansion of health insurance programmes in agricultural belts by providing a regulatory framework and support to community federated organisations, including self-help groups, with private insurance partners (an example is the Andhra Pradesh government's Aarogyashri Health Care Trust).[8] Such state-sponsored programmes have since expanded – partly through the momentum generated by the National Commission for Enterprises in the Unorganised Sector's (NCEUS) recommendations of a social security programme for informal workers (Government of India 2009). Although the demand-side initiatives have spurred access to certain innovations, whether or not innovations themselves increase is yet to be determined.

Conclusion

Generating innovations that respond to needs and to demand is complex since they have distinct institutional facets. Demand can be expressed through both market as well as non-market channels (e.g., families, communities, neighbourhoods, unions, or caste groups) and through mixed-market instruments (e.g., cooperatives or highly structured and regulated auctions). This is especially true in countries industrialising 'late' such as India, where supply and demand may have no simple or easily reconciled character and where rapid urbanisation and anaemic welfare systems may generate both supply and

[8] For a detailed discussion of structure, see ILO, Series: Social Security Extension Initiatives in South Asia, 'India: State Government Sponsored Community Health Insurance Scheme (Andhra Pradesh)'; http://www.ilo.org/public/english/region/asro/bangkok/events/sis/download/paper16.pdf, accessed: 18 August 2012.

demand dilemmas for innovation in essential areas like health, water, waste, and renewable energy technologies. Such uncertainty regarding innovation's connection to demand affects innovation in both macro- and micro-terms.

If we view demand in societies as being the part of the iceberg that is actually visible, with a considerable amount of potential demand submerged below sea level, then innovation policies must be geared towards elevating the iceberg so that a larger fraction of the iceberg is within sight. This is important not simply to generate growth and incomes, but also to find fulfilling ways for skills participation, for generating products and products relevant to various needs, and for accommodating the heterogeneity in production structure and demand that any society comprises. From an evolutionary standpoint, the elevation of the hidden part of the iceberg is not a simple lift and push upward. Rather, it requires coordination of multiple institutions over their different time horizons – from production, to demand, to delivery. In practice, this will require redesigning traditional industrial and innovation policies. This, of course, is easier said than done. The caveat that effective demand is only part of the story has already been stated.

The challenge for policy-making is not simply to be aware that demand exists or can be molded to national priorities, rather to see supply and demand as truly evolutionary. This requires an appreciation that price signals may not be the most dynamic means of signalling demand, and recognizing the structural constraints on switching policy paths. Early Indian nationalists were concerned with how elites might capture the priorities of imports toward higher-income consumer goods, but perhaps underestimated the complex, changing relationship between economic structure and institutional design· Yet, the political economy challenge has always been to ensure *transition* from one type of paradigm to another, especially when one developmental path comes often at the cost of another because planning processes under-recognize the scarcity and resources context (Srinivas and Sutz 2008). Not all technological systems and elements of learning are easily applied to solve problems of inequality, and a single design for policy results in 'Moon and the Ghetto' type problems where advanced capabilities – the ability to reach the moon – rarely translates to learning systems applied to local living problems such as the ghetto's education, crime, or health problems (Nelson 1977). Although India has had repeated national policy calls for bridging its scientific and technological capabilities with basic needs, the latest of such efforts to plan a mission to Mars, announced on the Independence Day 2012 celebrations by Prime Minister Manmohan Singh, has met with resistance from aid agencies, grassroots groups, parliamentary opposition parties, and Indian citizens.[9]

Finally, it must be noted that there is no *a priori* role for any actor in building or responding to demand in innovation. For instance, Indian private providers have been quick in telecoms markets but this was at least in part due to a complicated and unresponsive prior public sector involvement in landlines, and more recently profitable incentives for responding to unmet demand. However, these involvements have recently resulted in highly controversial and potentially criminal actions by regulators and private firms in establishing and responding to these new markets.[10] Similarly, while NGOs have been primary Indian proponents of prosthetics and other disability aids, the State and for-profit sector have played a more active role on this front in other countries. In the USA, for example, the American Disabilities Act and its urban effects as well as health insurance coverage for several aids (involving the private for-profit sector) have both been important mutually reinforcing policy supports for the prosthetics industry.

This reiterates the caveats issued by Chris Freeman and his colleagues, who generated much to benefit science policy research, and recognized that the need for a public policy that addresses the limitation of the competitive paradigm is greater than ever:

> The advance of science and technology must find its support and justification, nor merely in the expectation of competitive advantage, whether national or private, military or civil, but far more in its contribution to social welfare, conceived in a wider sense. To the extent that the competitive mechanism may contribute to the achievement of this wider goal, it may serve a useful purpose; but in so far as it fails to do so, it will require drastic modification (Freeman et al. 1971).

Innovation is generated by many different bundles of institutions of which formal institutions of S&T comprise only one part. Of these, the combinations

[9] For instance, see recent discussions on both financial and political implications in the Financial Times, http://www.ft.com/cms/s/0/ffcb6198-e693-11e1-ac5f-00144feab49a.html#axzz23zsqYGZM, accessed: 18 August 2012. The Christian Science Monitor, which covered the story overseas, stated: 'Indian scientists dismiss the criticism, saying that technology developed as part of the space program has resulted in spinoffs in other areas. [..] It is certainly not a question of misplaced priorities,' former ISRO chief U.R. Rao was quoted as saying by The Asian Age newspaper, http://www.csmonitor.com/World/Latest-News-Wires/2012/0815/India-s-PM-confirms-82-million-Mars-mission, accessed: 18 August 2012.

[10] The recent 2010–2012 '3G' scandal for auctioning telecoms licenses is a case in point.

that affect demand can exert powerful effects on the supply of vital technological innovations. This has been the story of urban innovations such as electricity, and health innovations such as vaccines. The brief examples discussed here and the arguments laid out yield two hypotheses that can be further tested through detailed case studies. First, technological innovations at micro-level are shaped, at least in part, by demand-led policies that shape the cognitive contents of the 'innovation set' for suppliers and create new potential product markets. Second, national welfare regimes and other public policies generate path dependencies, affecting innovation rates and domestic relevance. In short, public policy and economic and physical plans for technology and industry can better embed innovations to serve needs and facilitate the transformation of to-be-identified needs into effective demand through both market and non-market instruments.

Many of the other chapters of this book will further illustrate these hypotheses in varied ways. For instance, in education (chapter by Gita Surie) and energy (chapter by Ambuj Sagar), State policies led by demand considerations have affected innovation generation. Furthermore, the innovation rate and content have been directly influenced by welfare policies in agriculture (chapter by Carl Pray and Latha Nagarajan) and pharmaceuticals (chapter by Shyama V. Ramani and Samira Guennif). It will also be shown that the failure of State policy to satisfy needs in certain sectors like traditional medicine (chapter by Arijita Dutta) and sanitation (chapter by Shyama V. Ramani et al.) has led to other economic actors converting needs into effective demand.

References

Albuquerque, E. 2007. 'Inadequacy of Technology and Innovation Systems at the Periphery.' *Cambridge Journal of Economics* 31 (5): 669–90.

Arocena, R. and J. Sutz. 2000. 'Looking at National Systems of Innovation from the South.' *Industry and Innovation* 7 (1): 55–75.

Arora, S. and H. Romijn. 2009. 'Innovation for the Base of the Pyramid: Critical Perspectives from Development Studies on Heterogeneity and Participation.' UNU-MERIT Working Paper Series 2009–036. Maastricht: United Nations University.

Carlson, W. B. and M. E. Gorman. 1990. 'Understanding Invention as a Cognitive Process: The Case of Thomas Edison and Early Motion Pictures, 1888–91.' *Social Studies of Science* 20 (3): 387–430.

Chandrasekhar, C. P. 2009. 'Product Composition, Jobless Growth and the Role of Demand-side Influences.' Paper written for the ILO Sub-regional office, New Delhi.

Chintan Environmental Research and Action Group. 2003. *Space for Waste, Planning for the Informal Recycling Sector*. New Delhi: Chintan Environmental Research and Action Group. Accessed: 3 September 2013; http://www.chintan-india.org/documents/research_and_reports/chintan_study_space_for_waste.pdf.

Courvisanos, J. 2009. 'Political Aspects of Innovation.' *Research Policy* 38 (7): 1117–124.

Edquist, C. and L. Hommen. 1999. 'Systems of Innovation: Theory and Policy for the Demand Side.' *Technology in Society* 21 (1): 63–79.

Edquist, C. and B. A. Lundvall. 1993. 'Comparing the Danish and Swedish Systems of Innovation.' In *National Innovation Systems, a Comparative Analysis*, edited by R. R. Nelson, 265–98. Oxford and New York: Oxford University Press.

European Commission. 2005. 'Innovation and Public Procurement. Review of Issues at Stake, Study for the European Commission (No ENTR/03/24),' final report. Germany: Fraunhofer Institute for Systems and Innovation Research. Accessed: June 2012; http://cordis.europa.eu/innovation-policy/studies/full_study.pdf .

Fontana, R. and M. Guerzoni. 2008. 'Incentives and Uncertainty: An Empirical Analysis of the Impact of Demand on Innovation.' *Cambridge Journal of Economics* 32 (6): 927–46.

Freeman C., C. H. G. Oldham, C. M. Cooper, T. C. Sinclair, and B. G. Achilladelis. 1971. 'The Goals of R&D in the 1970s.' *Science Studies* 1 (3/4): 357–406.

Freeman, C. and C. Perez. 1988. 'Structural Crises of Adjustment, Business Cycles and Investment Behaviour.' In *Technical Change and Economic Theory*, edited by Dosi G. Freeman, R. Nelson, G. Silverberg, and L. Soete, 38–66. London: Pinter.

Fujita, K. 2003. 'Neo-industrial Tokyo: Urban Development and Globalization in Japan's State-centered Developmental Capitalism.' *Urban Studies* 40 (2): 249–81.

Government of India. 2009. *The Challenge of Employment in India: An Informal Economy Perspective.* Volume 1, Main Report. New Delhi: National Commission for Enterprises in the Un-organised Sector. Accessed: September, 2013; http://dcmsme.gov. in/The_Challenge_of_Employment_in_India.pdf.

Hård, M. and T. J. Misa, eds. 2008. Urban Machinery – *Inside Modern European Cities.* Cambridge, MA: MIT Press

Hodgson, G. M. 2008. 'An Institutional and Evolutionary Perspective on Health Economics.' *Cambridge Journal of Economics* 32 (2): 235–56.

Hughes, T. P. 1993a. *Networks of Power: Electrification in Western Society, 1880–1930.* Baltimore: Johns Hopkins Press.

———. 1993b. 'The City as Creator and Creation.' In *Berlin/New York: like and unlike: Essays on Architecture and Art From 1870 to the Present*, edited by Josef Paul Kleihues and Christina Rathgeber, 13–32. New York: Rizzoli.

Kalecki, M., 1968. Trend and business cycle reconsidered. Economic Journal 78 (2): 263–276.

Kannan, K. P. 1990. 'Secularism and People's Science Movement in India.' *Economic and Political Weekly* 25 (6): 311–13.

Lember, V., T. Kalvet, and R. Kattel. 2011. 'Urban Competitiveness and Public Procurement for Innovation.' *Urban Studies* 48 (7):1373–95.

Lotz, P. 1993. 'Demand as a Driving Force in Medical Innovation.' *International Journal of Technology Assessment in Health Care* 9 (2): 174–88.

Moran, M. 2000. 'Understanding the Welfare State: The Case of Health Care.' *British Journal of Politics and International Relations* 2 (2): 135–60.

Mowery, D. and N. Rosenberg. 1979. 'The Influence of Market Demand upon Innovation: A Critical Review of Some Recent Empirical Studies.' *Research Policy* 8 (2): 102–53.

Myers, S. and D. G. Marquis. 1969. *Successful Industrial Innovation*. Washington: National Science Foundation.

Nelson, R. 1977. *The Moon and the Ghetto: An Essay on Policy Analysis.* New York: W.W. Norton.

Nelson, R. R. 2005. 'On the Complexities and Limits of Market Organisation.' *Review of International Political Economy* 10 (4): 697–10.

Nelson, R. R. and Rosenberg, N. 1993. 'Technical Innovation and National Systems.' In *National Innovation Systems: A Comparative Analysis*, edited by Richard Nelson, 1(n dash)21. Oxford: Oxford University Press.

Papaioannou, Theo. 2011. 'Technological Innovation, Global Justice and Politics of Development.' *Progress in Development Studies* 11 (4): 321–38.

Porter, M. E. 1990. *The Competitive Advantage of Nations.* New York: Free Press.

Reddy, A. K. N. 2004. 'Science and Technology for Rural India.' *Current Science* 87 (7): 889–98.

Rosenberg, N. 1969. 'The Direction of Technological Change: Inducement Mechanisms and Focusing Devices.' *Economic Development and Cultural Change* 18 (1): 1–24.

Rothbarth, E. 1942. 'Review of Business Cycles: A Theoretical, Historical and Statistical Analysis of the Capitalist Process, by Joseph A. Schumpeter.' *The Economic Journal* 52 (206/207): 223–29

Ruttan, V. W. 1997. 'Induced Innovation, Evolutionary Theory and Path Dependence: Sources of Technical Change.' *The Economic Journal* 107 (September): 1520–29.

Sahu, S. K. 1998. *Technology Transfer, Dependence, and Self-Reliant Development in the Third World: The Pharmaceutical and Machine Tool Industries in India.* New York: Praeger Publishers.

Schmookler, J. 1966. *Invention and Economic Growth.* Cambridge, Mass: Harvard University Press.

Schumpeter, J.A. 1934. *The theory of economic development: an inquiry into profits, capital, credit, interest, and the business cycle.* Cambridge, MA: Harvard University Press.

Srinivas, S. 2012. *Market Menagerie: Health and Development in Late Industrial States.* Palo Alto, CA: Stanford University Press Economics and Finance Imprint.

———. 2006. 'Industrial Development and Innovation: Some Lessons from Vaccine Procurement.' *World Development* 34 (10): 1742–64.

Srinivas, S. and J. Sutz. 2008. 'Developing Countries and Innovation: Searching for a New Analytical Approach.' *Technology in Society* 30 (2): 129–40.

Srinivas, S. and S. Wallin. 2012. 'Labour Unions to Living Labs: The Locus of Innovation in 21st Century Welfare States.' n.p.: Mimeo.

Sutz, J. and R. Arocena. 2006. 'Integrating Innovation Policies with Social Policies: A Strategy to embed Science and Technology into Development Processes.' Strategic Commissioned Paper for IDRC Innovation, Policy, and Science Program Area. Ottawa: IDRC.

Von Hippel, E. 2005. *Democratizing Innovation.* Boston: MIT Press.

White, G. 1999. 'Rethinking Kalecki on the Trend and Cycle.' *Review of Political Economy* 11 (3): 341–53.

II
Sectoral studies

CHAPTER 4

Innovations in Indian Seed/Biotech Industry

Carl E. Pray and Latha Nagarajan

Introduction

Innovations from public sector research and development (R&D), which were spread by public sector seed corporations were an essential component of the Green Revolution. The modern varieties of rice, wheat, and coarse grains increased food production in India, helped millions of people come out of poverty, and made India self-sufficient in grain production. Now, however, rapid growth in per capita income and population growth in India are pushing up the demand for vegetables, fruits, and coarse grains to feed livestock. The demand for the basic food grains such as rice and wheat is growing too, if not as rapidly, and global prices for all agricultural commodities have increased. At the same time, the growth in yields of basic food grains in the most productive areas of India such as Punjab is slowing down. Fortunately, unlike in the early period of the Green Revolution, the private sector is now making major investments in R&D, and so, India no longer has to rely entirely on innovative new varieties from the public sector or from the international agricultural research institutes such as the International Rice Research Institute.

The objectives of this chapter are three-fold: (i) to assess the current innovation system in the seed and biotechnology sector; (ii) to identify the investments in public agricultural research and innovation; and (iii) to reflect on the policies supporting private research and innovations so that farmers can have the innovative new cultivars[1] that they need to meet the rapidly increasing

[1] In this chapter, 'cultivars' is a general term that refers both to varieties, which are bred to be genetically stable so that farmers can save and replant the seeds that they produce on their farms, as well as to hybrids, which are the seeds of two different varieties that will start separating into their original varieties if grown and replanted.

demands of Indian consumers. The second section of this chapter reviews the history of innovation in the Indian seed sector. Thereafter, in the following three sections, we describe the structure, research, and innovations of the current seed innovation system. The sixth section focuses on the impacts of innovations in this industry. The final section describes the policies and government investments currently at work to provide innovations, and it also identifies the components of the seed innovation system that need to be improved.

Evolution of the seed sector innovation in India

Technical innovations by the seed industry are of two broad types: new products and new production processes. In the seed industry, new products are primarily new plant varieties but also include new chemical seed treatments to protect against diseases, weeds, and insect pests, and new genetic traits such as insect resistance or improved grain quality. New production processes such as methods of making hybrid seed can reduce the cost of hybrid seed production. In addition, there have been some important changes in the research tools such as genetic engineering and molecular breeding, which can reduce the cost of innovation. These are also a type of process innovation. Institutional innovations include public–private partnerships, the shift from family ownership and management of the seed companies to professionals, the use of the stock market to raise funds, the entry of large Indian companies into the seed industry, and the entry of foreign companies, dedicated biotech companies, and private foundations. This chapter will primarily concentrate on new product innovations in the form of new varieties and new biotech traits because unlike other innovations, these are relatively easy to measure.

The science-based innovations from the seed sector can be divided into four periods. The first is the colonial period. During this period, the central and provincial governments of British India established a series of agricultural experiment stations and agricultural colleges that first imported crops and plant varieties from the other parts of the world. Starting at about the time of the First World War, these institutions produced new plant varieties using the newly rediscovered Mendelian genetics (Pray 1983). They produced high yielding sugar cane, wheat, and rice varieties that also had improved taste quality, and cotton varieties that also had longer fibre. These varieties spread extensively and had a major impact on Indian and Pakistani farmers' incomes (Pray 1984).

The second period of major innovation in the seed sector is the Green Revolution period. Major government investments were vital to the development of these innovations also. Building on what was left of the colonial system, the new Government of India developed a public agricultural research system with its headquarters in New Delhi, supported by a network of state agricultural universities (SAUs). During this period, which started in the mid-1960s, Indian scientists took the short, stiff-stemmed wheat and rice varieties from the International Agricultural Research Centres (IARCs), crossed them with local varieties to meet Indian taste and cooking requirements, and fitted them into Indian soil, pest, and climatic conditions. They produced varieties of rice and wheat, which fitted into all the major areas of India where there was good water control – areas neither too dry nor too wet. In addition, the public sector produced improved varieties of maize, sorghum, pearl millet, cotton, sugar cane, and other major crops. The private sector played a small role in innovation, importing and distributing improved vegetable and fruit varieties, and developing new varieties of some commercial crops such as tobacco.

The enormous impact of the Green Revolution rice and wheat and to a lesser extent maize, sorghum, and pearl millet on Indian food production, productivity, and prices has been documented in many studies (Evenson, Pray, and Rosegrant 1999; McKinsey and Evenson 2003). In addition, a number of scholars have shown that these varieties had a major impact on reducing poverty in India (Lipton 2001; Hazell and Ramasamy 1991).

The third major period of innovations is the hybrid period. It started in the 1980s and is the first period in which the private firms started to play a major role in developing new cultivars. All of the varieties that they created were hybrid varieties. These offered a higher yield per unit of land than varieties of the same crops, because of the scientific phenomena of heterosis or hybrid vigour. In addition, hybrids – unlike the Green Revolution wheat and rice varieties – had the characteristic that they could not be reproduced by farmers and had to be purchased every year. Hybrids are also difficult for other seed companies to reproduce, which meant that the company that developed them could prevent copying and protect their intellectual property. This made the production of new hybrids potentially attractive investments for the private sector, and in fact the private sector did invest.

In each crop, a system of producing the hybrid seed must be developed for commercialization. The first crop for which such a system was developed was maize in the USA between the first and second world wars. This system was introduced into India soon after its Independence from the British, by government scientists and the private sector, but it did not have much impact. In the 1960s and 1970s, many public sector institutions moved away from hybrid maize, because they did not think that small farmers would be able to afford to buy such expensive seeds every year. A few local and foreign firms continued to experiment with it and started selling hybrid maize in some regions.

At the same time, the public sector was developing systems for producing hybrid seed of some other crops. Hybrid sorghum and pearl millet production systems were developed by government scientists and International Crops Research Institute for the Semi-Arid Tropics (ICRISAT) scientists. India was the first country in the world to commercialize cotton hybrids. The first cotton hybrid, H-4, was intra-*hirsutum* (i.e., both parents are hirsutums) and was produced by Dr C. T. Patel in 1970 at the Surat Agricultural Experiment Station of the Gujarat Agricultural University. In 1972, an inter-specific (female – *hirsutum* × male – *barbadense*) cotton hybrid (Varalakshmi) was released by the University of Agricultural Sciences at Dharwad, Karnataka (Murugkar, Ramaswami, and Shelar 2007). At the same time, new crops were also introduced from outside India. For example, hybrid sunflower cultivars were introduced from the USA and Australia in the early 1980s.

In the late 1980s and 1990s, the private sector made major investments in plant breeding in all of these crops, and developed hybrids that took over much of the market for improved cotton, pearl millet, sorghum, and maize. Also, several public–private partnerships proliferated during the late 1990s, which led to the introduction of hybrids in field crops such as rice, wheat, and rapeseed. For instance, the International Rice Research Institute, along with some Indian government institutions and several private firms, adapted the Chinese hybrid rice system to India, and hybrid rapeseed systems were developed by both the public and the private sectors.

In the twenty-first century, hybrid rice has started to spread in different parts of India. Like the Green Revolution varieties, these hybrids have basically been yielding higher per unit of land than past varieties because they can increase grain or fibre production in response to fertiliser and good water conditions without falling over (lodging). The impact of the proprietary sorghum, pearl millet, and maize hybrids on the income of small farmers has been documented by Pray and his colleagues (Pray et al. 1991; Ramaswami and Pray 2002).

The fourth period of innovations that began at the beginning of the twenty-first century is the period of biotech innovations. This period started with the introduction of genetically engineered cotton hybrids that have the trait for resistance to one important group of insect pests. The trait is called Bt – short for *Bacillus thuringensis* – the gene from a soil bacterium, which gives the cotton hybrids resistance to the cotton bollworm. The gene was identified and then developed into a commercially useful trait, first by the American company Monsanto and then first brought into India in 2000 by an Indian company, without going through the government biosafety regulatory system. Around the same time, a joint venture between Monsanto and Maharashtra Hybrids Company (MAHYCO), one of the biggest Indian seed firms, was working on Bt trait through the Indian regulatory system since 1996. It was officially approved in 2001. The Bt trait was then licensed to 20 + seed companies in India that put it into their proprietary cotton hybrids. Recently, several new Bt genes have been approved – one from China and one from an Indian public research institute. Cotton hybrids with Bt genes now cover about 80 per cent of the India cotton area (Prabhakar 2008). They have increased yield per hectare and reduced pesticide use, and have helped increase the total area under cotton (Tripp 2009).

Overview of the current seed innovation system

The current innovation system in India contains a combination of public sector and private sector actors working together, sometimes in competition and sometimes in collaboration. Fig. 4.1 shows the basic structure of the Indian seed and biotech sector at present. The national-level institutes that come under the Indian Council of Agricultural Research (ICAR), such as the Indian Agricultural Research Institute in New Delhi or the Central Rice Research Institute in Cuttack, along with the 42 SAUs develop most of the new varieties of wheat, rice, pulses, and oilseeds through conventional plant breeding. The IARCs of the Consultative Group on International Agricultural Research (CGIAR) provided access to international germplasm and knowledge about key food crops, which encouraged the development of new research products in India. ICRISAT and 7 of the 15 CGIAR-supported centres have offices and research facilities in India. The international centres provided research support not only during the Green Revolution but are still a major source of new varieties and germplasm.

The new varieties or hybrids developed by public institutes are tested for traits such as yield per hectare, resistance to biotic and abiotic stresses, and their grain or oil quality by national and state regulators. If they are an improvement over the standard varieties, they are approved for dissemination by government agencies or the private sector. The research institutes provide small amounts of breeders' seed of the new varieties primarily to the public sector organisations such as the state seed corporations (SSCs) and extension services, who multiply

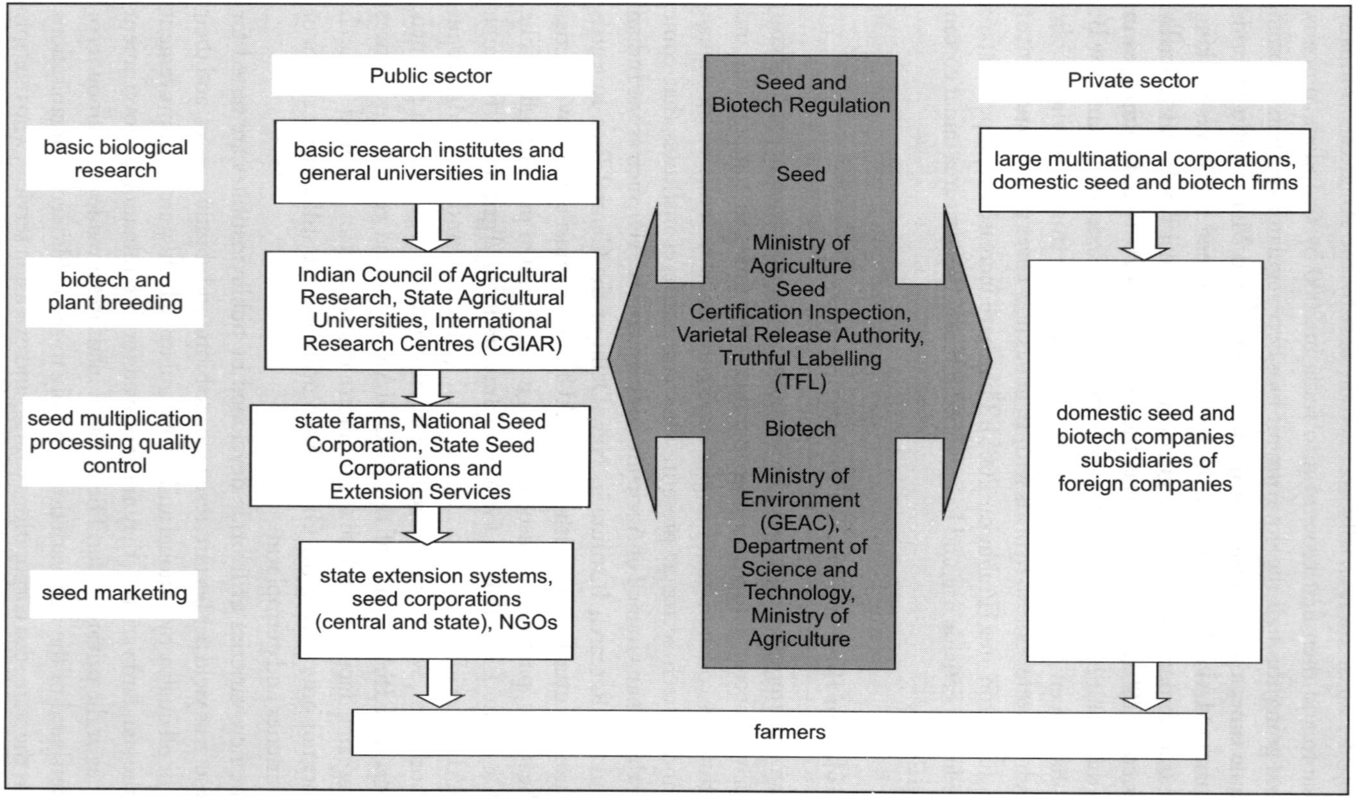

Fig. 4.1: Current structure of the Indian seed and biotech system
Source: Created by the author

the seed and market it through state extension services or through private seed dealers. Some of the breeders' seed may also be licensed to private firms who then multiply and distribute it.

The actual innovations from the private sector firms primarily take place at the same level as in the public sector – the plant breeders. The private sector almost exclusively produces hybrids and concentrates on the crops in which hybrid can be profitably grown – maize, cotton, sorghum, pearl millet, sunflowers, and recently, rice and rapeseed. Once the breeders are satisfied with the varieties that they have produced, they can enter it into the same government trials that the public sector does or they can sell without going through the trials as 'truthfully labelled' (TFL) seed.[2] Indian seed law until recently, allowed private companies to sell their seed either as TFL or as certified seed that has been tested by the government to ensure a certain level of quality (Tripp and Pal 2000). Currently, it is estimated that nearly 500 hybrids of field crops and vegetables are being marketed as TFL seeds, mostly by private firms.

Public sector innovation

The Indian national agricultural system, which is one of largest in the world, has contributed substantially towards crop improvement, and increased high yielding variety (HYV) adoption over the last four decades. The development of new agricultural institutions as well as increased investments into research during and after the Green Revolution period have had a major impact in terms of research outputs. These have given rise to more improved varieties whose adoption has played an essential role in attaining self-sufficiency in food. Besides increased investment in research and infrastructure, the government policies and the regulatory environment that was provided over the years have contributed enormously to the overall growth of Indian agriculture.

Research institutions

The Indian Agricultural Research System went through a series of innovations in its institutional structure since the 1920s (Pal and Byerlee 2006). The foremost was the creation of the Indian Agricultural Research System (now referred to as

[2] Well-established enterprises with reputations to protect may sell seed that has no official seed certification. Such seed is often described as 'truthfully labelled' as it bears a label describing minimum seed quality standards, self-certified, and not certified by an official agency (Tripp 2001).

the ICAR) in 1929, a semi-autonomous body to promote, guide, and coordinate agricultural research nationally. The post-British era saw another institutional innovation in the development of at least one agricultural university in each state. They were modelled after the USA land grant agricultural universities.

A second important institutional innovation was the establishment of the All India Coordinated Research Projects (AICRPs), initiated in 1957 under ICAR, to promote multidisciplinary and multi-institutional research and research collaboration between the SAUs and the national research institutes under ICAR. Under the AICRPs, the performance in yield potential, disease, and pest resistance, and nutritional value of the newly developed varieties (including hybrids) were evaluated. In 1961, the first maize hybrid was released for general cultivation, followed by hybrid varieties of sorghum, pearl millet, and non-hybrid HYVs of rice and wheat. The first intra-specific cotton hybrid was introduced in 1970 and since then it has paved the way for subsequent cotton hybrids in India. The current National Agricultural Research System (NARS) of India covers an array of disciplines and serves a wide range of agricultural industries, from dairy to biofuel to horticulture. One of its most successful areas of research has been plant breeding. In India, the seed industry growth can be attributed in part to the strength of the NARS that trained plant breeders and developed improved germplasm, in-bred lines, varieties, and hybrids that private firms continue to use as the basis for their breeding programmes.

There are now 42 state and central agricultural universities specializing in agriculture, animal, and horticulture sciences, fisheries, forestry, and home economics. They are supported by state governments, but their research is largely supported by ICAR. The ICAR has also established National Research Centre on Plant Biotechnology at the Indian Agricultural Research Institute (IARI) in New Delhi to conduct biotech research on stress tolerance for various field crops. The centre is working on gene isolation, marker-assisted selection, and developing transgenic plants for biotic and abiotic stress.

In addition to the traditional NARS, advances in the biological sciences led to new research institutions and new sources of money for agricultural research. For example, the Department of Biotechnology (DBT), Department of Science and Technology (DST), and Department of Scientific and Industrial Research (DSIR) under the Ministry of Science and Technology support and conduct agricultural research at their institutes and sometimes fund research in the ICAR/SAU system (Pal and Byerlee 2003). The DBT, established in 1986, promotes molecular biology and biotechnology research in agriculture by supporting seven Centres of Plant Molecular Biology (CPMBs) in different parts of the country (Pal and Byerlee 2003). The centres are functioning at Tamil Nadu Agricultural University (TNAU) in Coimbatore, Jawaharlal Nehru University (JNU) in New Delhi, Madurai Kamaraj University (MKU) in Madurai, National Botanical Research Institute (NBRI) in Lucknow, Osmania University (OU) in Hyderabad, Bose Institute (BI) in Kolkata, and University of Delhi South Campus (UDSC) in New Delhi. Other institutes doing some research related to agriculture under the DBT are the Indian Institute of Science (IISc) in Bengaluru, the Centre for Cellular and Molecular Biology (CCMB) in Hyderabad, and the Centre for DNA Fingerprinting and Diagnostics (CDFD), Hyderabad.

Research investment

India has one of the largest and most complex agricultural research systems in the world, with more than a century of organised application of science to agriculture, with annual economic rates of return to investment in research ranging from 35 to 155 per cent (Pal and Byerlee 2003; Evenson, Pray, and Rosegrant 1999). In India, most of the public research in agriculture is conducted through the ICAR institutes, research centres, and a network of SAUs. The central government of India supports the ICAR, the apex body of agricultural research, extension, and education in the country. Besides, the Government of India also allocates a part of the research funds to SAUs in the form of research schemes and grants (ICAR Budget Book 2005–06). In addition to central funding for research, the SAUs receive funding through their respective state governments.

Over the last four decades, there has been more than a ten-fold increase in public investments for agricultural research and education in India. In real terms, this suggests an increase from Rs 1,600 million in 1960–61 to Rs 22,000 million in 2004–05. The investment in research is led by crop sciences (27 per cent), followed by animal sciences (11 per cent) and horticultural research (7 per cent) (Mohapatra and Sahoo 2008). From 1996 to 2009, public research investment grew rapidly until 2000, levelled off through 2004, and then grew rapidly again till 2009 (Fig. 4.2). However, at least two-thirds of the agricultural research in India is still supported by the public sector through its vast network of research facilities and institutions (Pray and Nagarajan 2012).

The scientists working for the public institutions or agencies are engaged in triple functions of teaching, research, and extension activities. Pal and Byerlee

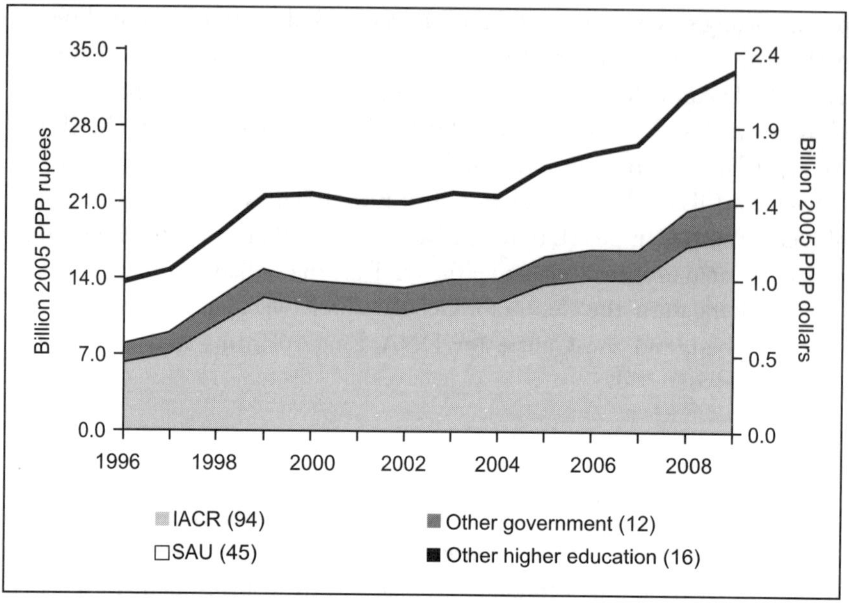

Fig. 4.2: Public sector R&D spending adjusted for inflation: 1996–2009
Source: Pal, Rahija, and Beintema (2012)

(2006) and Ramaswamy and Selvaraj (2007) estimated that the number of scientists working in the ICAR/ SAU system during the late 1980s was 4,189 at ICAR and 14,851 at the SAUs, totaling 19,040. The number of scientists remained steady at ICAR during the 1990s (4,092 in 1998) and increased marginally to 4609 in 2005–06 but decreased significantly at the SAUs (17,678 in 1992) and further declined by 24 per cent in the last decade (Ramaswamy and Selvaraj 2007). The number of full-time equivalent (FTE) scientists conducting research in the late 1990s was 2,999 within ICAR and 8,132 within the SAUs (Pal and Byerlee 2006). Agricultural extension played a major role in promoting Green Revolution technologies and the World Bank aided training and visit systems that proved effective in the most favourable regions of India (Birner and Anderson 2007). The training and visit system was phased out later in the early 1990s. After the training and visit system, a variety of innovative technology transfer initiatives were tried out across Indian states. These approaches included decentralisation; contracting; group extension; provision of extension by para-extension workers, producer cooperatives, or farmer-based organisations; the establishment of agro-clinics by private entrepreneurs with initial government support; public–private partnerships in financing and providing extension; and the establishment of Internet-based extension though village kiosks (e-*choupals*) set up by the private sector.

There is a general perception that after the training and visit approach was phased out in the 1990s, the existing extension system deteriorated, even though a variety of new approaches to provide and finance extension emerged (Sulaiman 2003; Sulaiman and Van den Ban 2001). The Agricultural Technology Management Agency (ATMA) – sponsored by the ICAR – has developed a model that embodies several of these reform elements. A national policy framework for agricultural extension was developed in 2000, to support state-related extension activities (Birner and Anderson 2007). India's Tenth Five-year Plan (2002–07) also emphasises on the need for revamping and modernising the extension systems and encouraging the private sector to take up extension services.

Innovations

A closer look at the activities of public sector research reveals that currently the innovation activities are focused on developing higher-yielding, open-pollinating varieties (OPVs) of wheat and rice. The seeds of these varieties have high volumes and can therefore be sold at low prices. Public sector research has an important but more limited presence in higher-valued hybrid cultivars of coarse grains and cotton, where the private sector plays a large role. The other crop for which the contribution of public sector plays a substantial role is pulses. Almost 95 per cent of the pulse varieties in India are developed by public research. The number of varieties released through ICAR and SAUs, especially for pulse crops, has increased three-fold from the early 1990s (ICAR 2008). A good measure of innovation by the public sector is the number of cultivars that the Department of Agriculture has notified or recognized as superior new cultivars during various periods. This is, however, an imperfect measure of innovation by the private sector, because while notification is required for all public cultivars, it is voluntary for private firms. Most companies prefer not to notify their cultivars, because this process takes a number of years, which means foregoing profits from sale of the new cultivar. According to this indicator, the rate of innovation holds steady from the 1980s to the 1990s, but then grows rapidly after 1999 (Table 4.1). This indicates that public research continues to be an important innovator.

Table 4.1: Trends in notified varieties of major field crops

Crop	*Number of notified varieties and hybrids by decade*		
	1980–89	*1990–99*	*2000–10*
Rice	198	188	303
Wheat	84	66	112
Maize	43	64	113
Pearl millet	38	45	51
Sorghum	55	49	55
Cotton	72	78	95
Total	490	490	729

Source: Pray and Nagarajan (2012)

The public seed sector confines its activities to (i) producing and selling seed of government-notified popular varieties from the government research programmes, (ii) certifying seeds from the government seed corporations and private firms as being notified varieties and TFL, and (iii) regulating seed distribution activities involving government subsidy programmes and emergencies. To facilitate the production and availability of quality seeds across the country, in 1963, the Government of India started a central seed organisation, the National Seeds Corporation (NSC) of India. This was followed by the establishment of SSCs in several states from 1969, with an aim to decentralise and specialize in crop seed production and distribution. With technical help from the Rockefeller Foundation, the NSC's first products were hybrid maize varieties. Later it began to produce seeds of cereals and cotton on a commercial scale. With the assistance of the World Bank, the Indian government established the National Seeds Programme (NSP) in 1967, to coordinate the efforts of the NSC and SSCs and private companies. Another aim of the NSP was to create new facilities and modify the existing infrastructural facilities for seed testing, research, and certification.

The public sector mainly produces and sells seeds of self-pollinating crops, such as rice and wheat and OPVs of sorghum, accounting for 60 per cent of sales and production of these types of crops during 2004–05 (GOI 2006). SSCs did play a large role in marketing hybrid seeds of cotton, maize, pearl millet, and sorghum in the past, but since 2000, the marketing of the seeds of these crops has been largely taken over by private firms.

Regulations and policies

Until the late 1980s, public agencies played the major role in crop variety development, seed multiplication, and their distribution activities. After

Independence, the seed industry was reserved for small-scale Indian firms. Large firms (domestic or foreign) were prohibited from operating in the seed industry, and the private importation of seeds for either commercial or research purposes was also restricted. One consequence of these policies was a seed supply system dominated by the public sector. During the liberalisation of the Indian economy in the late 1980s and 1990s, a series of regulatory and trade reforms affecting the seed sector were initiated. These reforms stimulated domestic and multinational private participation in this market.

The various regulatory policies and laws governing the Indian seed sector can be classified into three major groups, including (i) seed sector regulations and quality controls, (ii) phyto-sanitary regulations and laws, and (iii) policies related to the implementation of intellectual property rights. One of the major reasons for the success of private seed firms in the Indian seed sector is their ability to appropriate some of the benefits of their research via the use of hybrids.

With the recent enactment of the Protection of Plant Varieties and Farmers' Act (PPVFR), intellectual property protection has been extended to the rights of breeders of OPVs. PPVFR was passed in the Parliament in April 2001 and became effective from April 2007. In the first phase, it covered 14 major crops and few more have been added in 2008. The effect of plant varietal protection on Indian agriculture and the Indian seed sector is unknown as the act has just recently been implemented.

Private sector innovation

The Indian seed industry is the sixth largest in the world with the market size of US$1300 million (Prabhakar 2008).[3] There are over 250 private seed firms in India, marketing around 600 hybrids of field crops and vegetables as TFL seeds. The private sector in India mainly comprises about 20 major players (sales of more than Rs 500 million), several medium (sales between Rs 50 million and Rs 500 million) and a large number of small and/or unorganised players (sales turnover less than Rs 50 million) with local presence (Dravid 2008). Among the firms perhaps 60 firms or fewer have R&D programme to develop and commercialize new cultivars. A few of these firms have substantial research being conducted in biotechnology, while quite a few firms have recently made some investments in biotech labs. Another group of firms primarily imports varieties of vegetables that were developed elsewhere. A growing share of the

[3] The market size is denoted in terms of current market exchange rate value, in US dollars.

hybrid vegetable market is held by Indian and foreign-owned companies with strong research bases in India.

Evolution of the seed industry for major field crops

The entry of large Indian firms and multinationals into the Indian seed industry greatly increased the amount of research and innovation by the private sector. These firms entered the seed industry from the mid-1980s, after India reduced the barriers to the entry of foreign firms and large Indian companies and relaxed regulations on importing new agricultural technologies. The key reforms were: (i) seed and biotechnology industries were included in the Indian government's list of core industries in 1986, which allowed foreign-owned firms and large Indian conglomerates to enter the seed industry; (ii) the New Seed Industry Development Policy of 1988 allowed the seed firms to import commercial vegetable seeds with no quotas and commercial seeds of foreign varieties of coarse grains and oilseeds for only two years (after which the seed firms had to produce the seed within India), and made it easier to import germplasm for research purposes; and (iii) in 1991, regulations on technology transfer and foreign investment for the entire economy were reduced.

These reforms led to a dramatic increase in the number of private seed firms and the capital invested in the seed industry. Private companies' seed sales were five times higher in 2003–04 than they were in 1994–95, while the public sector almost doubled in size (see Table 4.2). As a result, the private sector increased its share of sales in the seed market from 35 per cent in 1994–95 to 60 per cent in 1998–99 and 67 per cent in 2003 along with the size of the market.[4] Even the small-scale 'unorganised' sector increased sales but its share fell from 25 per cent to 11 per cent.

The structure of the private seed industry changed dramatically since the mid-1990s. Many large Indian firms entered the seed industry including J.K. Industries, Southern Petrochemical Industries Corporation (SPIC), Khatau-Junker Ltd, Godrej, Dunlop, and Harrisons. In addition, large, well-established firms that were partially foreign-owned such as Hindustan Lever Ltd (HLL), ITC, ICI, and Sandoz also entered the seed industry. The other key groups of new entrants were the multinational seed companies such as Cargill, Pioneer, and Monsanto, which initially came in as joint ventures but later became wholly-owned subsidiaries.

[4] Personal communication with Mr Sadananda, Director-Research, Emergent Genetics India (Now Advanta), Hyderabad, Andhra Pradesh in 2003.

Table 4.2: Size of the Indian seed market (US$ million, PPP adjusted)

	1994–95		*1998–99*		*2003–04*	
	Market size (US$ million)	*% of market share*	*Market size (US$ million)*	*% of market share*	*Market size (US$ million)*	*% of market share*
Public sector	612	40	672	25	1,162	22
Private sector	540	35	1617	60	3,593	67
Private sector (unorganised)	386	25	404	15	581	11
Total	1538	100	2,694	100	5,336	100

Source: Chaturvedi (2005)

The twenty-first century really marks the globalisation of research and innovation in the Indian seed and biotech industry. This process has three components: first, there is a growing investment of foreign seed and biotech companies in research and a seed industry in India and they are using of India as a research base for developing technology for elsewhere in the tropics; second, there is the expansion of Indian firms into international markets through the purchase of foreign-based multinationals and seed firms in the USA, Europe, and South America; and third, there are research alliances and contract research between the Indian firms and the American, European, and Chinese research organisations.

Globalisation through the expansion of multinational research in India recently has been of three types. First, a few multinationals entered India initially. These include large firms such as French-based Limagrain, which initially had a joint venture with Avesthagen, and smaller firms like the Belgium Biotech Company Devgen, which brought the rice, sorghum, and millet businesses in India from Monsanto. Second, several of the major international seed/biotech firms that were already in India have set up strategic research facilities that conduct research in India for their global enterprise. For instance, Monsanto set up a basic research facility in Bengaluru a decade ago, which is primarily a part of their international research programme, although it spends perhaps a third or quarter of its resources on Indian research issues. In 2008, DuPont (Pioneer) began investing US$25 million in a biotech research centre, which will eventually house 300 scientists in Hyderabad and have a US$5 million per year budget.[5] Bayer Agro Sciences and its vegetable seed subsidiary Nunhems are

[5] Downloaded 24 March 2007 from http://www.dnaindia.com/report.asp?newsid=1084795.

an example of the third type of expansion of research. They are expanding their applied biotech and breeding research in India so that India can be a research base for hybrid rice and vegetable seeds for all of South and Southeast Asia.

The most important example of globalisation through the expansion of Indian seed firms into foreign markets is the 2006 purchase of the Dutch-based multinational Advanta Seeds by the Indian generic pesticide firm United Phosphorus (UP). Advanta had research facilities in Australia, Argentina, and Thailand as well as India where it has research facilities in Hyderabad and Bengaluru. In 2008, it continued to expand by purchasing US-based hybrid grain and forage sorghum supplier Garrison & Townsend, LP. UP already had a small seed subsidiary in India in 2006 and was a major international player with pesticide manufacturing plants in the UK and Argentina, and 70 per cent of its pesticide revenue was from outside India (UP press release 2006).

Globalisation also took place through the development networks of research alliances, contract research, and acquisitions between small and medium-sized Indian and foreign companies. One example is Krishidhan Seeds Ltd's recent announcement that it has signed a long-term agreement with the Dutch biotech company Proteios International BV 'for the development and application of molecular markers and DNA detection technologies on cotton.'[6]

These firms have dramatically increased expenditure in private sector R&D and innovation. The amount of R&D expenditure by private seed firms quadrupled in real 2005 US dollars from $1.3 million in 1985 to 4.9 million in 1995 and then soared to $88.6 million in 2009. Multinational firms accounted for 44 per cent of the R&D expenditure in 2009 (Pray and Nagarajan 2012).

One result of the growth of private research is that a growing number of hybrid varieties are being developed by the private sector, both in field crops and vegetables, starting from the early 1990s. New cultivars of crops such as cotton, maize, pearl millet, and sorghum, which are all hybrids in India, now primarily come from the private sector (Table 4.3). Since the liberalisation of the vegetable seed trade in 1988, vegetable cultivars primarily come from the private sector, which either imports or develops its own cultivars (Table 4.4).

With the growth of hybrids, the private sector has been able to capture a substantial share of market from the public sector, especially in commercial crops such as cotton and vegetables, and in a few field crops such as sorghum and millet (see Table 4.5).

Multinational firms such as Pioneer Hi-Bred, DeKalb, Pacific Seeds, and others have made major contributions to research and innovation in hybrid

[6] SeedQuest 2008; http://www.seedquest.com/News/releases/2008/july/23257.htm.

maize and sunflower. Somewhat surprisingly, foreign firms like Pioneer, Bayer, and Syngenta have also played an important role in developing and promoting hybrid rice, which is not an important market for them elsewhere in the world.

Table 4.3: Number of field crop varieties by public and private sector institutions in India, 2005–10

Crop	*Truthfully labelled private hybrids*	*Notified public varieties*
	2005–10	*2005–10*
Rice	79[a]	240[b]
Wheat	40	95
Maize	136	87
Pearl millet	97	48
Sorghum	75	46
Cotton	255	70
Total	603	346

Notes:
[a] Includes only actual hybrids released by the private sector (not open-pollinated varieties)
[b] Includes open-pollinated varieties (mostly) 48 of which are hybrids
Source: Pray and Nagarajan (2012) with data from MoA (2010). For public varieties and truthfully labelled varieties, information collected from individual firms' websites and surveys (34 firms)

Table 4.4: New vegetable hybrids in India, 1998–2005

Crop	*Public sector*	*Private sector*
Tomato	3	160
Eggplant	8	218
Chilli	2	73
Capsicum	1	31
Cauliflower	1	35
Cabbage	-	20
Okra	2	32
Watermelon	2	25
Cucumber	2	10
Gourd	6	80

Source: Kataria (2005)

The most important recent contribution of a foreign firm is Monsanto's introduction of Bt cotton, the first genetically modified (GM) crop to be introduced in India. This revolutionised the seed industry in India. It led to substantially increased income for farmers, increased cotton production, and large increase in profits of Monsanto and MAHYCO and also of a number of seed companies that licensed the technology from Monsanto or developed their own Bt technology. It also dramatically increased the industry's value, changed its structure, and increased industry investments in research. More than 150 Bt cotton hybrids, all developed by the private sector, containing any one of the four approved events are now available for cultivation, covering more than 80 per cent of the total cotton area in India (Pray and Nagarajan 2010).

Table 4.5: Contribution of the private sector in hybrid seed market: Per cent of sales volume (2008)

Crop	*% of hybrid in total seed sales*	
	Private	*Public*
Corn	98	2
Cotton	100	0
Millet	82	18
Sorghum	75	25
SSG	95	5
Sunflower	100	0
Rice[a]	90	10
Vegetables[b]	100	0

Notes:
[a] Less than 5 per cent of the rice seed planted in India is hybrid. The other 95 per cent includes varieties that are either from State Seed Corporations or are farmer-saved seed
[b] Includes chilli, tomato, watermelon, gourd, eggplant, okra
Source: National Seed Association of India (2008)

Who benefitted from the spread of these innovations?

The major innovations of the Green Revolution, hybrids, and biotechnology have all had important impacts on yields and productivity of farmers. The Green Revolution, which is the oldest, has undoubtedly had the most impact so far, but hybrids are increasing yields in regions and for crops that the Green Revolution originally neglected. Now biotechnology shows promise of taking care of some issues that the Green Revolution and hybrids have not taken care of – pest problems and rising pesticide use.

Green Revolution

Modern varieties of rice and wheat were the core of the Green Revolution. There is an extensive body of literature, which shows that they had a direct positive effect on the welfare of farmers and labourers in the areas of irrigation, good rainfall, and limited flooding. In those areas, both small and large farmers benefitted substantially. Between 1966 and 2000, wheat yields in developing countries increased by 208 per cent and rice yields, by 109 per cent (Pingali 2012). In addition, because the technology was labour dependent, farm labour in the regions of technology adoption and labour from regions in which there was surplus labour benefitted through new jobs. Poor and rich consumers benefitted because food prices would have been 35 per cent to 65 per cent higher in the absence of the Green Revolution (Hazell and Ramasamy 1991; Evenson and Gollin 2003; Pingali 2012).

Hybrids

Maize and cotton hybrids spread and increased yields in the favourable areas. The consumers who benefitted from maize included those who eat maize and who would have been willing to pay higher prices. This particularly helps the poor who cannot afford to buy wheat or rice. The animal feed industry and the commercial livestock industry also benefitted because their costs went down due to less expensive maize. Hybrid cotton farmers reaped benefits through higher productivity, and some of the benefits were passed on in the form of lower prices to cotton mills in India, and ultimately, perhaps consumers of cotton cloth, who got cheaper clothing.

Hybrid pearl millet, sorghum, and sunflower spread in the semi-arid tropics, which were not well served by the Green Revolution because of less reliable rainfall. The proprietary hybrids were responsible for a major increase in the yields of maize, sorghum, and pearl millet (Pray and Ramaswami 2001). However, they did not go into the most water-stressed environments, which continued to depend on land races of these crops. Consumers of millets and sorghums, who are some of the poorest consumers, benefitted from lower food prices. In addition, as with maize, some of the benefits were passed on to the animal feed industry.

Somewhat surprisingly, hybrid rice, which now covers about 5 per cent of the rice area, is spreading to some of the less favourable rice environments, which have not been well served by the Green Revolution. They are replacing older improved varieties or land races, but not Green Revolution varieties in eastern India. More than 80 per cent of the rice hybrids are cultivated extensively in eastern UP, Bihar, Jharkhand, and Chhattisgarh; 15 per cent in north western and western parts of India; and less than 5 per cent of the total rice acreage under hybrid rice cultivation is in southern Indian states (Viraktamath 2007). So far, 33 rice hybrids have been officially released for cultivation, of which 25 are from the public sector and eight from the private sector. In addition, several non-hybrid varieties of rice are being produced and marketed by private firms. The private sector is leading hybrid seed production activities and accounted for almost 95 per cent of the total seeds supplied (19,400 tonnes) in the year 2007–08 (Viraktamath 2007). More than 36 private firms have been involved in hybrid rice seed production and research activities.

Even though the hybrid rice was introduced in the late 1990s, its growth has been slow, occupying only 3.3 per cent (3.2 million/ha) of the total rice cultivated area (Yog Raj 2012). The major reason for the slow adoption was the lack of desired grain quality attributes in the initial hybrids released. Their yield and grain quality are not high enough to compete with the Modern Varieties (MVs) in the well-irrigated regions of India, but they are good enough to compete in lowland regions that have less reliable rainfall, such as Jharkhand, Bihar, West Bengal, and Orissa.

Gene revolution

So far, the main benefits from biotechnology have gone to cotton farmers who are producing higher yields and reducing insecticide use, and cotton mills and cloth consumers who pay lower prices. Both small and large cotton farmers have benefitted. Pray and Nagarajan (2010) showed that the farmers benefitted from the adoption of Bt cotton technology, through increased yields and reduced pesticide costs. Although Bt technology is targeted at reducing pesticide use and pest attacks rather than increased yield, there have been substantial increases in yields and total factor productivity levels. In spite of the higher seed costs of Bt cotton seeds, the cost savings due to reduced pesticide use offset the farmers' increased cost on seed. There are also additional spill-over benefits such as improved quality of life due to increased income, better health due to less exposure to pesticides, peace of mind due to better yield assurance, etc. (Dev and Rao 2007).

According to a study by Qaim and some other researchers (Qaim et al. 2006), 341 of cotton farmers, both adopters and non-adopters of Bt cotton from four states in India, namely, Maharashtra, Karnataka, Andhra Pradesh, and Tamil Nadu, indicated that compared to other conventional and non-Bt cotton varieties in their respective regions Bt cotton did exceedingly well in every aspect of its agronomic performance. Other studies by Barwale et al. (2004) and Bennett et al. (2004) also concluded that Bt cotton had significant positive impacts on average yields, and therefore, on the income of cotton growers.

Issues of the twenty-first century seed system

This section presents evidence of the components of an effective seed innovation system that is in place and seems to be working fairly well, namely, public sector research, liberalisation of trade and foreign direct investment, the strengthening of intellectual property rights, and the regulation of new varieties. The second part of this section discusses some of the weaknesses and limitations of the current innovation system and how they could be improved.

Strengthening public research

The evidence presented in this article indicates that the public sector remains an important source of innovation for the seed industry. Table 4.1 shows that the number of varieties and hybrids of major field crops produced by the public sector increased from 490 in the 1990s to 729 in the first decade of the twenty-first century. These innovations and others from the public sector research stations have contributed to continued growth in total factor productivity from 1985 to 2005. This study calculated a 42 per cent economic rate of return to public investments in agricultural R&D for that period. A breakdown of the returns to research by decade, in all of the major commodities, revealed positive rates of return to research in the more recent period from 1995 to 2005 too, which provides evidence that the productivity of the public research system has not decreased (Chand, Kumar, and Kumar 2012).

In addition to providing benefits directly to farmers, the public sector research has stimulated the private sector to do more research. In a recent study, Pray and Nagarajan (2013) used an econometric model to estimate the determinants of R&D expenditure by 35 Indian seed firms. The public sector research had a positive, statistically significant impact on R&D expenditure by these firms, and these firms are now an important source of innovations in agriculture.

The Indian government has recognized the importance of public research and is increasing the amount of money invested in research (see Fig. 4.2), but there are major concerns about the public system. One is that the total number of scientists declined by 17 per cent from 2000 to 2009. This decline is particularly acute at the SAUs, where the number of scientists dropped from a peak of 7,780 full-time scientists in 2000 to 6,158 in 2009. At Acharya N.G. Ranga Agricultural University (ANGRAU) and the University of Agricultural Sciences, Dharwad, total full-time researchers declined by more than 40 per cent during 2000–09 (Pal, Rahija, and Beintema 2012, 3). This might not be a problem if the increasing levels of expenditure allowed each scientist to be more productive. Unfortunately, it appears that much of the money at SAUs was spent on developing new campuses (10 were created in 2009) rather than strengthening research funding for scientists (Pal, Rahija, and Beintema 2012). The share of operating and capital costs in the total expenditure has been declining with inadequate operational back-up for scientists employed at different levels. This imbalance often lowers the scientists productivity of the scientists and degrades the quality of research output (Jha and Kumar 2006; Ramaswamy and Selvaraj 2007).

One way in which some parts of the public R&D system is trying to enhance its productivity is through public–private partnerships. The Indian seed industry is evolving into a science-driven stage in its development. Intense international competition coupled with increased costs of R&D has also prompted companies to seek alternative forms of partnerships to produce research products. Increased pressure on the public sector organisations to generate resources to undertake market-based research has also opened avenues for high value partnership with private sector in India and elsewhere (Dravid 2008).

The emerging public–private partnerships in India that are significant to Indian seed industry are mainly based on the development of finished products, for instance, the ICRISAT hybrid parents consortia. The public and private sectors also work together in the evaluation of hybrid parental lines – through AICRPs and SAUs. In the field of agricultural biotechnology, the partnerships are mainly in joint development of traits (e.g., JK-Agri-genetics and IIT, Kharagpur, developed and commercialized a Bt cotton event). Initiatives such as the World Bank funded National Agricultural Innovation Project (NAIP) are attempting to facilitate innovations through public research organisations in partnership with the farmers' groups, the private firms, and the civil society organisations.

In addition to the rise of the private sector research, another change in the Indian system has been the development of participatory breeding programmes by NGOs. One prominent example of this (or at least one of the best documented) is the rice breeding programme of the Gramin Vikas Trust. The programme started in 1995 with the support of Department For International Development – UK (DFID) funds and Center for Arid Zone Research (CAZR) supported research programme at Wales University in the UK. Under this programme, a multinational NGO plant breeding network was initiated, which has developed a number of popular rice varieties for upland and lowland regions where Green Revolution varieties did not spread. In addition to NGOs, some private research foundations have been active in technology generation in agriculture in the recent times. The M. S. Swaminathan Research Foundation, MAHYCO Research Foundation, and The Energy and Resources Institute (TERI) have developed considerable research capacity with a national presence and are working in close collaboration with the public research systems under ICAR/SAU/DBT. Besides, ICAR also promotes such NGO–private partnerships in the transfer of technology through their farm science centres (KVKs).

The policy implications are that there should be continued increase in funding research with more focus on increasing funding for scientists and recruiting more scientists – particularly at the SAUs. In addition, greater engagement of the public sector researchers with private firms, farmers, and NGOs could raise productivity.

Continuing liberalisation of agricultural input markets

Economic theories and empirical evidences suggest that some industry concentration increases R&D and innovation, but too much of it can reduce R&D and innovation (Aghion et al. 2005). Mergers and acquisitions among seed and biotech industry in the USA and Europe have increased global concentration in this industry (Fuglie et al. 2011) and this, in turn, could reduce competition in India. For example, Monsanto purchased Emergent Genetics, Seminis, and Delta and Pineland, all of which had Indian subsidiaries. In India, Monsanto then integrated them with some parts of the E.I.D.Parry. In addition, it purchased 26 per cent of MAHYCO. At the same time, Monsanto sold off the Stoneville cotton seed to Bayer in 2007 because of USA anti-monopoly requirements. It also sold its sorghum, pearl millet, sunflower, and rice seed businesses in India to the Belgium-based company Devgen in 2007, so that it could concentrate on its core crops: maize, cotton, soybean, canola, and vegetables.

Such high profile mergers and acquisitions have created concerns among Indian politicians and NGOs regarding too much concentration. The few quantitative studies that have looked at concentration have found that the entry of large Indian companies and multinationals into the seed industry has actually led to reduced concentration, rather than greater concentration. Murugkar, Ramaswami, and Shelar (2007) examine concentration in the cotton seed industry between 1996–97 and 2004–05. They find that it declined when measured either by the share of the top five firms or by the Herfendahl Index. In a recent study, Pray and Nagarajan (2012) calculated the concentration ratios in the commercial seed market[7] and the share of the market that was supplied by foreign firms. The four-firm concentration ratio in the private segment of the industry declined from almost 70 per cent in 1987, to only 36 per cent today. The share of foreign ownership in the private seed industry has increased from 10 per cent in 1987 to about 40 per cent today, which supports the argument that allowing the entry of multinationals has increased the competitiveness of the industry rather than increasing concentration.

The other type of concentration in the seed industry has been in the Bt gene business. It is the one part of the industry in which the Indian government officially ruled that a monopoly exists. Navbharat, a medium-sized seed companiy in Gujarat had a brief illegal monopoly on Bt cotton until it was discovered in 2001. Then Monsanto–MAHYCO Biotech (MMB), the joint venture of Monsanto and MAHYCO, was given a temporary monopoly by the biosafety regulatory system on the 'legal' Bt cotton industry until 2006 when Nath Seed and J.K.Seed were allowed to sell new Bt genes. In 2006, the Monopolies and Restrictive Trade Practices Commission (MRTPC) ruled that MMB had a monopoly and the prices of Bt seed were too high and should be controlled.

The question is did the temporary monopoly on Bt cotton harm farmers through higher prices or reduced innovation? The empirical studies on the Bt seed industry show that Bt seeds are more expensive than next hybrids, but they save money per unit of output or per unit of land by saving pesticide costs and increasing yields. Murugkar, Ramaswami, and Shelar (2007) find that farmers are willing to pay much higher prices for the private proprietary Bt cotton hybrid seed than for the public hybrid seed. They were also happy to get the seeds for lower prices when state governments forced the companies to greatly reduce the price of seed.

Regarding the impact on innovation, the high prices of Bt cotton hybrids led to a dramatic increase in investment by private firms in biotech research, which is now paying off in the form of new Bt genes and many new Bt cotton hybrids. This effort started with the firms Nath, JK Agri-genetics, Meta-Helix, and a consortium of companies lead by Nuziveedu, known as Swarna Bharat Biotechnics Private Limited (SBBPL), investing in developing their own Bt cotton hybrid programme. The JK Agri-genetics has developed a Bt cotton technology in collaboration with IIT (Kharagpur), and similarly, SBBPL entered into an agreement with the Lucknow-based NBRI for obtaining Bt cotton technology.

---

[7] This is the market share of the private sector seed market, which excludes sales by the State Seed Corporation and farmers and small traders.

Limiting the technology fee by imposing price controls might be a good way of transferring benefits from the owner of the technology to the consumers (in this case farmers) in the short-term (Pray and Nagarajan 2010). Yet the government-imposed price controls could have a negative impact on innovation in two ways. First, the companies, especially multinationals, may not invest in seeking approval for the release of new GM traits in India. For example, in China where there is virtually no protection of biotech intellectual property rights, Monsanto has not even attempted to introduce BG-II or any other traits since they introduced BG-I in 1997. Second, both Indian and multinational companies may defer the introduction of new technology and investments in research on traits /varieties that would benefit the Indian market in particular. For example, Meta-Helix had to cancel its plans to introduce a new type of Bt cotton in 2010 because they could not raise money to produce seed of the product. Their investors argued that with Bt cotton seed prices being held down by the government, Meta-Helix would not make enough profits to pay back their investment in seed production (Koshy 2010). Regarding research, Monsanto India state that they are unlikely to do the R&D of genes with resistance to sucking pests like thrips, leafhoppers, and aphids, despite large demand in India for such varieties and the potential benefits to farmers.[8]

Monopoly power in the seed and biotech industry is something that the government clearly needs to monitor. The evidence here suggests that the way the ruling was enforced – different price controls on Bt cotton seed by different states – had a dampening effect on innovation. The 2006 ruling by the MRTPC was one of the last before it was replaced by a new Competition Commission. Hopefully, the new commission can find better ways of fixing the problem than through price controls on innovation. These could include making it easier for more firms to enter the market and compete to bring prices down.

Intellectual property rights (IPR) legislation strengthening the Indian innovation system

Appropriability, the ability of inventors to capture some of the benefits from the technology they invent, is essential for private firms to invest in research and innovation. The changes that have taken place in intellectual property rights – implementation of the amended patent law in 2005 and the PPVFR Act – have been welcomed by much of the seed and biotech industry. Nagarajan, Pandey,

[8] Personal Communication with Monsanto India Ltd, January 2008.

and Smale (2007) noted that large and medium-sized companies favoured stronger plant variety protection systems than smaller firms, since stronger intellectual property rights could reduce the share of low quality seed materials (with fake brand names) and thus improve the overall quality of seeds available in the market by curbing the functioning of 'fly-by-night' operators. These changes in intellectual property rights seem to be having a modest impact on inducing more private sector R&D.

The primary method by which seed firms are able to protect the innovations that they have developed is through hybrid seeds. As mentioned earlier, the crops that can be protected this way are limited by the nature of their flowers, and research is needed to develop an efficient system of producing the hybrid seed. Maize, sorghum, and sunflower were some of the first crops in which the hybrid seed system became popular in the USA and elsewhere. Gradually, through public and private research, hybrid seed systems for pearl millet, rice, wheat, rapeseed, and canola have also been developed. Hybrid seed systems allow companies to limit the use of their varieties to those farmers or seed companies who are willing to pay for them. This price, which is higher than could be charged for a variety that is not protected, allows seed companies to pay for the research that was required to develop the technologies and still make a profit.

Trademarks have worked together with hybrids to strengthen appropriability. Companies can build up a reputation for providing high quality seed and new hybrids that are useful to farmers. This pays off if they can protect their seed with trademarks that competitors cannot copy. This also gives them the ability to introduce new hybrids as TFL, without needing government testing and approval to assure farmers that these are quality seeds. Finally, it gives them the ability to raise the price of innovative hybrids and prevent other firms from selling hybrids under their brand.

The biosafety regulatory system essentially provided MAHYCO-Monsanto Biotech Company (MMB) with temporary intellectual property rights on Bt cotton hybrids and set up a major barrier to entry of plant variety innovations produced through genetic engineering. The first trait to be approved in India in 2002, the Cry1Ac Bt gene for cotton, required six years of testing and possibly as much as US$1.6 million investment by Monsanto and MAHYCO (Pray, Bengali, and Ramaswami 2005). Subsequent traits have cost much less, because the regulatory system now has some experience, and has improved the efficiency of the process. However, no competing Bt genes were approved until 2005, and new GM varieties started to be planted extensively only from 2006. This gave MMB a monopoly on the trait except for the 'illegal' Bt seeds that were provided by NavBharat seed company in Gujarat. This allowed MMB to dramatically increase the price of cotton hybrids from about Rs 300/acre for non-Bt hybrids to Rs 1600/acre for Bt hybrids.

The Plant Variety Protection System started with 12 crop species from 21 May 2007. By November 2008, about 914 applications had been received for registration, out of which 207 were new varieties, 702 were existing varieties and 5 were farmers' varieties. The total number of applications received was 552 from ICAR, 337 from the private sector, 227 from the SAUs, and 5 from the farmers (Trivedi 2008). It is too early to say how important this system will be in increasing appropriability for private research firms. It will probably not be known until case laws start to define how much protection against copying can the law actually provide.

It is also too early to tell what the impact of the new plant breeders' rights legislation will be. Since most of the applications are from the public sector and on existing varieties, the number of applications gives little evidence of increased innovations. Thus, no predications can be made on the seed prices paid by farmers. However, it is clear that when there is strong protection of intellectual property, as there is for many hybrids, and as there was inadvertently with Bt cotton, farmers pay higher prices for the seed. In the case of Bt cotton, the increased revenue that farmers received nevertheless more than compensated for the higher seed prices (Pray et al. 1991).

Product patents were not allowed in chemicals (including pesticides, fertilisers, and veterinary medicines), agriculture, and biotechnology until 2005. Since then novel micro-organisms that are some of the main products of agricultural biotechnology are allowed to be patented in India. New plant varieties cannot be protected by product patents but can be protected using the plant variety protection. Almost all of the plant biotechnology patents so far were granted to multinational corporations (Pray and Nagarajan 2012). It appears that the ability to patent has had some impact in inducing foreign firms to conduct research in India, but there is no evidence yet on the impact of patents on innovation or prices of technology.

The policy implications are that the government should continue to strengthen patents and plant variety protection system. At the same time anti-monopoly authorities have to monitor developments in this sector to ensure that it remains competitive and does not hinder innovation.

Regulations on biotechnology

The seed, biotech, and fertiliser industries, in particular, expressed frustration with regulations for new technology approval. Bt cotton was the first GM crop approved for planting in India. The approval process took six years and cost

US$1.2 million (Pray et al. 2005). The first GM food crop put forward for approval was hybrid mustard, but Bayer Crop Science gave up after 10 years of trying to provide regulators with sufficient biosafety data (Pray, Bengali, and Ramaswami 2005).

We have argued in a previous section that the example of the profitability of Bt cotton was the major factor in the rapid growth in R&D in the seed industry in all hybrid crops in the first decade of this century. Modelling of Indian seed industry research also provides evidence that breakthroughs in biotech have induced more private R&D (Pray and Nagarajan 2013). The growth of R&D slowed down somewhat since 2006 and potential important innovations were not commercialized after the government ruled that MAHYCO-Monsanto had a monopoly on Bt cotton and forced all the seed companies to reduce prices of Bt cotton seed by about two-thirds (Pray and Nagarajan 2010).

Recent controversies about biosafety regulation have largely reduced the firms' interest in biotechnology research and innovation. In 2009, all regulatory authorities approved Bt eggplant as safe for consumers and effective for use by farmers, but the Ministry of Environment vetoed the approval in response to activists' concerns regarding genetically modified organisms (GMOs). Lawsuits to ban research on GM crops are being debated in the Indian Supreme Court. An Indian Agricultural Parliamentary Committee led by a member of Parliament from a Communist Party declared that Bt cotton was useless to farmers and GM technology should be effectively banned, and a proposed new independent biosafety regulatory authority is currently stuck in the Parliament. Finally, many state governments are refusing to permit the field trials of GM crops in their states.

Biotechnology has been and could in the future be an important stimulus to innovation in Indian agriculture. The Indian government and state governments need to develop a regulatory system that protects farmers and consumers from potential problems, but allows safe technology to be developed and make its way to farmers.

Subsidies and programmes to encourage private innovation

The Indian government and some state governments are investing in a number of programmes and tax incentives to promote private agricultural innovation. No one knows whether they would really pay off with important benefits to society or not. They offer an excellent research opportunity for economists and PhD students seeking a thesis topic. In this section, we have listed a number of programmes that are trying to support innovation in the seed and biotech area.

To begin with, for instance, there is the Small Business Innovation Research Initiative Scheme of the DBT, which financed Shriram Bioseed Genetics, Rasi Seeds, MAHYCO, and Nuziveedu TRANSGENIC RESEARCH for biotic and abiotic stresses. Public funding is now also available for private sector firms involved in high-risk technology development areas with global partnership. This is supported under the New Millennium Indian Technology Leadership Initiative (NMITLI) that aims to develop new management technologies for biotic and abiotic stress in rice and cotton crops.

The private sector has also initiated several programmes with multi-stakeholder partnership arrangements where IARCs or NARS operate programmes and support services to strengthen private sector investment in agricultural research. The Agri-Business Incubators (ABI) at ICRISAT is one such initiative, which facilitates the creation of agri-business enterprises through technology development and commercialization. The Biotech Consortium India Limited (BCIL) set up in 1990 also provides linkages between the R&D and industry to facilitate commercialization of products and processes.

The private and government-sponsored biotechnology/science/technology parks and incubators can provide a good platform for the promotion of start-up companies and to assist in the commercialization of their new technologies. Biotech parks and incubation centres have been established at Lucknow, UP, and Genome Valley, Hyderabad. Other places like Himachal Pradesh, Karnataka, and Kerala are setting up similar facilities.

In recent years, the growth in investment on R&D for products based on biotechnology has been enormous and encouraging for the firms involved. The biotechnology firms have attracted three sources of financial support, namely, offshore funds, Indian venture funds, and funding through government-sponsored Technology Development Board (TDB). The venture capital (VC) industry in India is also emerging as a vibrant sector to support IT, biotechnology, telecommunication, and food processing related industries.

The major initiatives undertaken by the public sector in creating infrastructure for promoting innovation through VC formation are: National Science and Technology Entrepreneurship Development Board (NSTEDB); S&T Entrepreneurship Park (STEP); Technology Business Incubators (TBI); Instrumentation Development Programme (IDP); Biotech Park and Biotechnology Incubation Center; Bio-IT Park; Software Technology Parks of India (STPI), and Food Parks. Besides, private financial and banking institutions such as ICICI are also active in setting up venture funds to promote start-up biotech firms and life sciences based on incubator model.

In addition, the Small Industries Development Bank of India (SIDBI) and the DBT have also set up a US$125 million biotechnology development fund in 1998 to encourage private–public partnerships in the small-scale sector, as well as to promote entrepreneurship in biotechnology. In the recent years, several state governments in India have also launched different initiatives to attract biotechnology industries. Another important development is the emergence of knowledge-based (biotech) clusters in major metropolitan cities such as Hyderabad, Bengaluru, Chennai, Delhi, and Pune.

Conclusion

The current seed and biotechnology innovation system in India has the essential components to develop and spread innovation that will increase the productivity of Indian agriculture. This consists of a large public sector agricultural R&D system with a set of national and state seed corporations that provide farmers with a steady stream of new varieties, particularly of wheat, rice, pulses, and oilseeds. The public system is complemented by a rapidly growing private sector seed industry that conducts research on hybrid crops such as maize, cotton, pearl millet, sorghum, hybrid rice, and hybrid vegetables. These companies supply proprietary hybrids that were developed in their research programmes, and hybrids and varieties that were developed by the public sector and other private firms.

This study suggests that the recent growth in Indian seed–biotech sector, which is most rapid in the private sector, can be attributed to three major government interventions. The first intervention was the growth of investment in public sector plant breeding and crop management research by the national government, state governments, and IARCs. In the early days of the new hybrids, these research institutions contributed OPVs, hybrid cultivars, and inbred lines to allow other researchers to develop new cultivars.

The second important intervention was government investments in seed production in government and private institutions. The Indian government, with the help of donors, made major investments in SSCs, which multiplied the seed of the Green Revolution cultivars of wheat, rice, maize, pearl millet, and sorghum. At the same time, the seed laws were written and enforced, which blocked large Indian companies and multinationals from entering the seed market, and allowed small private sector seed companies to profit and develop. Training for qualified personnel in both public and private institutions was provided by the government.

The third important intervention in this story was the liberalisation of this sector, starting in the mid-1980s. Instead of allowing SSCs to continue as regional monopolies, the governments opened the doors to investment by large Indian firms, which were excluded from this sector until 1986 and allowed foreign direct investment in the sector. This coupled with continuing investments in public plant breeding and public–private partnerships, such as the hybrid millet, sorghum, and pigeon pea research consortia of ICRISAT, provided private firms with a steady stream of inbred lines and germplasm, which they used to develop proprietary hybrids. These factors and a seed law that allowed TEL seed and did not require mandatory registration and testing by the government, have led to a vibrant and sustainable supply of seed of new cultivars that are resistant to important diseases and pests and tolerant of drought.

To strengthen this sector Indian and state governments should consider the following investments and policy options:

- Continue to increase funding for public sector research with more focus on increasing funding for scientists and recruiting more scientists – particularly at the SAUs. There needs to be opportunities for SAU scientists to work with more basic biology and engineering researchers inside and outside India. In addition, greater engagement of the public sector researchers with private firms, farmers, and NGOs could raise the productivity of the public research system.
- Continue liberalisation of foreign investment and trade in technology in the agricultural input sector. Monopoly power in the seed and biotech industry is something that the government clearly needs to monitor. The evidence here suggests that the way one ruling on monopoly in this sector was enforced – different price controls on Bt cotton seed by different states – had a dampening effect on innovation.
- Strengthen the implementation of patents and the plant variety protection system in the seed sector. At the same time anti-monopoly authorities have to monitor developments in this sector to ensure that it remains competitive and does not hinder innovation.
- Biotechnology has been and could in the future be an important stimulus to innovation in Indian agriculture. The Indian government and state governments need to develop a regulatory system that protects farmers and consumers from potential problems, but allows safe technology to be developed and make its way to farmers.
- The Indian government and some state governments are investing in a number of programmes, subsidies, and tax incentives to promote agricultural innovation. No one knows whether they would really pay off with important benefits to society or not. This would be an appropriate time for scholars and the government to assess the programmes to find out if they are really stimulating innovations that are sufficiently beneficial to society to justify their costs.

Acronyms

CGIAR	Consultative Group on International Agricultural Research
IARC	The International Agricultural Research Centres
ICAR	Indian Council of Agricultural Research
ICRISAT	International Crops Research Institute for the Semi-Arid Tropics
NSC	National Seeds Corporation
SAU	State Agricultural Universities
SSC	State Seed Corporation
TFL	Truthfully Labelled Seed

Select Bibliography

Aghion, P., N. Bloom, R. Blundell, R. Griffith, and P. Howitt. 2005. 'Competition and Innovation: An Inverted-U Relationship.' *The Quarterly Journal of Economics* 120 (May): 701–28.

Anand, A., U. Singh, and O. P. Dutta. 2003. 'Vegetable Seed Industry in India—Vibrant & Looking ahead.' Paper presented at The World Seed Congress, Bangalore.

Anand, N. and U. Singh. 2008. 'Industry Status and Future Trends: Vegetable Crops.' Paper published in the souvenir of Asian Seed Congress (APSA-NSAI Souvenir), Hyderabad.

Arora, S. R. 2008. 'Indian Seed Industry Opportunities and Challenges under Regulatory Systems.' Paper published in the souvenir of Asian Seed Congress (APSA-NSAI Souvenir), Hyderabad.

ASC. 2002. *Census of Scientific Manpower in Agriculture—2001–02*. New Delhi: National Center for Agricultural Economics and Policy Research.

Barwale, R. B., V. R. Gadwal, U. Zehr, and B. Zehr. 2004. 'Prospects for Bt. Cotton Technology in India.' *AgBioForum* 7 (1,2): 23–26.

Bennett. R. M., Y. Ismael, U. Kambhampati, and S. Morse. 2004. 'Economic Impact of Genetically Modified Cotton in India.' *Agbioforum* 7 (3): 96–100.

Birner, R. and J. R. Anderson. 2007. 'How to Make Agricultural Extension Demand-driven?: The Case of India's Agricultural Extension Policy.' Discussion Paper #729. Washington DC: International Food Policy Research Institute (IFPRI).

Bonwoo Koo, B., C. Nottenburg, and P. Pardey. 2004. 'Plants and Intellectual Property: An International Appraisal.' *Science* 306 (5700): 1295–97.

Chand, R., P. Kumar, and S. Kumar. 2012. 'Total Factor Productivity and Returns to Public Investment on Agricultural Research in India.' *Agricultural Economics Research Review* 25 (2): 181–94

Chaturvedi, S. 2007. 'Changing Structure of Indian Seed Industry: Policy Prospects and Emerging Lessons.' Paper presented at the South Asian Conference on Trade and Development entitled *Harnessing Gains from Trade: Domestic Challenges and Beyond.* 20 December 2007, New Delhi.

———. 2005. 'Dynamics of Biotechnology Research and Industry in India: Statistics, Perspectives and Key Policy Issues.' OECD Science, Technology and Industry Working Papers. Paris: OECD Publishing.

Dev, S. M. and N. C. Rao. 2007. 'Socioeconomic Impact of Bt. Cotton.' *CESS Monographic Series 3*. Hyderabad: Centre for Economic and Social Studies.

Dravid, P. S. 2008. 'Creation of Successful Public-Private Partnerships in Indian Seed Industry.' Paper published in the souvenir of Asian Seed Congress (APSA-NSAI Souvenir), Hyderabad.

Evenson, R. E. and D. Gollin. 2003. 'Assessing the Impact of the Green Revolution, 1960 to 2000.' *Science* 300 (5620): 758–62.

Evenson, R. E., C. Pray, and M. W. Rosegrant. 1999. 'Agricultural Research and Productivity Growth in India.' IFPRI Research Report No. 109. Washington DC: International Food Policy Research Institute.

Fuglie, Keith O., Paul W. Heisey, John L. King, Carl E. Pray, Kelly Day-Rubenstein, David Schimmelpfennig, Sun Ling Wang, and Rupa Karmarkar-Deshmukh. 2011. 'Research Investments and Market Structure in the Food Processing, Agricultural Input, and Biofuel Industries Worldwide.' Economic Research Report (ERR)-130. Washington DC: US Department of Agriculture, Economic Research Service (ERS-USDA).

Gadwal, V. R. 2003. 'The Indian Seed Industry: Its History, Current Status and Future.' *Current Science* 84 (3): 399–406.

Girdhar, M. 2006. 'Public-Private Partnership in Seed Development.' Presentation at National Seminar on *Public-private Partnership in Agriculture,* New Delhi.

Government of India. 2006. *Agricultural Statistics at a Glance*. New Delhi: Ministry of Agriculture.

Hazell, P. and Ramasamy, C. 1991. *The Green Revolution Reconsidered: The Impact of High-Yielding Rice Varieties in South India.* Baltimore: Johns Hopkins University Press.

ICAR (Indian Council of Agricultural Research). 2008. *ICAR Annual Report, 2008*. New Delhi: Indian Council of Agricultural Research.

ICAR Budget Book. 2005–06. *ICAR Budget Book, 2005–06*. New Delhi: Indian Council of Agricultural Research, Ministry of Agriculture and Cooperation.

Jha, D. and S. Kumar. 2006. 'Research Resource Allocation in Indian Agriculture.' Policy Paper 23, National Centre for Agricultural Economics and Policy Research. New Delhi: Indian Council of Agricultural Research (ICAR).

Kataria, A.S. 2005. *Development of F1 Hybrid in Vegetable Crops and the Availability in Public and Private Sector*. New Delhi: Seed Association of India.

Koshy, J. P. 2010. 'Biotech Companies Say Cotton Seed Price Cap Limiting Research.' *LiveMint& the Wall Street Journal.* Accessed: 4 May 2010; http://www.livemint.com/2010/05/04205303/Biotech-companies-say-cotton-s.html.

Lipton, M. 2001. 'Reviving Global Poverty Reduction: What Role of Genetically Modified Plants?' *Journal of International Development* 13 (7): 823–46.

Matuschke, I., R. R. Mishra, and M. Qaim. 2007. 'Adoption and Impact of Hybrid Wheat in India.' *World Development* 35 (8): 1422–35.

McKinsey, J. and R. E. Evenson. 2003. 'Crop Genetic Improvement Impacts on Indian Agriculture.' In *Crop Variety Improvement and Its Effect on Productivity: The Impact of International Agricultural Research*, edited by R.E. Evenson and Douglas Gollin, 387–408. Cambridge, UK: CABI Publishing.

MoA(Ministry of Agriculture). 2010. *Seed Net Portal.* Accessed: 2 December, 2010; http://seedvariety.dacnet.nic.in/AvarietyRpt.aspx.

Mohapatra, S. and B. K. Sahoo. 2008. 'Agricultural R&D in India: Trends, Patterns, Determinants and Policy Implication.' Paper presented at the International Conference on Issues in Public Policy and Sustainable Development, 26–28 March. New Delhi: Indira Gandhi National Open University.

Murugkar, M., B. Ramaswami, and M. Shelar. 2007. 'Competition and Monopoly in the Indian Cotton Seed Market.' *Economic and Political Weekly* 62 (37): 3781–89.

Nagarajan, L., P. Pandey, and M. Smale. 2007. 'Seed Systems and Millet Crops in Marginal Environments of India: Industry and Policy Perspectives.' *Quarterly Journal of International Agriculture* 46 (3): 263–88.

National Horticulture Board. 2008. *Annual Report.* New Delhi: National Horticulture Board.

Pal, S. and A. Singh. 1997. 'Agricultural Research and Extension in India: Institutional Structure and Investments.' Policy Paper 7. New Delhi: National Centre for Agricultural Economics and Policy Research.

Pal, S. and D. Byerlee. 2006. 'India: The Funding and Organization of Agricultural Research in India—Evolution and Emerging Policy Issues.' In *Agricultural R&D Policy in The Developing World: Too Little Too Late?* edited by J. M. Alston, P. G. Pardey, and R. Piggott. Washington DC: International Food Policy Research Institute.

Pal, S. and D. Byerlee. 2003. 'The Funding and Organization of Agricultural Research in India: Evolution and Emerging Policy Issues.' Policy Paper 16, National Centre for Agricultural Economics and Policy Research. New Delhi: Indian Council of Agricultural Research (ICAR).

Pal, S., M. A. Rahija, and N. M. Beintema. 2012. 'India: Recent Developments in Agricultural Research.' Agricultural Science and Technology Indicators (ASTI) Country Note. Washington, DC and New Delhi: International Food Policy Research Institute(IFPRI) and Indian Council of Agricultural Research (ICAR).

Pal, S. and R. Tripp. 2002. 'India's Seed Industry Reforms: Prospects and Issues.' *Indian Journal of Agricultural Economics* 57 (3): 443–57.

Pingali, P. 2012. 'Green Revolution: Impacts, Limits, and the Path Ahead.' *Proceedings of the National Academy of Sciences of the United States of America* 109 (31): 12302–308.

Prabhakar Rao, M. 2008. 'Indian Seed Industry Status and Future Trends: Field Crops.' Paper published in the souvenir of Asian Seed Congress (APSA-NSAI Souvenir), Hyderabad.

Pray, C. E. 1984. 'The Impact of Agricultural Research in British India.' *The Journal of Economic History* 49 (2): 429–40.

———. 1983. 'Underinvestment and the Demand for Agricultural Research: A Case Study of the Punjab.' *Food Research Institute Studies* 19 (1): 51–80.

Pray, C. E. and L. Nagarajan. 2012. 'Innovation and Research by Private Agribusiness in India.' IFPRI Discussion Paper 1181. Washington DC: IFPRI.

———. 2013. 'Role of Biotechnology in Stimulating Agribusiness R&D Investment in India.' *Agbioforum* 16 (2): 104–11.

Pray, C. E. and L. Nagarajan. 2010 'Price Controls and Biotechnology Innovation: Are State Government Policies Reducing Research and Innovation by the Ag Biotech Industry in India?' *AgBioForum* 13 (4): 297–307

Pray, C. E. and B. Ramaswami. 2001. 'Liberalization's Impact on the Indian Seed Industry: Competition, Research, and Impact on Farmers.' *International Food and Agribusiness Management Review* 2 (3): 407–20.

Pray, C. E., B. Ramaswami, and T. Kelley. 2001. 'The Impact of Economic Reforms on R&D by the Indian Seed Industry.' *Food Policy* 26 (6): 587–98.

Pray, C. E., S. Ribeiro, R. A. E. Mueller, and P. Rao. 1991. 'Private Research and Public Benefit: The Private Seed Industry for Sorghum and Pearl Millet in India.' *Research Policy* 20 (1): 315–24.

Pray, C. E., P. Bengali, and B. Ramaswami. 2005. 'Costs and Benefits of Biosafety Regulation in India: A Preliminary Assessment.' *Quarterly Journal of International Agriculture* 44 (3): 267–89.

Qaim, M., A. Subramanian, G. Naik, and D. Zilberman. 2006. 'Adoption of Bt. Cotton and Impact Variability: Insights from India.' *Review of Agricultural Economics* 28 (1): 48–58.

Rabobank International. 2001. 'Indian Seed Industry: On the Threshold of Consolidation.' *Food and Agribusiness Research*. The Netherlands: Rabobank International.

Ramanna, A. and M. Smale. 2004. 'Rights and Access to Plant Genetic Resources under India's New Law.' *Development Policy Review* 22 (4): 423–42.

Ramaswami, B. and C. E. Pray. 2002. 'Dissemination of Private Hybrids and Crop Yields in the Semi-Arid Tropics of India.' *Indian Journal of Agricultural Economics* 57 (1): 39–51.

Ramaswami, B., M. Murugkar, and M. Shelar. 2009. 'Product Proliferation in India's Cotton Seed Market: Are There Too Many Varieties?' *Journal of Agricultural & Food Industrial Organization* 7 (1): 1–15.

Ramaswamy C. and K. N. Selvaraj. 2007. 'Prioritizing Agricultural Research and Extension.' INRM Policy Brief No. 15. New Delhi: Asian Development Bank.

Rangnekar, Dwijen. 2004. 'Demanding Stronger Protection for Geographical Indications: The Relationship between Local Knowledge, Information and Reputation.' Discussion Papers 11. Netherlands: United Nations University, Institute for New Technologies.

Rao, S. 2008. 'Economic Impact of Price Controls and Regulations: Case of GM Cotton in India.' Unpublished MS dissertation, Department of Agricultural, Food and Resource Economics, Rutgers, The State University of New Jersey, New Jersey.

Revathy, E. and R. V. Ramana Murthy. 2005. 'Changing Seed Policy, Law and Regulation: An Appraisal of the Emerging Seed Markets in Andhra Pradesh.' GAPS Series, Working Paper: 6. Hyderabad: Centre for Economic and Social Studies.

SeedQuest. 2008. 'Seed Company Krishidhan and Dutch Biotech Company Proteios International Sign Agreement for the Development and Application of Molecular

Markers and DNA Detection Technologies on Cotton.' *SeedQuest.* Accessed: June 2012; http://www.seedquest.com/News/releases/2008/july/23257.htm.

Sethi, P. S. 2009. 'Biofuels in India.' Paper presented at the International Forum for Sustainable Asia and the Pacific. 26–27 June, Japan.

Spielman, David J., F. Harwich, and Klaus von Grebmer. 2007. 'Sharing Science, Building Bridges, and Enhancing Impact: Public-Private Partnerships in the CGIAR,' IFPRI Discussion Paper 708. Washington DC: International Food Policy Research Institute (IFPRI) Accessed: June 2012; http://www.ifpri.org/publication/sharing-science-building-bridges-and-enhancing-impact.

Sulaiman, R. V. 2003. 'Innovations in Agricultural Extension in India.' *SD Dimensions.* Rome: Sustainable Development Department (SD), Food and Agriculture Organization (FAO) Accessed: June 2012; http://www.fao.org/sd/2003/KN0603a2_en.htm.

Sulaiman, R. V. and A. W. Van den Ban. 2001. 'Funding and Delivering Agricultural Extension in India.' Paper presented at the 15th European Seminar on Extension Education. Netherlands: Wageningen.

Tripp, R. 2009. *Biotechnology and Agricultural Development: Transgenic Cotton, Rural Institutions and Resource-Poor Farmers.* London: Routledge.

———. 2001. *Seed Provision and Agricultural Development: The Institutions of Rural Change.* London: Overseas Development Institute (ODI).

Tripp, R. and S. Pal. 2000. 'Information and Agricultural Input Markets: Pearl Millet Seed in Rajasthan.' *Journal of International Development* 12 (1): 133–44.

Trivedi, R. K. 2008. 'Current Situation of Plant Variety Protection Legislation in India.' Presentation at APSA-UPOV Seminar. 9 November, Hyderabad.

USDA-FAS. 2005. 'Gain Report on India Planting Seeds Annual, 2005.' GAIN Report Number: IN5118, Global Agricultural Information Network, Foreign Agricultural Service. Washington DC: USDA.

Verma, P. 2008. 'Opportunities and Challenges of Using Biotechnology in the Seed Industry.' Paper published in the souvenir of Asian Seed Congress (APSA-NSAI Souvenir), Hyderabad.

Viraktamath, B. C. 2007. 'Hybrid Rice Seed Production in India.' In *DRR—Perspective Plan: Vision 2025.* Hyderabad: Directorate of Rice Research. Accessed: June 2012; http://drricar.org/Vision_1%20Final.pdf.

Virk, D. S., K. A. Steele, D. N. Singh, S. C. Prasad, J. S. Gangwar, V. Singh, and J. R. Witcombe. 2004. 'An International Partnership for the Breeding and Delivery of Drought-Tolerant Rice Varieties by Market-Oriented Plant Breeding and Marker Assisted Selection.' In *Resilient Crops for Water-Limited Environments: Proceedings of a workshop held at Cuernavaca, Mexico, 24–28 May 2004*, edited by D. Poland, M. Sawkins, J. M. Ribaut, and D. Hoisington, 30–31. Mexico DF: CIMMYT.

Yog, Raj. 2012. 'Hybrid Rice Breeding – Private Sector Perspective.' Presented at the 6th Hybrid Rice Symposium. 10–12 September, Hyderabad.

CHAPTER 5

National Innovation Systems and the Indian Software Industry Development

Nagesh Kumar

Introduction

The recent rise of India as a competitive exporter of software services has attracted worldwide interest. India's performance in this field is notable as it is leading the developing countries in this activity, which is as highly knowledge intensive as R&D is. The companies from across the world are rushing to India to outsource their software requirements, to set up software development centres to take advantage of the pool of talent available, and also to recruit experienced software personnel. The governments in different developed and developing countries are opening their doors to invite Indian software engineers. The factors contributing to the Indian success are being debated. It is argued here that the Indian success owes largely to the substantial investments made by the government in building the national innovation system (NIS) over the past 50 years. Being an activity very similar to R&D, investments made in creating NIS have also facilitated software development activity.

In this context, this chapter examines the patterns of the technology upgrading and innovative activities of the Indian software industry and analyses the role that NIS has played in general and in particular to the industry in determining these. Such analysis may have useful policy lessons for other developing countries.

The structure of the chapter is as follows. The following section provides an introduction to the software industry in India and discusses its significance in relation to macro aggregates to put them in a perspective. The next section reviews the performance of the Indian software industry with respect to the indicators of technological competence, maturity, productivity improvement, and international competitiveness. It also discusses the extent and nature of MNE (Multinational Enterprise) involvement in the industry. The subsequent section identifies the different elements of NIS as evolved over the past 50 years. The elements of NIS specific to the IT industry that have contributed to the Indian success are discussed in the next section. These elements include government policies to create infrastructure for the development of human resource in science, technology, and engineering, create institutions of excellence and public-funded R&D institutional infrastructure, and provide communication infrastructure to software firms, among others. The following section briefly summarizes the patterns of clustering of the activity around certain centres such as Mumbai, Bengaluru, Delhi, Hyderabad, among others. The patterns of clustering of the software activity around major centres of excellence and other publicly funded institutions highlight the role of NIS in explaining the success story. This is further documented with a more detailed case study of the software development cluster in Bengaluru. Finally, the last section concludes the chapter with a recapitulation of the findings and some remarks on the policy lessons from Indian experience on the relevance of NIS.

Indian software industry development in a perspective

The rise of the IT software and services industry (henceforth software industry) during the 1990s represents one of the most spectacular achievements for the Indian economy. As shown in Table 5.1 and Fig. 5.1, the Indian software industry has grown at a phenomenal compound annual rate of over 50 per cent during the 1990s from a total revenue of US$197 million in 1989–90 to a US$5.7 billion industry by 1999–2000. Furthermore, the industry has earned 70 per cent of its revenue (totaling US$4 billion) from exports. The industry body NASSCOM, or National Association of Software and Service Companies, expects the industry to grow to US$8.7 billion in the year 2000–01.

Table 5.1: Indian software industry revenues and exports (US$ million)

Year	*Total*	*Domestic*	*Exports*
1989–90	197	97	100
1994–95	835	350	485
1995–96	1,224	490	734
1996–97	1,755	670	1,085
1997–98	2,700	950	1,750
1998–99	3,900	1,250	2,650
1999–2000	5,700	1,700	4,000
2000–2001^{P}	8,750	2,450	6,300
Compound annual growth rate 1994–99	56.3%	46.05%	60.71%

Note: P – projected

Source: Kumar (2000b) based on Hanna (1994), Heeks (1996) and NASSCOM (2000a,b)

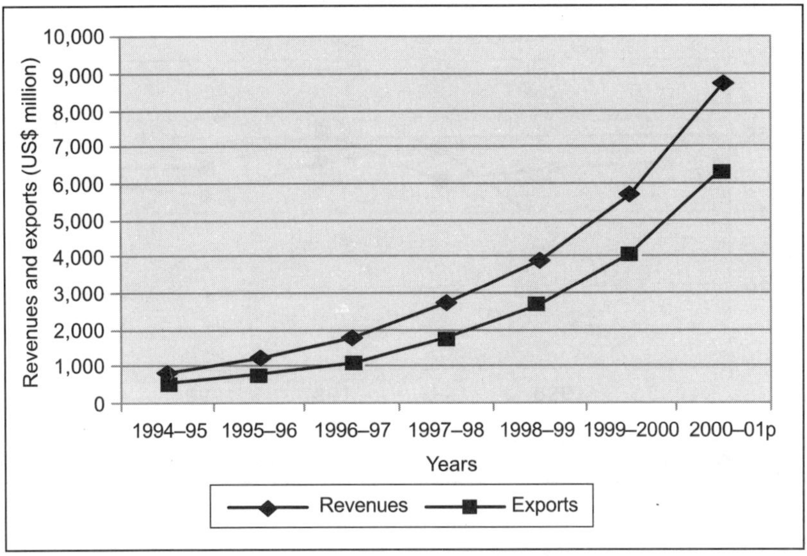

Fig. 5.1: Revenues and exports of Indian software industry
Source: Table 5.1

The growth rates of revenues and India's software exports in US$ terms are summarized in Table 5.2 and Fig. 5.2. It is clear that except for chairing 1996–97, the annual export growth rate has been higher than 50 per cent in all other years. The growth rate of overall revenue has been lower than that of exports except for 1996–97 on account of a somewhat slower rate of growth of the domestic software market. However, the growth rates of overall revenues and exports have tended to converge over the past couple of years as a result of the growth rates of the domestic software market also picking up.

Table 5.2: Growth rates of magnitudes of software revenue and exports

	1994–95	*1995–96*	*1996–97*	*1997–98*	*1998–99*	*1999–2000*
Growth rate of revenue in US$		34.61	56.14	53.85	44.44	48.65
Growth rate of exports in US$	54.46	51.34	47.82	61.29	51.43	50.94

Source: Based on Table 5.1

If the above growth performance is put in an international perspective, the Indian software industry would account for roughly 1–1.5 per cent of the US$400 billion global software industry. However, India's share in the global market for customised software that is outsourced across borders is significant at 18.5 per cent in 1999 compared to 11.9 per cent in 1991 (NASSCOM 2000a).

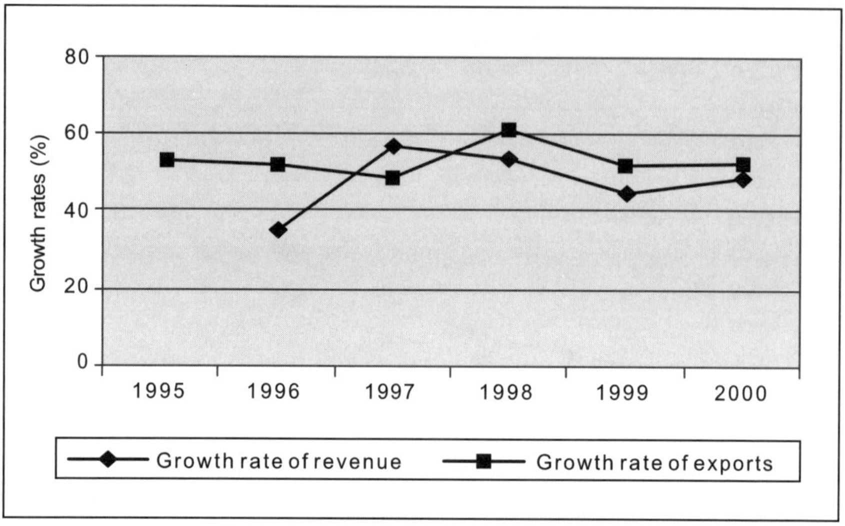

Fig. 5.2: Growth rates of software revenue and exports

Source: Based on Table 5.2

The growth rate of the Indian software industry has been substantially higher than the global software industry. Apparently, India is the only country in the world to register a growth rate of around 50 per cent in the software industry.

The software industry still accounts for a rather marginal share of India's gross national product (GNP) but it has been rising fast. In 1998–99 the industry accounted for just 1 per cent of India's GNP. In the following year, its proportion had gone up to 1.4 per cent. Even with just 1 per cent share of GNP, the sector contributed 7.2 per cent of the growth rate of the economy (Table 5.3). Given the rates at which it is growing, it is bound to emerge as an important sector of the Indian economy in the future. The McKinsey Report projected a 7.5 per cent share of the sector in the overall gross domestic product (GDP) by 2008. The growth performance described above has endeared the Indian software companies to the investors. Their stocks are among the most sought after at the Indian stock exchanges and command hefty premia. The market capitalisation of IT industry had peaked to US$95 billion in February 2000 and had been estimated at US$55 billion in June 2000 compared to just US$4 billion in January 1999. The share of software exports to India's exports of goods and services has risen from negligible in 1990, to over 8 per cent 1999–2000 as is clear from Table 5.3. The NASSCOM–McKinsey Report (1999) and NTITSD project that by 2008, software and services exports would account for 35 per cent of India's exports.

Table 5.3: Software industry in relation to India's macro parameters (in billion US$)

	1998–99	*1999–2000*	*Incremental*
India's GNP	379.7	404.7	25
Indian software industry revenues	3.9	5.7	1.8
Share of software industry in the GNP	1.027	1.41	7.2
India's exports of goods and services	45.76	49.7	3.94
Exports of software services	2.65	4.0	1.35
Share of software	5.79	8.05	34.26

Note: GNP – gross national product
Source: Kumar (2000b)

Nature of technological upgrading and innovation in Indian software industry

Although the magnitudes of exports of software and services from India have grown rapidly over the past decade, the general perception is that these exports comprise low-value services. That perception emanates from the fact that in the early years, the bulk of the software export activity of Indian enterprises consisted in lending their software professionals to their clients to deliver their services 'on-site.' It was considered to be a rather lower level of skill intensity compared to software product designing and development and has been derisively termed as 'body-shopping' (Heeks 1996). However, the Indian software industry has since come of age in terms of capabilities, sophistication, range of expertise, and worldwide reach. In what follows, we briefly review the performance of the industry in terms of certain indicators of growing capability.

Moving away from body-shopping

As observed earlier, the bulk of the software export activity of Indian enterprises in the early period comprised body-shopping or on-site. The advantage of Indian enterprises in the on-site work emanated largely from the lower salaries of Indian software professionals compared to those available in developed countries. However, Indian companies have progressively demonstrated their technological and project management skills by successfully completing turnkey projects for large companies. As a result, the proportion of on-site exports has begun to come down in India's software exports from 90 per cent in 1988 to 49 per cent by 1999–2000, as shown in Table 5.4. An increasing proportion of India's software is developed 'off- shore' at the home bases of exporters in India and exported. The 'off-shore' development, as it is called in the industry, has been partly facilitated by the improved communication links in the software technology parks (STPs) set up by the government that allow teams of professionals at vendors' and clients' ends to be in constant touch on a real time basis, and by the growing visa restrictions in the USA and Europe. It also turns out to be cheaper for the clients. Besides, the 12-hour time lag between India and the USA virtually doubles the working time per day and hence cuts the development life cycle by half. The share of off-shore development peaked in 1997–98 at 55.5 per cent and has since declined to 44.4 per cent in 1999 and to 49 per cent in 2000. This recent rise in the share of on-site development has been attributed to rising focus on e-commerce related jobs in the recent years, which require a greater presence of software developers at the clients' sites.[1]

Table 5.4: Locational division of labour in Indian software development (in %)

Location of work	*1988*	*1995*	*1997–98*	*1998–99*	*1999–2000*
On-site (at client's site abroad)	90	66	43.8	54.4	49.5
Off-shore (at vendor's site in India)	10	33	55.5	44.4	49.4

Source: Kumar (2000c) based on NASSCOM; Heeks (1996); *Dataquest* (various issues)

Increasing focus on high-value consulting and packaged software

So far, the Indian software enterprises have generally focused on services that are considered to be low value adding. Having got themselves established as suppliers of these services, Indian companies are now making a conscious effort to increase exports of high-end consulting with the development of domain expertise and export of packaged software. Infosys, for instance, is focusing on the export of end-to-end services.[2] As Indian software enterprises establish their credentials and competence, they are consciously seeking fair value for their work.[3] This, however, may be applicable for leading companies such as TCS, Infosys, HCL Technologies, WIPRO, and Satyam Computer Services, which are now providing higher-end programming solutions to their clients. Infosys has successfully renegotiated its per man-hour charges with its clients. It reportedly commands US$90 per hour.[4]

Over the past few years, Indian companies have also managed to develop and launch a number of proprietary software products. A niche market has

[1] *Dataquest*, 15 July 2000: 228.

[2] Kumar (2000d).

[3] *Economic Times*, 25 April 2000.

[4] *Dataquest*, 15 July 2000: 235.

been created in banking, financial, and accounting software. These include, for instance, I-Flex that has been used by over 240 financial institutions in 69 countries. Polaris has developed a proprietary retail banking software Polaris Point and is tying up with Bull, France, for its marketing in Europe.[5] Banking solutions from Infosys (Financle, Bankaway, and Payaway) have been adpated by 22 domestic and 16 overseas banks across 12 countries.[6] TCS has launched packaged software for banking insurance, securities, accounting, and healthcare industries. It is currently developing industry-specific software products for several manufacturing industries such as cement, steel, chemicals, petrochemicals, and refineries in collaboration with industry firms. TCS is targeting 30 per cent of its revenue from the branded products within the next three years up from 5 per cent at present.[7] TCS has also launched its branded integrated suite of software tools Mastercraft, which is claimed to have been received well in the USA and Europe and carries a price tag of US$150,000.[8] WIPRO Technologies has recently launched two branded products, viz., Teleprodigy, a billing system for ISPs, and WebSecure, an Internet security package. It is focusing on global brand building and plans to come up with a branded product every year.[9] NIIT and Pentamedia are developing multimedia products on CD-ROMs in large numbers. A number of even smaller software companies have developed packaged software that are sold in domestic market. For example, Tally, a popular accounting package for small and medium enterprises, which is being used by 50,000 companies and has been approved by the Accountants' professional bodies in India and the UK, has been developed by a smaller highly specialized software company. Despite these efforts, the share of products and packages in the Indian exports of software is still low at 7.9 per cent (NASSCOM 2000a). However, given the high entry barriers in the packages market, the entry of Indian companies in their exports is, nevertheless, significant. Furthermore, success in markets such as those of software products is a cumulative process. Once the image or brand value of India as a reliable supplier improves with growing exports of products and services, given the current trend, and as leading companies augment the requisite scale of operations, global reach, and financial clout for sustaining large marketing efforts, it would be easier for these companies to make a serious dent in the markets for products.

[5] *Business Standard*, 25 May 2000.

[6] Kumar (2000d); *Economic Times*, 13 July 2000.

[7] *Economic Times*, 6 June 2000; *Business Standard*, 8 June 2000.

[8] Kumar (2000d).

[9] *Times of India*, 23 July 2000; *Economic Times*, 23 July 2000.

Broad and expanding supply base

An interesting feature of the Indian software industry is the relatively large and growing number of companies participating in the development and export activity. One indicator of the supply base is the membership of the industry body, namely, NASSCOM, which that has grown from just 38 members in 1988 to 572 members as on December 1999. NASSCOM members are generally medium and large companies (those with 20 employees or more). There are numerous small and informal sector enterprises as well, which have displayed considerable dynamism (see Kumar 2000a). As expected, larger firms do account for a disproportionate share of revenue and exports, with the top 25 companies accounting for 61 per cent share and the top five, 29 per cent share of exports in 1999–2000 (NASSCOM 2000b). However, the vendor concentration is not as high as prevalent in many other industries.

Locally anchored capability

There is a qualitative difference between the export success of a country based on subsidiaries of foreign-based MNEs and another based on indigenous enterprises in terms of local anchoring of capabilities. In terms of this, the Indian export success is primarily driven by local enterprise, resources, and talent. The role played by MNEs in software development in India is quite limited. Although all the major software companies have established development bases in India, their overall share in India's exports of software is rather small. MNEs do not figure among the top seven software companies in India, ranked either on the basis of overall sales or exports. Among the top 20 software companies too, no more than six are MNE affiliates or joint ventures. Seventy-nine of the 572 member companies of NASSCOM are reported as foreign subsidiaries. Some of these are actually subsidiaries of companies promoted by non-resident Indians in the USA such as Mastech, CBSI, IMR, and Syntel, rather than being associated with USA MNEs (Arora et al. 2000). Some others were Indian companies to begin with but have been subsequently taken over by foreign companies such as Hinditron, which has been taken over by TAIB Bank E.C., Bahrain, or IIS Infotech, which has been taken over by the FI Group of UK. The foreign subsidiaries include software development centres of software MNEs and also subsidiaries of other MNEs that develop software for their parents' applications. The latter include subsidiaries of financial services companies such as Citicorp, Deutsche Bank, Churchill Insurance, and Phoenix Life Mutual; and telecommunication MNEs such as Hughes,

Motorola, among others. In addition, MNEs have set up 16 joint ventures with local enterprises such as British Aerospace with Hindustan Aeronautics, Bell South with Telecommunication Corporation of India (TCIL), and British Telecom with the Mahindra Group, among others. In all, 95 companies have controlling foreign participation.

Table 5.5 summarizes the shares of these 79 foreign subsidiaries in the total sales and exports of the software industry. It would appear that MNE subsidiaries have a higher degree of export orientation with 94 per cent of their earnings coming from exports compared to local companies. This is because often, they are exclusively catering to the demand of their parents. However, collectively, they account for less than 19 per cent share of exports in 1998–99.

Table 5.5: Share of foreign subsidiaries in India's software industry

Share of 79 foreign subsidiaries in	*1997–98*	*1998–99*
Total revenue	12.27	13.7
Total exports	16.77	18.66

Source: Kumar (2000d)

Another aspect of the role of foreign direct investment (FDI) and MNE subsidiaries in the development of Indian software industry is apparent from the pattern of their entry. As Table 5.6 shows, the bulk of the entries took place from 1994 by which time India's potential as a base for software development was already established and not the other way round.

Table 5.6: Time profile of entry of MNEs in Indian software industry

Period	*Entries of MNEs as subsidiaries or joint ventures*
Upto 1987	11
1988–90	14
1991–93	15
1994–96	39
1997–99	16

Source: Kumar (2000d)

Increasing international orientation of Indian companies

The Indian software exporting companies themselves are sufficiently global in their outlook. As many as 212 Indian software companies have set up 509 overseas offices or subsidiaries. Of these 509 offices, 266 had been set up in

North America, 122 in Europe, 59 in Asia, excluding India, 25 in Australia and New Zealand, 25 in Africa, and 12 in Latin America (NASSCOM 2000a). A few leading companies have established extensive networks of offices and subsidiaries all over the world to tap opportunities in different markets similar to the operations of a multinational corporation. These firms include TCS, HCL Technologies, Infosys Technologies, and NIIT.[10] Four Indian companies have got themselves listed on American stock exchanges and more are planning such moves.

International quality accreditations and process maturity levels

International orientation and the increasing professionalism of Indian software enterprises have prompted them to align their processes with global best practices and obtain international certifications. For instance, 170 Indian companies have obtained the International Standards Organization 9000 (ISO 9000) certification by June 2000. Fifty more companies are expecting to get their ISO 9000 certifications by March 2001 (NASSCOM 2000a). Furthermore, as many as 38 Indian companies have received SEI-CMM, or Software Engineering Institute, USA's, Capability Maturity Model, Certification at Levels 3 or above. India's lead in high maturity levels is quite clear now with 29 of 31 non-USA companies, which have been certified at high maturity levels, namely, Levels 4 and 5 in terms of SEI-CMM, being Indian. Of the 31 companies certified at Level 5 world-wide, 16 that are outside the USA are in India (Table 5.7). Level 5 represents the Optimising Level of process maturity and is the highest stage to be reached. Outside the USA, Australia and Israel have one organisation each qualifying Level 4. This shows that Indian software enterprises, especially the leading ones, have strived to attain excellence in their professionalism and best practices.

Table 5.7: High-maturity organisations

Level	*Number of organisations certified worldwide, as on 4 May 2000*	*Of which non -USA*	*Of which in India*
Level 4	45	15	13
Level 5	31	16	16
Total high maturity	76	31	29

Source: Based on Software Engineering Institute (2000)

[10] See Kumar (2000d) for detailed case studies.

Geographical reach

Indian software services are exported worldwide. However, the bulk (62 per cent) of export is concentrated in the North America, mainly the USA, which is also the largest market for software. Europe accounts for 23.5 per cent of India's exports, and the Asia-Pacific for a further 10 per cent (Table 5.8). Language also contributes to a high concentration in the USA.

Table 5.8: Geographical distribution of Indian software exports, 1999–2000

Region	*Share of India's exports*	*Share of global software market*
North America	62.0	45.1
Europe	23.5	27.6
Japan	3.5	16.9
South East Asia	3.5	2.8
Australia and New Zealand	1.5	2.1
West Asia	1.5	1.8
Rest of the world	4.5	3.7

Source: On the basis of NASSCOM (2000b), and OECD (1997)

Range of domain expertise and applications

Indian companies have developed domain expertise in a wide range of domains and industries. Banking, insurance, and finance have emerged as areas in which they have developed particular expertise and have even launched packaged software (Table 5.9). They are also able to undertake a variety of tasks as listed in Table 5.10.

Table 5.9: Major domain specializations of Indian software companies

Domains/ Sector	*Number of companies offering expertise*
Banking, insurance, stock exchange, financial accounting	247
Manufacturing, retail, trading and distribution	331
Transport/airlines/railways/ports	157
Web applications/online information services	295
Engineering, electronics, design automation/robotics	224
Medical and health	163
Education, training/entertainment	115
Telecommunications	174

Source: Kumar (2000d)

Table 5.10: Major areas of specialization of Indian software companies

Areas	*Number of companies offering expertise*
Web technologies/intranet/internet/e-commerce	319
Euro currency solutions	132
Software product development	286
Software maintenance and migration	233
RDBMS/data warehousing/datamining	215
ERP/MRP solutions	200
GIS/imaging	55
System integration/networking	192
Business process consultancy/reengineering	168

Source: Kumar (2000d)

An evidence of the growing ability and expertise of Indian software companies was provided by their ability to manage transition from Y2K-related projects successfully. In 1998–99, 16.5 per cent of the export earnings of Indian companies were derived from Y2K-related projects.[11] During 1996–99, Indian companies are reported to have earned US$2.5 billion from Y2K projects (NASSCOM 2000a). Hence, it was widely expected that the loss of these projects with the turn of the century would lead to a decline in the growth rates of exports. However, it is quite clear by now that the Y2K transition has been managed successfully. The industry has, in fact, clocked a 51 per cent rise in exports in 1999–2000, which included the duration of three months post-Y2K. All the leading companies have announced a 50–100 per cent rise in their revenues and export earnings for the first quarter of the accounting year 2000–01. This transition has been managed because of their ability to quickly diversify into Internet and e-commerce related technologies and applications that are now booming. Evidently web-based revenue increased its proportion from 4.8 per cent to 15.6 per cent over the period 1998–99 to 1999–2000.[12] In fact, Y2K provided Indian companies with an opportunity to know new potential clients and display their competence. According to surveys conducted by CMU, the expertise levels of Indian companies on UNIX and Windows NT platforms is considered to be on par with the USA firms. Indian companies have also grown in their ability to handle larger and more complex projects than in the past (Arora et al. 2000). Now 300–500 man-year projects are not a rarity any more.[13]

[11] See *Dataquest*, 15 July 2000.

[12] *Dataquest*, 15 July 2000.

[13] *Dataquest*, 15 July 1999.

Increasing research and development thrust

In tune with their growing internationalisation and product focus, the R&D thrust of leading Indian companies is increasing. TCS has a well-equipped Tata Research Development and Design Centre (TRDDC) in Pune. TCS has also set up a centre specializing on Linux at Columbus, Indiana, on networking at Riverside, CA, and on business components at Milwaukee, Wisconsin. Other leading firms such as Infosys, WIPRO, and Computer Maintenance Corporation (CMC) are also known to have in-house R&D laboratories. NIIT has recently set up a Centre for Research in Cognitive Systems (CRCS) at IIT Delhi. NASSCOM (2000b) finds that R&D spending in the industry has increased from 2.5 per cent of total spending on R&D in 1997–98 to about 3.4 per cent during 1999–2000. Although the proportion of R&D spending has increased, it is small compared to international norms in the industry. A bigger R&D thrust will be necessary for Indian software enterprises to establish themselves as innovators and developers of new products and technologies rather than just providers of coding and programming services. Such an evolution of Indian software enterprises is important if they are to realize the target of generating revenues equal to US$8 billion from export of products.

Moving up the value chain? Evidence from recent trends in enterprise performance

An important issue concerning the technological upgrading, international competitiveness, and the ability to move up the value chain of Indian software enterprises is whether the enterprises are able to constantly improve their productivity, improve the efficiency in the use of other inputs, and hence, reduce the unit cost of production, and improve profit margins. An attempt has been made to examine the performance of Indian software enterprises in terms of these indicators with the help of an exclusive panel data set covering a sample of 66 Indian software companies with a fair representation of small, medium, and large companies for the period 1994–99 [see Kumar (2000c) for more details]. The patterns emerging from this analysis are summarized subsequently.

Domestic value addition

Software companies spend quite a substantial proportion of their foreign exchange earnings in the process of delivering these services; for instance, on travel and overseas stay of personnel. The movement in proportion of foreign exchange utilisation per unit of exports over time could also indicate some trends in terms of domestic value addition. Table 5.11 and Fig. 5.3 summarize the patterns of foreign exchange utilisation by sample companies per unit of exports. It is apparent that foreign exchange utilisation per unit of exports by the industry firms at large has declined in a steady manner since 1996, from 62 per cent in 1996 to 48 per cent in 1999. This suggests that the proportion of net exports in total exports has gone up from 1996, from 38 per cent of total exports to 52 per cent of total exports. It is a healthy sign and suggests that the proportion of domestic value added has gone up in the Indian software export activity. To some extent, it reflects the increasing importance of off-shore development, and hence, savings of foreign exchange on travel, etc. In terms of size classes, smaller firms have a significantly higher proportion of foreign exchange utilisation per unit of exports compared to other two groups. It tends to indicate the economies of scale inherent in foreign activity such as maintaining a foreign office, foreign travel, etc. That makes smaller firms least efficient from the point of view of foreign exchange utilisation. Medium-scale firms also had a slightly higher foreign exchange utilisation compared to larger firms till 1998. However, in 1999, their foreign exchange utilisation has come down sharply as a proportion of their exports, making them the most efficient in terms of domestic value addition (Table 5.11 and Fig. 5.3).

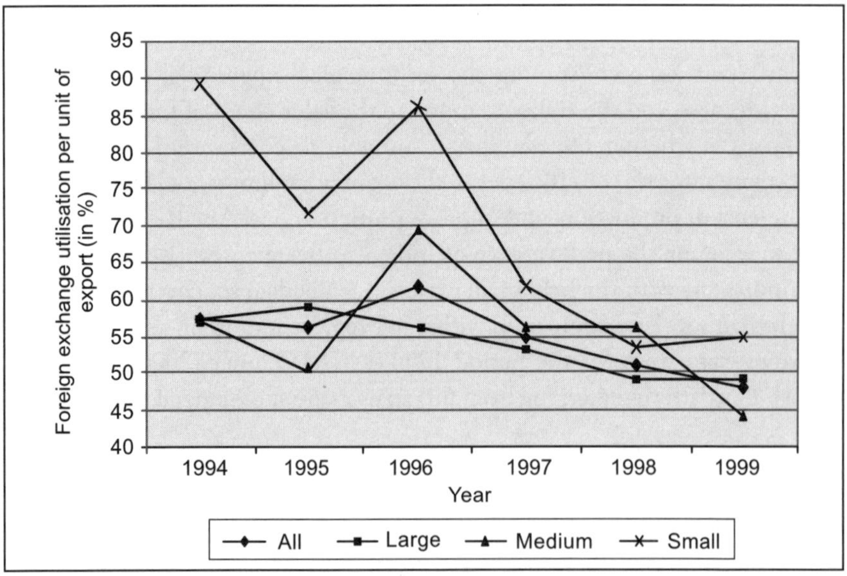

Fig. 5.3: Foreign exchange utilisation per unit of exports
Source: Kumar (2000c)

Table 5.11: Foreign exchange utilisation per unit of exports (in %)

	1994	*1995*	*1996*	*1997*	*1998*	*1999*
Full sample	57.41	56.39	61.66	54.65	51.11	48.01
Large firms	57.26	59.21	56.31	53.19	49.01	49.04
Medium-sized firms	56.96	50.00	69.35	56.36	56.29	44.18
Small firms	89.22	71.83	86.27	61.66	53.38	54.95

Source: Author's own calculations

Labour productivity

Productivity performance in the software industry has to be judged with respect to the key resource in the industry, that is, human resources. Labour productivity has been measured in terms of revenue per unit of wage bill to take care of possible differences in the quality of manpower and to capture the overall efficiency in the use of labour, keeping in mind rapidly rising salaries of the work force. Table 5.12 summarizes the patterns with respect to productivity measured in terms of revenue per unit of wage bill. The productivity measured this way declines over the period 1994–97 for the full sample as well as for different groups of firms. Since 1997, however, Indian software industry is able to improve the productivity even after taking care of rising wage cost. It would appear that over the past few years, the Indian software industry has made an effort to improve the efficiency in the use of its key resource, that is, manpower. This effort might have been prompted by the rising costs and growing scarcity of trained manpower. Among the groups, smaller firms had the highest ratio of revenue per unit of wage bill in 1994, probably owing to lower rates of employee compensation. However, the tight labour market conditions that have prevailed in the subsequent years due to increasing competition for knowledge workers have led to a converging trend in productivity levels of larger and smaller firms. The productivity of medium-scale firms had roughly converged to the levels of larger firms in 1996. Since then, however, it has declined and has diverged from that of larger firms. Thus, among the size classes, larger firms have best levels of efficiency in the utilisation of the key resource in the industry, that is, manpower (Table 5.12 and Fig. 5.4).

Table 5.12: Trends in labour productivity (in %)

	1994	*1995*	*1996*	*1997*	*1998*	*1999*
Full sample	6.21	5.52	5.33	4.46	4.81	4.58
Large firms	7.36	5.69	5.27	4.76	5.31	5.03
Medium-sized firms	4.15	4.68	5.22	3.71	3.80	3.72
Small firms	10.06	9.43	7.45	6.07	4.69	4.35

Source: Author's own calculations

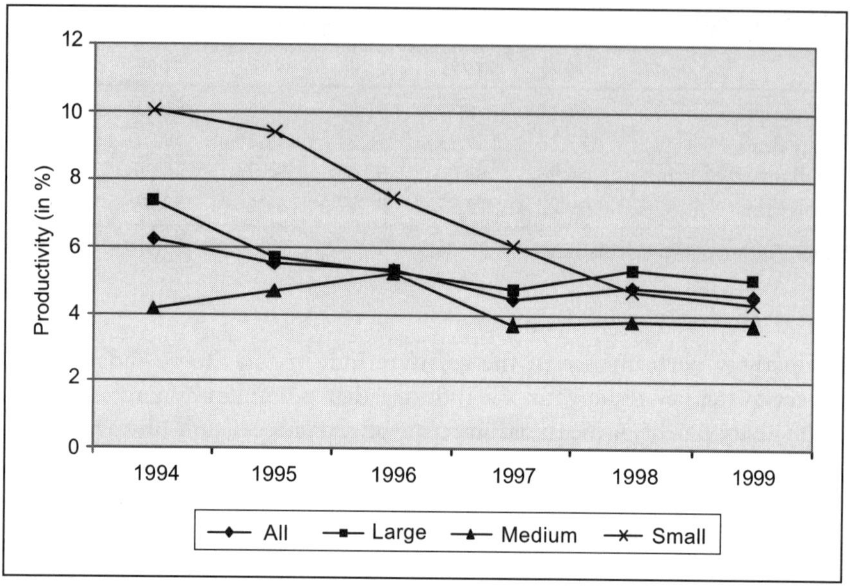

Fig. 5.4: Trends in labour productivity (in %)

Source: Kumar (2000c)

Unit cost of production

The unit cost of production could be another measure of efficiency in overall resource use by companies. It has been measured by dividing the overall cost of production (a sum of material and energy costs, wages, other operating expenses, and depreciation) by sales revenue. Table 5.13 summarizes the pattern with respect to unit cost. In general, the trend is of rise in the cost of production over the years covered. In large parts, it reflects the rising cost of main item of cost that is, wage bill. However, the costs have been relatively stable during 1996–97 and also during 1998–99. In part, this reflects the growing efficiency in the labour use since 1998, which has been observed above (Table 5.13 and Fig. 5.5).

Table 5.13: Trends in unit cost of production (in %)

	1994	*1995*	*1996*	*1997*	*1998*	*1999*
Full sample	46.10	48.60	50.77	50.22	54.35	54.15
Large firms	43.52	47.29	48.22	47.71	53.98	54.63
Medium-sized firms	52.88	53.27	56.07	54.75	55.72	53.12
Small firms	49.12	45.66	57.22	57.28	52.92	52.56

Source: Author's own calculations

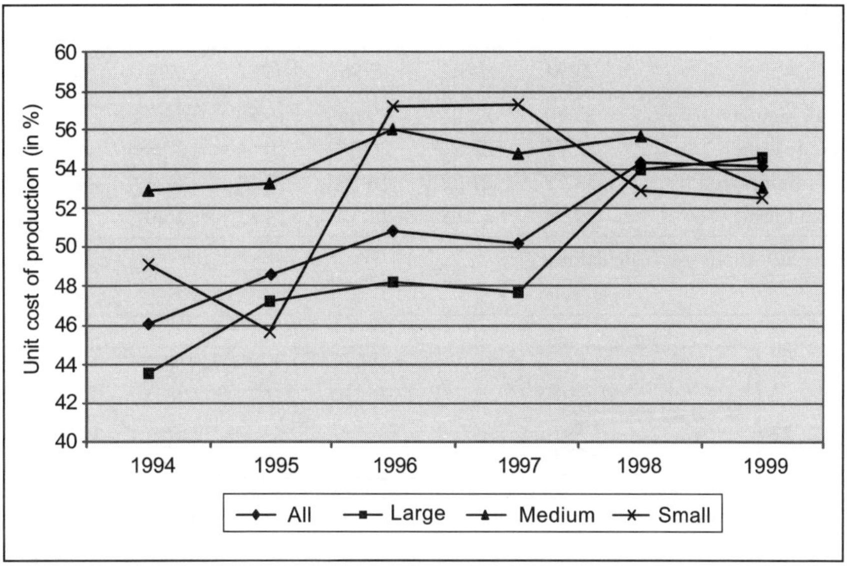

Fig. 5.5: Trends in unit cost of production (in %)
Source: Kumar (2000c)

Profit margins

Besides indicating the overall efficiency, profit margins of firms competing on the basis of costs are generally under pressure because of rising competition. On the other hand, firms moving up the value chain may be able to improve their margins. Table 5.14 summarizes the trends pertaining to profit margins for the sample firms. Here profit margins have been measured in terms of proportion of profit before tax in revenue. Like some of the other performance indicators, the profit margins also show a shift in 1996. In the period 1994–96, the margins show a declining trend and a rising trend from 1996. It may have something to do with the effort of the industry towards improving the efficiency in resource use and rising export-orientation since 1996. Among the groups, smaller firms reveal lower margins during the period. However, they have improved their margins sharply from 1996. In fact, in 1999, profit margins of smaller firms have exceeded those of the other two groups. It is to be seen whether this performance will be sustained in the future. Medium-sized firms also have somewhat lower margins. However, the margins of these firms have tended to converge with those of larger firms (Table 5.14 and Fig. 5.6).

Table 5.14: Trends in profit margins (in %)

	1994	*1995*	*1996*	*1997*	*1998*	*1999*
Full sample	21.70	22.03	19.66	19.88	19.74	21.69
Large firms	22.71	22.24	20.55	20.64	20.89	21.82
Medium-sized firms	18.69	21.22	18.66	19.13	17.33	20.86
Small firms	22.98	22.88	12.52	14.72	16.07	24.11

Source: Author's own calculations

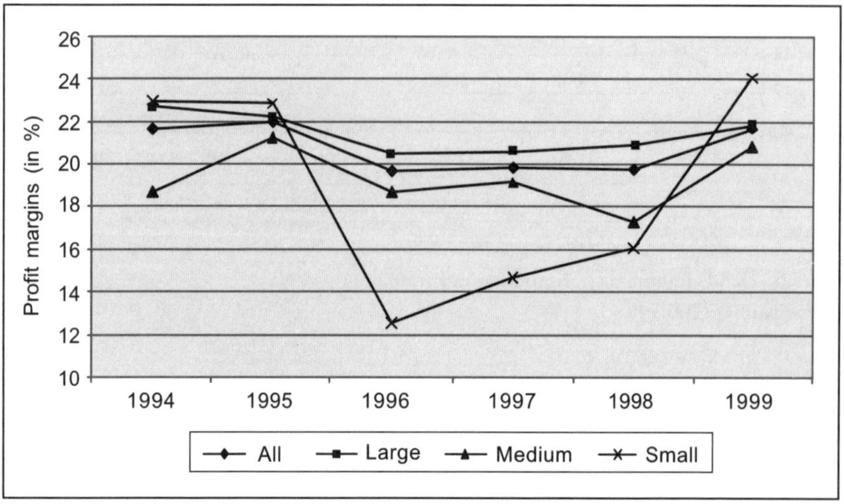

Fig. 5.6: Trends in profit margins (in %)
Source: Kumar (2000c)

The overall conclusion emerging from the above analysis of enterprise performance suggests that 1996 (that is, the financial year 1995–96) was a year of transition for the industry. The trend with respect to a number of indicators shows a change with this year. This includes rise in export-orientation, rising net exports, rise in profit margins, and decline in unit cost of production. These trends got reflected in terms of improved productivity since 1997. Small-scale firms are quite different from medium and large-sized firms with a greater focus on domestic markets. However, they are fast catching up with the larger ones. Finally, the analysis also indicates the presence of economies of scale in the utilisation of foreign exchange and human resources. Hence, some consolidation of the industry may lead to improvement of overall competitiveness.

National innovation system in India: Generic skill formation and institution building

The term NIS has been recently incorporated into the literature to define the institutional set-up that results in technological inventions and capability (Lundvall 1992; Nelson 1993). NIS is defined as a set of institutions whose interactions determine the innovative performance of the national firms. The Government of India has consciously invested considerable resources in building different aspects of NIS over the past five decades to accelerate the pace of local generation of technology. These steps include the efforts to develop infrastructure for the human resource development, institutional framework and infrastructure for technological development, government's direct participation in scientific and technological development, incentives for promotion of in-house R&D activity in industry, enactment of new Patent Act, and encouragement to utilise indigenous technology. We briefly discuss below the major initiatives taken in each respect.

Infrastructure for human resource development

The evolution of the software industry in India is largely due to the substantial investments made by the government over the years in building infrastructure for human resources development, particularly in engineering and science and technology (S&T). The Scientific Policy Resolution of 1958 recognized the need for the creation of S&T manpower and placed emphasis on training and encouraging the scientific personnel. As a result, a large number of institutions training S&T personnel were set up in the country. By 1999, India had set up 237 universities (including Deemed Universities) and 10,555 colleges with a total enrolment of nearly 7 million students, of which 2 million were in technical disciplines. This system of higher technical education now produces more than 250,000 scientists and technologists per year. These include over 10,000 doctorates awarded in various disciplines (DST 1999). In particular, for the engineering education that is relevant for software development, by 1997 India had set up 607 institutions for degree qualification in engineering and another 1,135 institutions granting engineering diplomas (Table 5.15). These institutions produced 73,936 engineering degree holders and 93,121 diploma holders in engineering. Among these, the six Indian Institutes of Technology (IITs) and Indian Institute of Science (IISc), Bengaluru, are known for their excellence and are the preferred sources of manpower for the software companies. About 75 per cent of the professionals employed by software companies are engineering degree holders, according to NASSCOM surveys. India has been ranked third in the world (after Israel and France) in terms of the quality of scientists and engineers by the World Economic Forum (1999). India also enjoys a relatively high rank (17th compared to overall 52nd rank) in terms of quality of mathematics and science education in the world competitiveness rankings (ibid). The government also took the initiative in the early 1980s to start graduate and postgraduate courses in computer applications (BCA and MCA) at various universities.

Table 5.15: Engineering training institutions and students in India

Year	*Institutions*		*Students*			
	Degree level	*Diploma level*	*Degree level*		*Diploma level*	
			Admission	*Pass-outs*	*Admission*	*Pass-outs*
1951	53	89	4,788	2,893	6,216	2,626
1961	111	209	15,497	7,026	26,525	10,349
1971	134	301	18,207	18,223	33,154	17,699
1981	171	363	34,835	19,012	61,114	35,487
1986	248	680	36,328	29,291	94,728	50,390
1991	351	910	70,481	44,724	117,835	65,325
1997	607	1135	138,450	73,936[a]	186,155	93,121[a]

Notes: [a]Estimated
Source: IAMR (2000)

Scientific and technological infrastructure

Since the early post-Independence period, the government has attached a high priority to building S&T infrastructure as a part of the planned development strategy. A three-tiered structure of policy-making bodies on S&T has evolved over time to guide S&T programmes. This includes a Cabinet Committee on Science and Technology (CCST), the Science Advisory Committee to the Cabinet (SACC), and the Committee of Secretaries on S&T (CoSST) (Government of India 1999). A separate Ministry of Science and Technology was created in 1985. The public funded S&T infrastructure has been created over the years in the framework of the research councils, viz., the Indian Council of Agricultural Research (ICAR), Council of Scientific and Industrial Research (CSIR), Indian Council of Medical Research (ICMR), and other scientific departments such as the Department of Atomic Energy (DAE), Department of Space, Department of Electronics, Department of Science and Technology (DST), Department of Scientific and Industrial Research (DSIR), Department of Biotechnology (DBT), Department of Agricultural Research and Extension, Department of Environment, Department of Space, and Defence Research and Development Organization. The infrastructure covers now a network of about 140 specialized laboratories and institutions providing a variety of services such as testing and measurement, fabrication and pilot plant facilities, consultancy, etc., to technology development programmes of other institutions, besides their own research programmes. These laboratories have proved to be the largest source of experienced scientists/technologists for in-house R&D activities in the industry (Desai 1980).

Direct involvement in technology development

The government itself is the largest spender on technology generation in India. Over 63.3 per cent of India's total expenditure on R&D and related S&T activities is still spent by the central government with another 8.7 per cent by the state governments. The government-funded S&T activity is conducted in the framework of the Five-year Plans. The government expenditure on R&D is concentrated in the areas of strategic importance to the economy as a whole, such as defence, communication, meteorology and remote sensing (space research), electronics, harnessing atomic energy, agriculture and livestock, medicine and health, non-conventional energy, environment, and oceanography (DST 1999). In terms of proportion, 64 per cent of the central government's expenditure on S&T is accounted for by three sectors, viz., defence, space, and atomic energy. Agricultural research accounts for nearly 10 per cent. The rest is spent by other scientific activities such as industrial research (CSIR), environment, biotechnology, health research, electronics, ocean development, non-conventional energy, etc. The CSIR laboratories undertake research programmes in different major branches of industry, which cover chemicals and bio-chemicals, physical metallurgical and mechanical engineering, machine tools, electronics, aeronautics, drugs, leather, post-harvest technology, food, rural technologies, astro-physics, etc. (DSIR 1999).

Promotion and support of corporate R&D activity

Besides the creation of S&T infrastructure, the government has encouraged industries to take up in-house R&D activity through other policy instruments. In 1974, a scheme for recognition of in-house R&D establishments of industrial units was started. The recognized R&D units get facilities for import equipment, raw material, samples, prototypes, etc., for their R&D work under Open General

License, without any ceiling. Full income tax relief on the in-house R&D expenditure by the company related to the business of the company is permitted. R&D expenditure in government-approved in-house R&D centres is allowed a weighted relief of 125 per cent since 1998 for companies engaged in the business of manufacture or production of drugs, pharmaceuticals, electronic equipment, computers, telecommunication equipment, computers, and chemicals. In the Budget 2000, the weighted relief was raised to 150 per cent. Starting with 106 recognized R&D units in 1973, the scheme now covers 1,222 units as of the end of December 1998 (DSIR 1999). Of the 1,224 in-house R&D units, 135 belong to the public sector undertakings. These in-house R&D units are encouraged to work towards import substitution, export promotion, process/product development and design improvement, development of new process/product technologies, increased efficiency in use of resources and fuels, recycling of wastes, etc., in the respective areas of the industrial units. Sometimes foreign collaboration approvals/extensions are granted with the understanding that importer would undertake R&D activity to absorb the technology. Technology Absorption and Adaptation Scheme (TAAS) of DSIR aims to provide a catalytic support for accelerated absorption and adaptation of imported technologies by the industrial units. It was made mandatory to highlight the efforts taken towards absorption of technology imports in a separate chapter of the annual report of all the importing firms. In addition, industry research associations have been set up to take up work on common problems. In 1988, the DSIR launched a scheme of granting recognition to Scientific and Industrial Research Organizations (SIROs). At present, there are 159 SIROs recognized by the DSIR. The SIROs have employed qualified scientists and researchers and also established good infrastructural facilities for research. In 1996–97, the government proposed to provide a five-year tax holiday to approved companies whose main objective is scientific and industrial research. It is provided to all new and existing companies, which are accorded approval before 1 April 1998. Besides, the government has introduced a system of allowing accelerated depreciation in respect of blocks of assets and rationalised the rate structure by reducing the number of rates as also by providing for depreciation at higher rates.

Funding of R&D projects in industry

Over the years, a number of programmes for directly supporting R&D activity in the industry have been started by different scientific agencies of the Indian government. DSIR, for instance, operates a Programme aimed at Technological

Self Reliance (PATSER) to support R&D projects in Industry. About 100 R&D and design and engineering projects have been supported by the end of 1998. Some of these projects involve collaboration with public-funded R&D institutions. DST is funding several industrial R&D programmes such as Home Grown Technology Projects, Drugs and Pharmaceuticals Research Programme, Instrument Development Programme, and Advanced Materials development Programme. DBT has been promoting and financing various aspects of biotechnology R&D activity undertaken by industry and other institutions. A Technology Development Fund has been created out of a collection of a 5 per cent cess imposed on the technology import payments, which is used to help the indigenously developed technologies reach the stage of commercial production. A Technology Development Board has been constituted in 1995 to utilise the Fund by providing grants, loans or equity capital for the purpose of promoting indigenous technology development and application.

Incentives for utilisation of indigenous R&D

The government has promoted the National Research Development Corporation (NRDC) with the specific responsibility of transferring technology from R&D laboratories to industry. NRDC commercializes the technologies developed with government support, undertakes further work towards upscaling the laboratory know-how, setting up pilot plant, etc., and even provides risk finance to development projects. In addition, the utilisation of indigenous R&D is sought to be promoted by various other incentives. All goods manufactured by a wholly Indian owned company are exempted from excise duty for a period of three years, provided these are patented in any two countries from amongst India, USA, Japan, and any one country of the EU. A higher rate (40 per cent) of investment allowance and depreciation is applicable to plant and machinery installed (since 1987) for manufacture of goods based on indigenous technology. The indigenous technology-based products are exempt from the provisions of industrial licensing. Royalties earned by Indian companies abroad through export of indigenous technologies are completely free of tax, and those earned within the country are given a 40 per cent rebate.

Furthermore, to inculcate technological entrepreneurship in the country, the public sector financial institutions such as IDBI, ICICI, and IFCI have set up venture capital funding companies to assist new generation of techno-preneurs. Private venture capital funds and angel investors have been allowed to operate in India as per the SEBI regulations.

DSIR has set up Technology Business Incubation Centres at the research institutions to facilitate speedier transfer of know-how developed. IITs and other technology institutions are setting up industrial consultancy and extension centres to facilitate the utilisation of domestic R&D and encourage technology entrepreneurship among their alumni. DST has set up S&T Entrepreneurial Parks at different locations. These Parks provide infrastructural facilities to techno-entrepreneurs to start their business activity expeditiously.

A critical assessment of the Indian national innovation system

The Indian government policies in building local technological capability and promoting enterprise-level technological effort have many achievements to their credit. These include building up of one of the largest stocks of scientifically and technical manpower in the world. Indian talent has been widely sought after by the IT industry worldwide. Investments made by the government over the years in S&T development have led to the creation of a vast network of institutional infrastructure. There are pockets of excellence in this system, which have attracted MNEs from the Western countries to seek tie-ups with some public-funded laboratories in the form of contract R&D [see Kumar (1999) for illustrative examples]. A number of innovations of the public-funded laboratories have also successfully gone into the production system such as Swaraj Tractor, indigenous optical glass, innovations for dyes and tanning agents, indigenous electronic-grade silicon, modular digital switching technology, fluidised bed combustion technology for thermal power equipment industry, indigenous super computer param, and ability to build satellites and launch them, among others. Indian companies have also been able to export technology-intensive goods as well as technology itself to a number of countries. Indian companies have emerged as the most competitive suppliers of a large number of generic drugs and computer software in the world. India is fast emerging as a preferred host of overseas R&D activity by MNEs with a number of them having invested in global or regional R&D facilities in India or R&D joint ventures with Indian companies [see Kumar (1999) for illustrative examples].[14] However, the Indian technological effort, especially with respect to industry, has been sub-optimal as would become clear from the following. The bulk of

[14] Kumar (2001) provides empirical evidence on the relevance of NIS for relocation of R&D activity of MNEs.

the industrial production continues to be dependent on imported technology; international competitiveness of Indian knowledge-intensive goods has been poor; and India's share of international patenting has been marginal.

Human resources development

Despite massive expansion of the educational system, both the quantitative and qualitative achievements have been short of the targets. The most conspicuous failure of the system has been its inability to universalise elementary education. Barely 52 per cent of the population is literate according to the 1991 census. Even in case of higher education, where a rapid growth in numbers is noted, less than 5 per cent of the relevant age group population is enrolled in universities and colleges. A mismatch between the demand and supply of skilled manpower has led to growing shortages of experienced IT manpower side by side with growing educated unemployment. The quality of higher education system in engineering and technology also seems to have suffered over the years as is evident from the declining number of engineering doctorates awarded by Indian universities and other institutions, from 511 in 1982 and 629 in 1990 to 298 in 1996–97. As compared to this, the number of Indian nationals obtaining doctorates in science and engineering has grown from 371 in 1980 to 1,276 in 1996 (Kumar and Agarwal 2000b). This trend may be indicative of falling standards of excellence of research in Indian institutions of higher learning in technical disciplines. To some extent, the declining standards may be related to the declining levels of funding provided to the educational system by the government. For instance, Panchamukhi (2000) has observed that the average annual growth rate of per pupil expenditure on education during the 1990s has been just 1.2 per cent for primary education (compared to 6.9 per cent during the second half of 1980s), nil for secondary education (compared to 5.0 per cent during the second half of 1980s), and has been –4.4 per cent for higher education.

Scale of science and technology activity and its composition

As a result of increasing involvement of the government in S&T activity, the national expenditure on R&D as a proportion of GNP has risen steadily from 0.17 per cent in 1958–59 till 1987–88 when it peaked at nearly 1 per cent level of GNP. Since then, it has declined gradually to reach a level of 0.66 per cent in 1996–97. The decline owes partly to the budget constraints faced by the government agencies as a part of structural adjustment programme in the 1990s.

The other disturbing trend is the relatively low proportion of national R&D expenditure being spent by industry. As observed earlier, only 28 per cent of India's national R&D expenditure is actually spent in the corporate sector or industry comprising both public sector and private sector enterprises. Despite so many promotional and support measures towards the corporate R&D undertaken by the government over the years, industry's share continues to be low compared to other countries, e.g., 66 per cent in Japan, 70 per cent in the USA, and 30 per cent in South Korea. Thus, the bulk of the national R&D activity is conducted in isolation with the production system, and hence, the latter is unable to benefit from it. Furthermore, the bulk of the S&T expenditure by the government is spent on non-industrial strategic sectors such as defence, space, and atomic energy, which generate fewer spillovers with the production system compared to industrial R&D activity. Industrial sector not just accounts for a small proportion of national S&T activity, but studies have shown that R&D expenditures by the industry in real terms have also declined over the 1990s (Basant 2000). Finally, the linkages between the publicly funded R&D institutions, higher education system, and the industry have been very weak, leading to poor levels of diffusion and commercialization of the results of innovations.

Indicators of science and technology output

Indicators of S&T output are scientific publications, patent holdings, and technology-intensive exports. India's record in respect of these indicators has been hardly encouraging. According to NSF (2000, A429), the ratio of scientific and technical articles to GDP for India was 6 in 1997. As per this, India ranked 50th in the world. This ratio, although low, is comparable with some of the more developed countries than India, such as Turkey (6.1) and is higher than countries such as Brazil (4.8), Mexico (3.6), Malaysia (1.8), Thailand (1.0), and China (2.5). However, India's share of scientific publications in engineering and technology has come down from 3.37 on average during 1986–88 to 2.68 during 1995–97.

Share of indigenous innovators in patent grants is one indicator of the success of indigenous R&D effort. In terms of this indicator, the Indian experience of the past two decades has not been encouraging. The bulk of patents granted in India are still obtained by foreigners, and the share of residents in total number of patents in force has stagnated over the 1990s at around 21 per cent.

Patents obtained by Indians at the US Patents Office could also be indicative of the success of local technological effort in the country. In terms of this indicator, India's share is negligible at 0.5 per cent in the late 1990s. However, it is encouraging to note that it has risen sharply over the 1990s from 0.0186 per cent during the late 1980s to 0.05338 per cent during the latest period. This is no mean achievement. However, it may partly reflect the growing patent consciousness among R&D institutions, which has been cultivated recently by joint efforts of CSIR, DSIR, TIFAC, and industry bodies to secure protection on their intellectual property.

Productivity growth and technological upgrading of export structure

The relative neglect of R&D activity by industry coupled with poor linkages with the publicly funded innovation system has meant that the contribution of total factor productivity growth to output growth is negligible in India (Das 1999). The growing technological capability of a nation's enterprises helps them to upgrade their product profile and allows them to compete in high value adding activities over time. India's export structure does not provide an evidence of moving up the value chain in terms of growing technological complexity of export profile. Lall (1999) has found India's export structure to be dominated by simple and undifferentiated products, where the main comparative advantage lies in cheap labour, low levels of skills, and simpler technologies. India's modest high technology exports consist mainly of pharmaceuticals and a few low-end electronics goods and software services. This suggests 'a narrow base of capabilities, with electronics production being particularly weak... At this time, there is little capacity in India to expand high technology manufactured exports' (Lall 1999). In comparison to India, Southeast and East Asian countries have rapidly diversified their export structure in favour of technologically advanced goods. For instance, share of technologically advanced goods (differentiated and science-based goods) in India's manufactured exports rose marginally to about 8 per cent by the mid-1990s from 5.6 per cent in the mid-1970s. In China, this proportion increased from 8.8 per cent to 23 per cent over the period 1987–95, and in Malaysia, from 12 per cent to 57 per cent over the period 1980 to 1995 (Pigato et al. 1997). Therefore, India has not been able to channel its technological resources effectively to upgrade its export structure and for moving up the value chain.

To sum up, therefore, the above analysis suggests a relative failure of technological effort of Indian corporate sector to take-off in a significant manner. However, the cumulative investments made by the Indian government in building the NIS over the past 50 years have succeeded in creating an extensive infrastructure and a base for innovative and skill-intensive activities. This includes the largest and expanding mass of technically trained manpower, even though they may be constituting a rather small proportion of the country's population, and a network of centres of internationally renowned excellence in specific sciences such as IISc, IITs, and national laboratories, among others. This base has attracted a large number of foreign MNEs to come to India to set up R&D facilities, contract research, or R&D joint ventures in different areas. It is this base that has made India a successful participant in the international division of labour in software development. A further evidence of the role NIS has played in contributing to India's success for software development activity emerges from the pattern of clustering or agglomeration of software development activity around the centres of excellence in the cities of Mumbai, Bengaluru, New Delhi, and Hyderabad.

National innovation systems for IT and software development

Within the overall pattern of development of NIS, the government has recognized the critical importance of electronics industry and computing for national development in view of their 'pervasive' applications and has consciously strived to build local institutional infrastructure for the development of local capabilities since the early 1960s (Electronics Commission of India 1975). As will be seen subsequently these early initiatives have provided a base for rapid development of IT software industry in the 1980s and 1990s.

Institutional infrastructure for policy analysis and implementation

A series of government committees have contributed to the evolution of the NIS in the sector. The early initiatives include Bhabha Committee of 1963, Electronics Committee Chaired by Dr V.A. Sarabhai in 1966, and the National Conference on Electronics of March 1970. Significantly, the National Conference had recognized the potential of India to emerge as a force in software exports way back in 1970 and had put emphasis on manpower development (Electronics Commission of India 1975). As a follow-up of their recommendations, a separate Department of Electronics (DoE) was set up within the government to coordinate and implement policies for the development of electronics industries, including computer software, in 1970. In 1971, the government constituted the Electronics Commission as a policy formulation body with a heavy emphasis on R&D and technology development. An Information, Planning and Analysis Group (IPAG) was also created at the Commission. In 1973, a Technology Development Council was set up to assist the Electronics Commission on the recommendation of the National Seminar on R&D Policy in Electronics (Electronics Commission of India 1975). The Electronics Commission had prepared a 10-year perspective plan for the development of electronics for implementation during the Fifth and Sixth Five-year Plans. This plan covered a number of significant measures for development of a base for software industry development. The DoE has spearheaded a number of programmes of human resources development for software engineers, technological and communication infrastructure for software development, and other facilitating measures over the past three decades in tune with the recommendations of the Electronics Commission's perspective plan, as will be summarized subsequently. More recently, recognizing the potential of IT-related industries and software for India's development, the Prime Minister had appointed a National Taskforce on Information Technology and Software Development (NTITSD) in May 1998 under the chairmanship of the Deputy Chairman, Planning Commission. NTITSD submitted its first report outlining a National IT Plan comprising 108 recommendations in early July 1998 (Government of India, NTITSD 1998). These recommendations have since been notified by the Government in the Gazette of India dated 25 July 1998. NTITSD has set before the country an ambitious target of US$50 billion software export by 2008. The National IT Plan is a comprehensive set of recommendations dealing with various aspects of IT and software industry and the infrastructure dealing with it. DoE was upgraded into a full-fledged Ministry of Information Technology (MIT) in October 1999 to coordinate the promotional role of the government in the industry.[15]

Supply of trained manpower for software development

The National Conference on Electronics in 1970 had estimated the need for about 300 MTechs and 50 PhDs in computer science and technology and had recommended launching of specialized Masters-level programmes at the IITs and other major institutions, besides making proficiency in computer programming mandatory for undergraduates of IITs and science postgraduates of all major universities in the country (Electronics Commission of India 1975). As a follow-up of these recommendations, MTech (2-year postgraduate course) and BTech (4-year graduate course) in computer science were started in

[15] See Government of India, MIT (2000a) and http://www.mit.gov.in for programmes of the Ministry.

1974 and 1977, respectively, at the IITs with DoE support. In 1982, two new courses, viz., a three-year Master of Computer Applications and a Diploma of Computer Applications were started besides expansion of MTech/BTech courses, as a follow-up of the Rajaraman Committee of 1978. These facilities were further expanded and new polytechnic diplomas were started in 1984 further to Computer Manpower Development Programme launched in 1983. In 1984, the Sampath Committee reviewed the training needs, and in 1985, a Standing Committee on Computer Education was set up to plan further actions. The new courses introduced under the Computer Manpower Development Programme, supported by the DoE, in about 400 institutions had produced some 15,000 software personnel by 1996 (Heeks 1996). The DoE's support has not only been restricted to financial grants but has also involved curricula development. Besides the courses started at the educational institutions, a number of enterprises and other institutions promoted by DoE have also been providing training in software development. These include National Centre for Software Technology (NCST) and Centre for Development of Advanced Computing (C-DAC) offering advanced software engineering courses and CMC Ltd., ETTDC, NIC offering routine software application training.

Besides these, the government has permitted private investment in IT training since the early 1980s. By now, 80 private companies are operating some 4,000 training centres, offering 1000 training courses throughout the length and breadth of the country through networks of franchises.[16] These privately run centres offer diplomas of various duration, ranging from short-term specialized courses to longer-term basic courses. However, the quality of the training imparted by these institutions has been uneven. The DoE has stepped in to provide accreditation of their courses as a step towards standardisation of these courses. A scheme called DOEACC was started in 1990 to provide accreditation to specified level of courses, viz., O – foundation course, A – Advanced Diploma, B – MCA Level, and C – MTech Level. By January 2000, a total of 699 institutes had been accredited by DOEACC Society. The Society conducts examinations for all the four levels twice a year and grants certificates /diplomas (Government of India, MIT 2000a).

The demand for software personnel, especially engineering graduates, has grown rapidly over the past few years due to the expansion of the software development activity in India as well as the growing brain drain. In view of this, easing the supply of IT professionals is one of the most important challenges

[16] *Dataquest*, 31 May 2000, 15 June 2000

faced by the country. In a recent survey, 57 per cent of the firms interviewed indicated manpower and skill shortage as the major problem (Arora et al. 2000). The NTITSD has made a number of recommendations dealing with augmenting the quality and quality of trained manpower for software industry. In tune with these recommendations, the capacity of the higher education system in engineering in the country is being currently expanded, besides new institutions being set up (Kumar 2000d, for details).

Infrastructure for technology development in software

The DoE, as the government arm, has played an instrumental role in creating necessary infrastructure for the development of a base for software industry development. In the 1970s, the DoE was setting up Regional Computer Centres functioning like public utilities, attached to educational institutions that were following the recommendations of the National Conference on Electronics of 1970. These centres were set up at Mumbai, Delhi, Bengaluru, Kolkata, Kanpur, and Hyderabad. CMC was set up in 1976 for maintaining these and the computer centres. Over time, CMC has grown into a full-fledged software company with strong R&D capability. NIC was set up to facilitate automation and networking of government offices at the centre, state, and district levels.

Since the late 1980s, the DoE has concentrated on providing data communication and networking infrastructure to the educational and research community and to the software industry. This infrastructure has played a critical role in the development of the industry in the 1990s. The Education and Research Network (ERNET) project was initiated in 1986 with participation of NCST Mumbai, IISc Bangalore, and five IITs, and support of DoE and the United Nations Development Programme (UNDP) with the objective of enhancing the national capability in the area of computer communication by progressively setting up a nation-wide computer network for the education and research community. ERNET has evolved into a separate institution now, providing networking services to over 80,000 users in 750 academic and research institutions with its dedicated satellite data transfer backbone.

STPs of India is a scheme that DoE has initiated in 1990 to provide high-speed dedicated data communication links and other infrastructural facilities to the software industry. Eventually, STPI has also been spun off as a separate society. It has set up STPs at 15 locations across the country. Software units located in these parks are provided high-speed data communication facilities through dedicated earth stations set up by STPI. STPIs are located at Bengaluru, Hyderabad, Thiruvanathapuram, Noida, Bhubaneshwar, Gandhinagar, Mohali,

Jaipur, Navi Mumbai, Mysore, Coimbatore, Manipal, Pune, Vishakhapatnam, Guwahati, and Chennai. The data communication links are of international standards (64 kbps) and are provided at internationally competitive tariff. These have helped the Indian software exporter firms to interact with their clients on a real time basis and have helped to obviate the constraints posed by poor telecommunication infrastructure in the country. Besides the data communication facilities, STPIs have played a promotional role for the industry by providing business incubation facilities, single-window clearance, among others. The importance of the STPI scheme for India's success in software exports is clear from the fact that STPI units account for nearly 60 per cent of India's total software exports (Government of India, MIT 2000a).

R&D capability building

The DoE has put heavy emphasis on R&D activity relating to, among other fields, development of computer software by supporting R&D activity in the area at different institutions such as TIFR, IITs, IISc, select universities (such as Jadavpur University), ISI, and CSIR laboratories since the early 1970s. The Technology Development Council has been supporting R&D projects since its inception in 1973 (Government of India, MIT 2000a). These programmes of technology development have led to building up of capabilities and have provided experienced manpower for the rapid development of the industry in the coming years. For instance, the capabilities built in the process of early work on data communication at TIFR, started in the late 1970s and anchored at the DoE, supported the NCST, set up in erstwhile Bombay in 1984, proved instrumental for the development of country-wide networks and Internet in the country in 1980s and 1990s. The National Aeronautical Laboratory (NAL) also developed a super computer Flosolver. The government S&T agencies have set up a parallel Super-computer Education and Research Centre (SERC) and Department of Computer Science and Automation at IISc, which have provided high end expertise and manpower to the industry in software. Besides NCST, DoE has also set up another institution for technology development in the 1980s, viz., C – DAC. C – DAC has developed India's first super computer Param and has developed software for Indian languages' script. The Electronics Research and Development Centre (ER&DC) is another new R&D institution set up by the DoE. ER&DC has research facilities at Thiruvanathapuram and Kolkata and is acquiring another unit at Noida near Delhi. The government has also stimulated and supported R&D activity of industry through tax incentives and direct funding on a limited scale by DoE. For instance, DoE and now MIT have been operating a programme called Funding R&D in Electronics Industry (FRIEND) for funding industrial R&D.

Procurement policies and other promotional measures

The government has also helped in the development of software industry by generating large and complex assignments that have given confidence to the local industry firms. These include automation of railways reservation and bank automation, among others (Heeks 1996).

The government has also assisted the industry firms in standardisation and quality control, for instance in their effort to obtain ISO 9000 certification. The smaller and medium enterprises have been assisted in their overseas expansion by subsidies for their participation in industry fairs such as CebIT, venture capital, and tax breaks for export activity.

To sum up, the government has played a critical role in India's software industry success by ensuring skill supply with timely initiatives and investments since the early 1970s, and building necessary networking and other institutional infrastructure for technology development and capability building. The World Bank (1992) and Heeks (1996) also underline the importance of the government's intervention in this sphere.

Besides government intervention, a few other factors have contributed to the Indian success story. These include English language capability of Indian engineers and a cultural tradition of logic and mathematics. Furthermore, for once, the pool of expatriate Indian engineers migrating from the country over the years has also helped the country and has proved to be an asset as seen subsequently.

Making a virtue out of brain drain

The evolution of export-oriented software industry in the country has been helped to some extent by the sizeable expatriate Indian community of engineers in the Silicon Valley. The early emphasis on higher education in engineering and technology in India generated a supply of trained manpower that exceeded the demand. A large number of trained engineers found career opportunities in the West, especially in the USA. The brain drain has been a significant phenomenon, especially in respect of highly trained manpower. According to a study, 58.5 per cent of graduates in computer science from IIT, Madras, during the period 1964–86, migrated abroad.[17] The export-orientation of the Indian software industry has benefited from the presence of a substantial number of non-resident Indian engineers working in USA MNEs. Lateef (1997) and Arora et al. (2000) observe that some of them have played an important, though yet to be documented, role in facilitating the contacts between the buyers in the USA and the potential suppliers in India. Non-resident Indians in the software industry in the USA have also invested back home in subsidiaries that develop software for their USA operations. These include investments in subsidiaries of Mastech, CBS Inc., and IMR, among others. The rise of the software industry has also prompted a number of non-resident Indians to return to the country to start software ventures. According to some estimates, the rate of returning of professionals increased from 2 per cent in 1991 to 8 to 10 per cent in the late 1990s, with several senior software professionals returning to India to set up their own companies here. Apparently in Hyderabad alone, about 100 companies have been set up by returning software professionals (*Express India*, 2000).

While the early round of brain drain may have helped the country in the evolution of the export-oriented software industry, the rapid growth of the software industry worldwide has created fresh possibilities of a further and potentially sizeable new wave of brain drain. The recent emergence of India as a centre of software development has focused attention on the country as a potential source of trained manpower for IT industries to meet the growing shortages in other countries. According to IDC, in the USA, half of all IT jobs now go 'begging.' In Western Europe, the shortfall of IT workers has risen from 6 per cent in 1998 to 14 per cent in 2000. It is expected to rise to 23 per cent by 2002 (IDC 2000). As a result, a large number of countries have been planning to import software engineers from India. These include Germany's offer of 20,000 green cards for software workers; Japan seeking 10,000 IT workers from India over the next three years, apart from projections as: Ireland, 32,000 by 2005; France, 10,000; Italy, 8000; and Korea another 10,000.[18] There are other countries like Belgium, Syria, Iran, Singapore, and Spain, which have also shown interest in importing Indian talent, although precise figures are not indicated.[19] The British government has recently enacted a special fast track work permit system to allow IT workforce into the country to meet an estimated demand for about 150,000 professionals. This system will reduce the

[17] Cited in IAMR (2000).

[18] See Kumar (2000d).

[19] *ET* 3 June 2000, 17 June 2000; *TOI*, 23 July 2000.

time needed for issuing a work permit for IT professionals from six to eight weeks to only two weeks.[20] In the USA too, the gap between demand and supply of IT manpower has been increasing and this shortfall has been bridged with imports from India among other countries.[21] Due to growing pressure from the industry, the administration has recently raised the annual cap on H1B visas from 115,000 to 195,000 for the next three years. This trend raises the prospect of a further rise in the rates of brain drain of software manpower from India. It is not clear yet, as to how many of Indian personnel will leave the country to take advantage of these opportunities. However, an outflow will certainly add to the growing scarcity of software talent, thereby pushing up the salaries of software professionals in India. This, in turn, may have an adverse bearing on the external competitiveness of software exports from India. Thus, it has the potential of adversely affecting the domestic software industry and its export performance. Much will depend on the ability of the country to increase the supply of software manpower quickly.

National innovation system and patterns of spatial agglomeration of software industry

Software industry development in different parts of the world is characterized by a strong tendency of clustering because of agglomeration economies. In India, the software industry developed initially in Mumbai. Subsequently, especially after the entry of Texas Instruments in the mid-1980s, Bengaluru emerged as a centre of software industry development. Besides, Bengaluru and Mumbai, Delhi along with its suburbs, namely, Noida and Gurgaon, has emerged as the third most popular concentration of software units (See Table 5.16). Hyderabad and Chennai have started to provide alternative location in the South after the saturation of Bengaluru in terms of available infrastructure and scarcity of space. The state government's promotional role has also contributed to the emergence of Hyderabad as the fourth most important centre of concentration of software companies. The top five cities together account for 80.5 per cent of the 600 top companies. But other cities such as Kolkata, Pune, Thiruvanantpuram, Ahmedabad, and Bhubaneswar are coming up as increasingly popular locations. One important factor is the availability of high-speed data communication links and built-up space in the STPs.

---

[20] *BS*, 2 June 2000.

[21] ESA (2000): Chapter 5.

Table 5.16: Patterns of clustering of top 600 software companies

City	*Number of company headquarters located*	*Percentage share*
Mumbai	131	21.83
Bengaluru	122	20.33
Delhi and around	111	18.50
Hyderabad	64	10.67
Chennai	55	9.16
Kolkata	25	4.16
Pune	23	3.83
Thiruvanathapuram	14	2.33
Others	55	9.16

Source: Adapted from NASSCOM (2000a)

The patterns of concentration of software development industry in and around select cities do corroborate the key importance of NIS for the activity. Table 5.17 shows that the cities of high concentration of software development, viz., Mumbai, Bengaluru, Delhi, and Hyderabad have shared a disproportionate share of national innovative infrastructure, skills base, and other resources for technology development. Because of significant agglomeration economies present in skills and knowledge-intensive activities such as software development, this disproportionate share of national innovative infrastructure has crowded in the software development activity to these cities.

The presence of higher educational enterprises in engineering and technology, thus, ensuring the supply of engineering manpower, centres of excellence, a nucleus of local software industry companies, and clustering of high technology enterprises, together with telecommunication infrastructure in the form of STPs in these cities has facilitated agglomeration of software industry in these cities. It will become clear from a case study of Bengaluru presented in the Box 1.

Box 1

Bengaluru and the software industry: A case study

The city of Bengaluru has emerged as a favoured location for the software companies from within the country and outside MNEs that have set up development centres in the country so much so that it is referred to as the Silicon Valley of India. Besides hosting headquarters of 122 companies, software centres of a number of other companies are located in Bengaluru. Most of the foreign software majors that have set up development centres in

Table 5.17: Illustrative science and technology infrastructure in four cities of concentration of software industry

Type of NIS infrastructure	*Mumbai*	*Bengaluru*	*Delhi and around*	*Hyderabad*
Institutions of higher technical education and excellence	IIT - B; Bombay University; SNDT Women's University; Bajaj Institute of Management; and several other engineering and management institutes	IISc; University Visvesraya College of Engineering; SKSJ Technology Institute; and 28 private engineering colleges; Indian Institute of Management-B	IIT- D; Delhi College of Engineering; Delhi University Department of Computer Sciences; Roorkee University of Engineering (within 200 kms.); J.N. University; Jamia Milia Islamia Engineering College; FMS; IIFT; plus several private institutions.	J.N. Technological University; Hyderabad University; Osmania University; Kakatiya University;
Public-funded research laboratories and institutions	TIFR; NCST; BARC; UDCT; SAMEER	ISRO; NAL; CMTI; Electronics and Radar Development Establishment Aeronautical Development Establishment; Gas Turbine Research Establishment; Centre for Aeronautical Systems Studies and Analysis;ER&DCI	NIC; NPL; Institute for Systems Studies and Analysis; SPL; C-DOT;	National Remote Sensing Agency; RRL; NGRI; IICT; Defence Electronic Research Laboratory; DRDL
Local software champions	TCS; PCS; Tata Infotech; Mastek; L&T ITL; APTECH; COSL; Datamatics; Silverline	Infosys Technologies Ltd.; WIPRO Information Technologies	HCL Technologies; NIIT Ltd; CMC Ltd;	Satyam Computer Services Ltd
High-speed data communication facilities	Earth Station of STPI	Earth Station of STPI	Earth Station of STPI	Earth Station of STPI
High technology enterprises	L&T; Godrej; Tata group; and a large number of engineering and electronics enterprises	ITI; BEL; HAL	Central Electronics Ltd; NRDC; EIL; RITES; ETTDC; ET&T; RITES; TCIL	ECIL; BHEL

Source: Compiled by the author from different sources

India have also preferred Bengaluru. These include Texas Instruments (TI), which entered the country in 1986, followed by Hewlett-Packard, Novell, and Oracle, among others. This pattern of clustering of software industry in Bengaluru is to be explained in terms of a high concentration of the national innovation system (NIS) infrastructure located in the city. Bengaluru has been an important centre for engineering and technical education since the early twentieth century. The Indian Institute of Science (IISc) was set up with an endowment from Tata industrial family in 1909. IISc has emerged as a centre of excellence in different branches of engineering and technology. A polytechnic was founded in the city in 1920s by M. Viswesvarayya, a prime minister of the princely state of Mysore. Since the Independence, many government research establishments and high-tech industries, besides educational centres, have been set up in the city. The infrastructure provides a ready availability of a pool of trained manpower, high-end computer programming facilities and expertise, and other infrastructure for technological development besides, agglomeration economies because of this concentration. Besides, the mild climate has also helped (see also Lateef 1997; Holmstrom 1998).

Supply of engineering and managerial skills

As observed earlier, Bengaluru has a history of higher education in engineering and technology. Besides the IISc that offers integrated engineering degree programmes right from graduation to doctoral level, there are many engineering colleges in the city, providing degrees and diplomas in different streams of the subject. According to a city website, the city has 30 engineering colleges out of 77 located in the Karnataka state, including private colleges. These colleges in Bengaluru and at other places in the state ensure a steady supply of trained engineering manpower for the software houses. Besides, the location of one of the premier management institute in the city Indian Institute of Management- Bangalore (IIM-B) ensures supply of trained management skills.

Technology development infrastructure

Bangalore has a high concentration of high technology institutions and research laboratories of different public-funded scientific and technological establishments, which have developed expertise in computer programming and software engineering. Hence, their presence provides benefits of agglomeration that can be substantial in the skill-intensive activities such as software development. These institutions include:

1. *Indian Institute of Science*: It is a centre of excellence in various areas of engineering. In computer software, it has a Supercomputer Education and Research Centre (SERC) and a Department of Computer Science and Automation (CSA). SERC has been conducting industrial consultancy for a number of companies. Its clients include Digital Equipment, Texas Instruments, HP, Tata Elxi, Wipro, and BHEL, among others. CSA conducts Masters programmes in computer science, Internet science, systems science, and automation. IBM and Infosys have instituted fellowships at the CSA. Its faculty includes 21 PhDs in computer science.
2. Public-funded R&D organisations concerning IT industry or using computer software: Bengaluru hosts a number of CSIR national laboratories, viz., National Aerospace Laboratory (NAL), Centre for Mathematical Modelling and Computer Simulation (C - MMACS), Central Machine Tools Institute (CMTI); Indian Space Research Organization (ISRO)'s Satellite Centre and Telemetry Tracking and Command Network (ISTRAC); Defence Research and Development Organization (DRDO) 's Centre for Artificial Intelligence & Robotics (CAIR), and Centre for Airborne Systems Studies and Analysis, and Electronics and Radar Development Establishment; Microwave Tube Research and Development Centre; Gas Turbine Research Establishment; and Central Power Research Institute. Some of these institutes, even though having a different focus than primarily on software development, have built strong capabilities in computers, electronics, and software engineering. NAL, for instance, has developed India's second parallel super-computer, Flosolver. NAL has attracted research contracts worth $25 million from global IT leaders such as IBM, Hitachi, and Boeing among others. CMTI has spearheaded India's advance in CNC machine tools that require considerable software skills. Besides these specialized R&D institutions, Bengaluru also hosts a number of other scientific and technology institutes such as Raman Research Institute and J.N. Nehru Centre for Advanced Scientific Research, both funded by the Ministry of Science and Technology; Indian Institute of Astro-Physics; and National Institute of Advanced Studies. The presence of all these highly specialized high

technology institutions in the city gives it a critical mass of expertise in nearly every domain of engineering that the software industry can tap.

High technology enterprises

A number of public sector electronics, communication, and aerospace enterprises have been set up in Bengaluru since the 1950s. These include Bharat Electronics Limited, Indian Telephone Industries (ITI) Limited, and Hindustan Aeronautics Limited (HAL) started in 1950s and 1960s, and Hindistan Machine Tools Ltd (HMT) and Bharat Heavy Electricals Limited started in 1970s. Activities of these enterprises had numerous backward and forward linkages with the software development activity. HAL's capability in software engineering for aerospace industry has attracted British Aerospace (BA) to form a joint venture with it, viz., BAeHAL Software Limited in 1993, headquartered in Bengaluru. BAeHAL has developed an aircraft logistics package, CALSS, besides other products and services (NASSCOM 1999).

Finally, some of India's more successful national software enterprises such as WIPRO and Infosys Technologies have also originated in Bengaluru, while others like Tata Consultancy Services (TCS) have established their development centres there. The presence of national champions in Bengaluru has also crowded in other software enterprises from within and outside the country.

Source: The author's compilation based on information from different sources

Concluding remarks

This chapter has discussed some issues concerning the emergence of India as a significant exporter of software and services in the world. Starting from a rather negligible base, software industry in India has come up rapidly and has become a significant contributor of the country's export earnings and growth of output. The industry has gradually evolved in terms of capability and sophistication, and range of activities and geographical reach from export of low value adding services that were delivered on-site at the clients' locations. Another characteristic feature of India's development is that it is driven largely by local entrepreneurship, talent, and resources.

It has been shown that India's success owes largely to the cumulative investments made by the government over the past five decades in building what is now termed as NIS. These include resources in the development of a system of higher education in engineering and technical disciplines, creation of an institutional infrastructure for S&T policy-making and implementation, and building centres of excellence and numerous other institutions for technology development, among other initiatives. The Indian government recognized the potential of the country in computer software way back in the early 1970s and started building necessary infrastructure for its fruition, in particular, for training of manpower. The government also facilitated technological capability building with investments in public-funded R&D institutions and supporting their projects by creating computing facilities, and developing infrastructure for data transfer and networking. The patterns of clustering of the software development activity and in particular the case study of Bengaluru provide a further evidence to the contention that public-funded technological infrastructure has crowded in the investments from private sector in skill-intensive activities such as software development. It would appear from this that investments made by governments in NIS have substantial positive externalities.

Select Bibliography

Arora, A., V. S. Arunachalam, Jai Asundi, and Ronald Fernandes. 2000. *The Indian Software Services Industry*. Pittsburg: Carnegie Mellon University (mimeo).

Basant, R. 2000. 'Corporate Response to Economic Reforms.' In *Indian Economy under Reforms: An Assessment of Economic and Social Impact*, edited by Nagesh Kumar, 45–92. New Delhi: Bookwell.

Das, D. 1999. 'Productivity Growth in Indian Manufacturing: An Application of Domer Aggregation.' Paper presented at the Workshop on *Measurement of Productivity in India,* 9–10 July. New Delhi: Institute for Human Development.

Desai, Ashok V. 1980. 'The Origin and Direction of Industrial R&D in India.' *Research Policy* 9 (1): 74–96.

DSIR (Department of Scientific and Industrial Research). 1998. *Research and Development in Industry: An Overview.* New Delhi: DSIR.

———. 1999. *Annual Report 1998–99*. New Delhi: DSIR.

DST. 1999. *Research and Development Statistics 1996–9.* New Delhi: Department of Science and Technology.

Electronics Commission of India. 1975. *Perspective Report on Electronics in India.* New Delhi: Electronics Commission.

Government of India NTITSD. 1998. *IT Action Plan* 3 Volumes. New Delhi: National Task force on Information Technology and Software Development. Accessed: June 2001; http://it-taskforce.nic.in/.

Government of India. 1999. *Annual Report.* New Delhi: Department of Education, Ministry of Human Resources Development.

Government of India, Planning Commission. 1999. *Ninth Five Year Plan 1997–2002: Development Goals, Strategy and Policies*, Volumes1 and II. New Delhi: Planning Commission.

———. 2000. *Economic Survey 1999–2000.* New Delhi: Ministry of Finance.

Government of India, MIT (Ministry of Information Technology). 2000a. *Annual Report 1999–2000.* New Delhi: MIT.

———. 2000b. 'Action Taken Report of the National Task Force on Information Technology and Software Development.' *IT Action Plan: Part I.* New Delhi: MIT.

Hanna, Nagy. 1994. 'Exploring Information Technology for Development: A Case Study of India.'World Bank Discussion Paper No. 246, Washington, DC: The World Bank.

Heeks, R. 1996. *India's Software Industry: State Policy, Liberalization and Industrial Development.* New Delhi: Sage Publications.

———. 1998. 'The Uneven Profile of Indian Software Exports.' *Development Informatics* Working Paper Series No. 3, IDPM. Manchester, UK: University of Manchester.

Holmstrom, Mark. 1998. 'Bangalore as an Industrial District: Flexible Specialization in a Labour Surplus Economy?' In *Decentralized Production in India: Industrial Districts, Flexible Specialization and Employment,* edited by Philippe Cadene and Mark Holmstrom, 169–229. New Delhi: Sage Publications.

IAMR (Institute of Applied Manpower Research). 2000. *Manpower Profile, India,Yearbook 2000.* New Delhi: IAMR.

Kumar, Nagesh. 2001. 'Determinants of Location of Overseas R&D Activity of Multinational Enterprises: The Case of US and Japanese Corporations.' *Research Policy* 30 (1): 159–74.

———. 2000a. 'New Technology Based Small Service Enterprises and Employment: The Case of Software and Related Services Industry in India.' Paper presented at the National Seminar on *Strategic Approach to Job Creation*, February 2000, organized by ILO-SAAT and Ministry of Labour, at SurajKund, Haryana.

———. 2000b. *India's Software and Services Exports in a Perspective.* New Delhi: Research and Information System for Developing Countries, April (mimeo).

———. 2000c. 'Firm Size, Export Performance, and Productivity Growth: Recent Trends in Enterprise Performance in Indian Software Industry.' New Delhi: Research and Information System for Developing Countries, June (mimeo).

———. 2000d. 'Developing Countries in International Division of Labour in Software and Service Industry: Lessons from Indian Experience.' Background Paper for *World Employment Report 2001*. Geneva: ILO.

———. 1999. 'Multinational Enterprises, Overseas R&D Activity and Global Technological Order.' RIS Occasional Paper #56. New Delhi: Research and Information System.

Kumar, Nagesh and Aradhna Agarwal. 2000a. 'Liberalization, Outward Orientation and In- house R & D Activity of Multinational and Local Firms: A Quantitative

Exploration for Indian Manufacturing.' RIS Discussion Paper #7. New Delhi: Research and Information System.

———. 2000b. 'National Innovation System and Corporate R&D Activity in India: Some Explorations into the Effectiveness of Promotional Policies.' New Delhi: Research and Information System for Developing Countries (mimeo).

Lall, Sanjaya. 1999. 'India's Manufactured Exports: Comparative Structure and Prospects.' *World Development* 27 (10): 1769–86.

Lateef, Asma. 1997. 'Linking up with the Global Economy: A Case Study of the Bangalore Software Industry.' NIOP, DP/96/97. Geneva: International Institute for Labour Studies.

Lundvall, Bengt-Ake. ed. 1992. *National Systems of Innovation: Towards a Theory of Innovation and Interactive Learning.* London: Pinter.

NASSCOM (National Association of Software and Service Companies). 2000a. *The I.T. Software and Services Industry in India: Strategic Review 2000.* New Delhi: NASSCOM.

———. 2000b. *Indian Software Export Grows by 57% in 1999–2000: NASSCOM Report.* Mumbai: NASSCOM.

———. 1999. *Indian I.T. Software and Services Directory.* New Delhi: NASSCOM.

NASSCOM and McKinsey. 1999. *The Indian I.T. Strategy Summit: Beyond Y2K: Building a Dominant Position for India in IT software and Services.* New Delhi: McKinsey & Co. for National Association of Software and Service Companies.

Nelson, R. R. 1993. *National Innovation Systems: A Comparative Analysis.* Oxford: Oxford University Press.

NSF (National Science Foundation). 2000. *Science and Engineering Indicators 2000.* Washington, DC: NSF.

OECD. 1997. *IT Outlook 1997.* Paris: OECD.

Panchamukhi, P. R. 2000. 'Social Impact of Economic Reforms in India: A Critical Appraisal' *Economic and Political Weekly* 35 (10): 836–47.

Pigato, Miria, Caroline Farah, Ken Itakura, Kwang Jun, Will Martin, Kim Murrell, and T. G. Srinivasan. 1997. *South Asia's Integration into the World Economy.* Washington, DC: The World Bank.

Software Engineering Institute. 2000. *High Maturity Organizations.* Pittsburg, SEI, Carnegie Mellon University. Accessed: June 2001; www.sei.cmu.edu/~highmatorg.html.

Ramadorai, S. 2011. *The TCS story ... and beyond.* New Delhi: Penguin Books.

The World Bank. 1992. *India: An Information Technology Development Strategy.* Washington, DC: The World Bank.

World Economic Forum. 1999. *World Competitiveness Report 1999.* Geneva: World Economic Forum.

CHAPTER 6

The Secret behind India's Success in Pharmaceuticals

Seizing the Right Windows of Opportunity

*Shyama V. Ramani and Samira Guennif**

Introduction

When India attained its Independence in 1947, its pharmaceutical industry was of a very modest size with market sales of about US$28.5 million (Ahmad 1988). Western multinational companies (MNCs) held about 80 per cent of the market with the remainder being served by several Indian-owned companies operating on a much smaller scale. No Indian company had manufacturing capabilities in either bulk drugs or formulations. There was heavy dependence on imported drugs, which were marketed directly by the MNCs established in India and the local agents of MNCs that did not have an Indian presence. These MNCs mainly formulated their drugs in India, importing the bulk drugs from their home countries; their contention being that the locally available bulk drugs were not of the desired quality. In the process, not only were the technological spillovers and knowledge transfer absolutely minimal but Indian drug prices were also among the highest in the world (Greene 2007).

Given such history, it is indeed noteworthy that despite the erosion of scientific and technological entrepreneurship over nearly a thousand years of quasi-foreign rule, great progress has been made in the 60 years following Independence. A comprehensive survey of the Indian pharmaceutical sector notes that by 2005, India ranked 13th in terms of value and 4th in terms of value

* We gratefully acknowledge comments made during the UNU-MERIT workshop on 'The Indian system of innovation and access to medicines' held on 29–31 July 2010 at New Delhi.

of pharmaceuticals produced in the world with market sales of US$9.4 billion (CCI 2007). In 2007, about 70 per cent of the market was controlled by about 250 firms of a fragmented industry comprising about 20,000 manufacturing units (CCI 2007). Of these, foreign drug-makers accounted for about 15 per cent of the market by 2010 (Silverman 2011). By 2010, the pharmaceutical industry in India met around 70 per cent of the country's demand for bulk drugs, drug intermediates, pharmaceutical formulations, chemicals, tablets, capsules, orals, and injectibles.[1] Market competition remains fierce with low prices that are unmatched in most parts of the world (Greene 2007).

How did India catch-up from a state of nearly no-capabilities in pharmaceuticals to being a leading industrial power of the developed world? This is the question we will attempt to answer in this chapter using the 'national system of innovation' (NSI) framework. The NSI approach was spearheaded by the seminal work of Lundvall (1992), Nelson (1993), and Freeman (1995). Starting from the premise that the creation of technological and industrial competence in any knowledge-intensive sector is a collective and cumulative process, the NSI builds upon the rationale of actors and institutions within the country, involved in the creation, adoption, and diffusion of new technologies. The principal actors of the NSI are the State, public laboratories, firms, financial organisations, and consumers. The actions of these actors determine and are determined by institutions, i.e., State policies, regulations, and even economic, social, and cultural norms.[2] In the last three decades, the NSI approach has emerged as a useful framework to organise historical evidence and study the 'catching-up' processes of 'late-comer' countries with respect to the accumulation of industrial capabilities. It has also inspired the notion of a sectoral system of innovation (SSI), incorporating sectoral specificities in an innovation system (Malerba 2002).

Taking off with the discovery of antibiotics, the pharmaceutical industry emerged as a distinct sector in the developed world in the last century, when the R&D activity by firms greatly increased in order to find solutions to the problems and demands created by the two world wars. Since then, it has continued to be a highly R&D-intensive industry, with technological competencies and innovation creation being important determinants of firm survival and growth. Such trends were reinforced with the emergence and

[1] http://www.pharmaceutical-drug-manufacturers.com/pharmaceutical-industry/.

[2] 'Institutions are set of common habits, routines, established practices, rules or laws that regulate the relations and interactions between individuals, groups and organisations. They are the rules of the games'. (Edquist 2001, 5)

integration of biotechnology during the 1980s[3] and the commercial success of radical innovations such as recombinant insulin, human growth hormone, interferon, tissue plasminogen activator (TPA), erythroprotein (EPO), etc. (Achilladelis and Antonakis 2001; Malerba and Orsenigo 2002).[4] Major innovators of the pharmaceutical industry are found only in five countries, namely, USA, Germany, Switzerland, the UK, and France, where they are greatly supported by favourable national policies in terms of public and private research (McKelvey et al. 2004). Nevertheless, some emerging economies have developed significant industrial capabilities over the last 50 years.

Pharmaceutical products of today broadly fall under one of the three types: drugs, vaccines, and diagnostics. For the purpose of this chapter, we have focused only on drugs and refer to industrial capabilities as the capacity of domestic firms to satisfy national and international demand in terms of the quantity, quality, and variety of drugs locally produced. We have further considered industrial capabilities to include two components: manufacturing capabilities and innovation capabilities.

The manufacturing of drugs involves three main operations, and the associated capabilities are different in terms of technological complexity. The least complex step is the 'formulation' of drugs, which refers to the processing and packing of the basic ingredients called 'bulk drugs' into a consumable form such as a tablet, capsule, syrup, injection, plaster, etc. The production of 'bulk drug' containing the therapeutic molecule in powder or liquid is a more complex process requiring a higher level of scientific and technological capabilities. But the most complex step is to produce the core component of bulk drugs termed as the 'active pharmaceutical ingredients (API)'.

In defining innovation capabilities, we distinguish between 're-engineering skills' and 'new drug discovery skills.' Usually a late-comer country firm starts by building re-engineering skills, i.e., by developing new processes to produce

---

[3] Biotechnology refers to techniques that involve manipulation or change in the genetic patrimony of living organisms. It emerged from the advances in the life sciences from the mid-1970s.

[4] For instance, between 1980 and 1990, the radical innovations to be commercialized were: Insulin (by Eli Lilly in 1982); human growth hormone (by Genentech in 1985, Eli Lily in 1987, Novo-Nordisk in 1988, Ares-Serono in 1988, Biotechnology General in 1989); alpha 2 interferon (by Schering Plough in 1986, Hoffman La Roche in 1986); monoclonal orthoclone OKT3 (by Ortho biotech in 1986), tissue plasminogen activator TPA (by Genentech in 1987), erythroprotein EPO (by Amgen in 1989), and hepatitis B vaccine (by Smith Kline and Beecham in 1989).

the copies of existing drugs. Once a firm learns to manufacture bulk drugs and eventually API, it can envisage investing in the development of 'new drug discovery capabilities'. Capabilities in new drug discovery can take the form of integration of biotechnology and/or research capabilities in one or more of the steps in the new drug discovery process. To date, no developing country firm has patented a new chemical entity.

Furthermore, developing country firms have to build up complementary competencies that go beyond technology, if they want to commercialize drugs. The regulatory procedure to commercialize a copy of a branded drug is relatively simple. Firms just need to submit proof of their chemical and therapeutic equivalence with the proprietary drug; though some additional information and technical support need to be provided to the regulatory authority to enter the market, requirements that vary from country to country. On the other hand, to commercialize a new drug, data has to be generated on preclinical tests on animals followed by a three-phase series of clinical testing on human beings, at the end of which a new drug application can be made with the concerned regulatory authority. At present, very few developing countries have a patent or regulatory bureaucracy that can deal with an application for the commercialization of a new proprietary drug, especially if it involves biotechnology.

The rest of the chapter is organised as follows. The second section describes the Indian case study. The third section proposes some calculated speculations about the future. Finally, the last section contains concluding remarks.

Evolution of the pharmaceutical sector in India[5]

First attempts to reduce dominance of foreign firms (1947–70)

In order to reduce the dependence on imports and Western MNCs, at least for vitally needed antibiotics, the Government of India made large investments to establish public sector enterprises (Singh 1985). The most important among these were 'Hindustan Antibiotics Limited' (in 1954) and 'Indian Drugs and Pharmaceuticals Limited' (in 1961). The move was useful and timely, but it was not a comprehensive response to the country's healthcare needs.

Inspired by the economic growth models of Russia and China throughout the 1960s and 1970s, India, like most other developing countries, adopted

[5] Some parts of this section draw upon the previous articles by Guennif and Ramani (2012); Reid and Ramani (2010); Athreye, Kale, and Ramani (2009); Ramani and Maria (2005); Ramani (2002); and Ramani and Venkataramani (2001).

inward-looking trade and investment policies. The objective was to minimise dependence on imports and develop an industrial base to serve the needs of its citizens while promoting market competition and curbing monopolistic and oligopolistic tendencies. The ensuing 'import substitution' policy took the form of a complex system of price controls, high import duties, and export subsidies.[6] In addition, under the guise of controlling profiteering by the private sector, the Indian government practised the *License Raj*, i.e., 'rule by license' under which any firm wanting to expand its manufacturing base, export or import, had to get a license from the government to proceed.

The system of price controls was particularly stringent on pharmaceuticals as compared to other sectors, in the interests of social welfare. In the wake of the Sino-Indian War in 1962, in order to ensure accessibility to drugs to serve wartime needs, the 'Drug display of Prices Order' (DPO) and the 'Drug Price and Control Order' (DPCO) were passed under the 'Defence of India Act' in 1962 and 1963, respectively. Over time, these two acts were merged into one, and price control was introduced in 1970 for a long list of 'notified' drugs that were deemed essential (Kaushal 2007). The objective was to curb profit margins and promote access to drugs.

This system of regulation was continually opposed by both MNCs and fledgling Indian companies in the pharmaceutical sector. They argued that while high import duties were responsible for pushing up the domestic prices, price ceilings were discouraging the flow of investment into the industry by depressing the earnings of companies. Besides, the *License Raj* promoted public corruption and diverted the efforts of firms towards securing fiscal and tariff concessions, permits, and licenses from the government rather than seeking support for R&D in any form.

Against this backdrop, India still adhered to a very tight system of Intellectual Property Right (IPR). At the time of Independence in 1947, India's IPR system was defined by the 'Indian Patents and Designs Act of 1911,' which itself was based on the British 'Patent Act of 1852'. Under this regime, patent holders were allowed exclusive rights to make, sell, and use both new processes and products for 14 years from the date of filing in India. Re-engineering of branded drugs was not allowed, and almost all patents of branded drugs were held by MNCs.

Thus, after 20 years of the *License Raj* and an import substitution policy, 80 per cent of the market share was still held by foreign controlled firms in 1970. Indian firms had capabilities only in formulations. Prices of drugs remained

[6] See Burton (1998) for a survey of import substitution policies adopted by India and other developing countries.

among the highest in the world, partially due to import duties, but mostly because firms were focused on brand competition and promotional activities (Lall 1974a, b). Indian consumers suffered from a shortage of essential drugs and a crisis in terms of healthcare provision. MNCs, on the other hand, fared well: they were in India 'not only the most profitable among manufacturing firms in the country generally but also among all types of foreign controlled enterprises, including those in non-manufacturing sectors' (Lall 1974b,163).

Development of re-engineering capabilities and conquest of internal markets (1970-90)

There were two possible solutions to the healthcare emergency at the beginning of the 1970s. Either medicine could be imported in large quantities as essential commodities or incentives could be provided for the development of the local pharmaceutical industry. The Indian government opted for the latter solution.

By the mid-1950s itself, it had been brought to the attention of the Indian government that most of the developed countries had put in place a strong IPR system with full product and process patents only after having acquired a certain level of technological competence in knowledge-intensive sectors and a good competitive position in the international market in targeted fields. Accordingly, in 1957, the Indian government appointed Justice Ayyangar to investigate this matter, and his report submitted in 1959 recommended that only process patents be recognized for essential commodities such as food and drugs. But, it was after a little more than a decade that his recommendation saw the light of day in the form of the new 'Patent Act of 1970,' which came into force in 1972. It is to be noted that even in 1972, developed countries such as Sweden, Switzerland, Spain, Italy, and Japan did not allow for product patents in pharmaceuticals, thus revealing the retard of the Indian government in opting for this much-practised option of industrial development in other parts of the world.

The Indian Patent Act of 1970 essentially constituted a 'narrowing' of the IPR regime with provisions for commercializing independently developed copies of branded drugs, if the production process was significantly different from that used to manufacture the branded product. Thereby, the incentives for Indian firms to become second innovators were increased. At the time, neither MNCs nor scholars expected the industrial organisations to change much; it was essentially viewed as a possibility for the public sector to accumulate technological capabilities and serve the low-income communities directly (Lall 1974b; Redwood 1994).

However, Indian firms correctly sized-up the *window of opportunity* opened by the new process patent regime. They did not patent new processes in the Indian patent office *en masse*, but experimented with new methods of production compatible with local resources and constraints, since the original method could not be copied. Thus, they developed technological capabilities required for 'duplicative imitation' and 'creative imitation' (Kale and Little 2007). Leading Indian pharmaceutical firms began to invest in building re-engineering capabilities and started producing essential drugs – slashing market prices heavily.

When some firms made huge profits via the creation and commercialization of lower priced generics, a secondary bandwagon effect was triggered within the market. In 'winner takes all' tournaments, often the incremental technological innovations continued, with a second or third innovator improving upon earlier re-engineered products and grabbing the market share yet again, lowering the prices even further, and increasing consumer welfare. As a consequence, Indian drug producers faced continual gales of Schumpeterian technological competition in which only the most diversified or the most technologically competent firms could survive, and the Indian pharmaceutical market became very dynamic and competitive.

In 1970, eight of the top 10 firms in the Indian market were MNCs, but by 1995, only four of the top 10 firms were MNCs (Athreye, Kale, and Ramani 2009). The share of MNCs in the Indian market was 68 per cent in 1970, 60 per cent by 1978, and 50 per cent by 1980 (Chaudhuri 2005). Indian firms even entered into production contracts with the original MNC inventors, permitting them too to enjoy lower costs and a greater mark-up. By the mid-1980s, leading Indian pharmaceutical firms were producing both bulk drugs and formulations for the domestic market. By the end of the 1980s, India was exporting bulk drugs and final therapeutics, supplying many parts of the developing and developed world at lower prices and edging towards a positive trade balance. But most important of all, with the increase in market supply, the Indian public healthcare system was finally able to stand up on its feet, and the proportion of the poor with access to basic drugs increased.

To conclude, the change in the IPR regime coupled with the dynamic response of local firms to acquire capabilities in all stages of drugs production led to a sharp reduction in import dependence and MNC domination. But, this would not have been possible had India not been equipped already with scientific capabilities in the form of public laboratories skilled in creating new processes, and universities producing large numbers of science graduates.

The demand side also supported the new trajectory as Indian consumers revealed themselves to be extremely price sensitive.

Development of regulation-handling capabilities and assault on international markets

The 1990s saw a number of extreme changes in the Indian regulatory environment, which influenced the accumulation of technological capabilities in almost all sectors, including pharmaceuticals. In 1991, the economy was liberalised and the pharmaceutical sector was de-licensed, it was no longer necessary to get a license from the concerned ministries to expand the manufacturing base, or export or import goods. The price control regime, DPCO, was modified and 50 per cent of the drugs were removed from price control in 1995 and only 76 drugs (26 per cent) remained under price control by 2004 (OPPI 2004). A new regulatory agency, 'The National Pharmaceutical Pricing Authority,' became functional in 1997 to coordinate the regulation of prices. Hot on the heels of liberalisation, India became a member of the World Trade Organization (WTO) in 1995 and thereby changed its regulatory framework to comply with Agreement on Trade-Related Aspects of Intellectual Property Rights (TRIPS). Between 1994, when TRIPS was ratified, and 2005, when it came into effect, three amendments to the patent law of 1970 were passed in the Indian Parliament in 1999, 2002, and 2005 successively to make it TRIPS compliant.

Economic liberalisation had a remarkable effect on the growth of the pharmaceutical industry, with more firms entering the market, and the established ones increasing their manufacturing base. Production, exports, and imports shot up, and the industry grew rapidly in the 1990s, with an average annual growth rate of 15 per cent for bulk drugs and 20 per cent for formulations (OPPI 2001).

Interestingly, the rise of exports was partly due to the foray of Indian firms into regulated markets of Western countries, with the principal target being the USA. During the 1980s, American policy-makers had become sensitive to the need for improving access to medicines and curbing the growth of health expenditures in the USA. With this objective, the Hatch-Waxman Act was passed in 1984 to stimulate the market for generics, and as a result, generics manufacturers no longer had to go through a lengthy period of extensive clinical trials. Demonstration of bio-equivalence was sufficient to acquire marketing approval for a generic drug. Ironically, the concerns that prompted the Hatch-Waxman Act were quite similar to those that had triggered the Indian Patent Act of 1970.

Indian firms with foresight like Ranbaxy recognized that the Hatch-Waxman Act in the USA in combination with the liberal economic policies in India was opening up new windows of opportunity. Such firms recognized that in order to sell generics in regulated markets, they had to upgrade their regulation handling capabilities, i.e., initiate routines to document the entire production process under specific formats. To enter a Western market, say that of the USA, Indian firms had to upgrade their 'Drug Master Files' procedure by increasing the comprehensiveness of the details supplied on the manufacturing and distribution process, to satisfy the requirements of the US Food and Drug Administration (FDA). The generics producer also had to prove that its manufacturing methods conformed to current 'good manufacturing practices' (GMP), as defined in the US Code of Federal Regulations. Then it could apply for an ANDA, or 'Abbreviated New Drug Application', under four types of filings termed paragraphs, the last even permitting entry of generics before patent expiration. Ranbaxy was the first Indian firm to use the ANDA filing route to enter the US generics market, leading others to follow. It used the steady but low return Paragraph I to Paragraph III approach of ANDA filings, whereby the generic manufacturer entered the market only after expiry of the product patent to secure a niche in the US antibiotics segment.[7] Such building of regulatory handling capabilities resulted in India having the largest number of manufacturing units validated by the FDA outside of the USA by 2007: India had 75; Italy, 55; Spain, 25; and China, 27 (Tribune des droits humains 2007).

In addition to exporting medicines to unregulated Southern markets, from 2000 onwards Indian firms began to get supply contracts from international agencies[8] supporting public health programmes in developing countries, thus

[7] On the other hand, Dr Reddy's Laboratories adopted the aggressive strategy of Paragraph IV filings, which involves invalidating existing patents or producing non-infringing process through a costly process of litigation, a high-risk strategy due to the litigation costs involved and the 180-day market exclusivity that the firm wins on a successful challenge. Though Dr Reddy's Laboratories got six-month exclusivity for selling Fluoxentine 40 mg capsules in the USA, it also received a severe setback when it lost the AmVaz case to Pfizer. Thus, the two early entrants differed quite markedly in their propensity to take risks. Other Indian firms then began to follow the example set by Ranbaxy and Dr Reddy's Laboratories, but the low-risk strategy of Paragraph I–III applications is the more commonly followed model. Such entry into Western regulated markets meant that the Indian pharmaceutical firms also had to set up litigation teams and invest in paying for the corresponding dispute settlement procedures.

[8] The WHO, the US President's Emergency Plan, AIDS Relief, the Global fund to fight AIDS, tuberculosis and malaria, the Bill and Melinda Gates Foundation, the Bill Clinton Foundation, etc.

responding effectively to yet another 'windows of opportunity.' Again, Indian firms had to learn to comply with a pre-qualification process of product-selection and the leading pharmaceutical firms adopted 'GMP' even when it was not required within India. The success of the Indian firms is illustrated by the fact that out of the 190 antiretrovirals pre-qualified by the WHO to treat HIV/AIDS, 108 are produced by firms such as Ranbaxy, Cipla or Matrix Laboratories (WHO 2010). As a result of being recognized as suppliers of quality drugs at low cost, Indian firms have been supplying more than 80 per cent of donor-funded AIDS medicines to developing countries in the last five years (Waning, Diedrichsen and Moon 2010).

Transition into biotechnology

Modern biotechnology pertains to a set of techniques that involve manipulation or change of the genetic patrimony of living organisms. Since the 1980s, modern biotechnology has been integrated in a number of industries such as pharmaceuticals, chemicals, agribusiness, agriculture, and environment. The biopharmaceutical sector refers to pharmaceutical firms that have integrated modern biotechnology in their research, production, or marketing activities. Though India was first among the developing countries to recognize the potential of biotechnology, in terms of industrial capabilities, the build-up has not been as successful as initially expected.

In India, the push to develop the biotechnology sectors came from reputed scientists who had been brought into the administration. In 1982, Dr S. Varadarajan, then secretary of the Department of Science and Technology, headed the National Biotechnology Board (NBTB). The objectives of the NBTB were to: (i) identify priority areas in biotechnology, (ii) identify infrastructural needs, and (iii) implement a coordinated programme to realize certain national objectives. To this end, a number of pilot programmes were proposed in the Sixth (1981–85) and Seventh (1986–1990) Five-year Plans but objective (iii) was not reached. In 1986, the NBTB was replaced by a separate government department called the Department of Biotechnology (DBT). It set out to implement the objectives of the earlier body such as the development of scientific competence in selected non-capital-intensive disciplines (genetic engineering, vaccines, food production, edible oils). The establishment of DBT served as a signal that the government considered biotechnology to be a priority area for development. It was welcomed by academics, national laboratories as well as industrialists.

The first target was to create a core of researchers competent in biotechnology. Grants were given to the network of research institutions and university departments to undertake biotechnology-related projects. Grants were also provided to selected teaching and research institutes partially supported by the government, such as the Indian Institute of Sciences, Indian Institutes of Technology, All India Institute of Medical Sciences, National Chemical Laboratory, Tata Institute of Energy Research, Tata Institute of Fundamental Research, etc. The DBT also participated in the creation of new institutions such as the National Institute of Immunology, Centre for Cellular and Molecular Biology, National Facility for Animal Tissue and Cell Culture, and International Centre for Genetic Engineering (in collaboration with UNIDO).

From the period of initiation of biotechnology, the pharmaceutical industry did not figure high in the thinking of the National Board (NBTB), and the non-association of any competent scientist or industrialist from the pharmaceutical sector in its deliberations had its own consequences. Subsequently, no grants were available from the DBT for the modest R&D establishments that were being set up by some pharmaceutical majors. The thrust of government strategy was on agriculture rather than healthcare, because of the former's intrinsic importance to the economy and the existence of a good record of indigenous research accomplishment. The meagre research output of pharmaceutical enterprises and the minor role of pharmaceutical-related research in the large government-supported research establishment had inevitable impact on the resources made available to the pharmaceutical industry. There were also no attempt from the industry itself to make the point that a determined effort should be made for developing new drugs through biotechnological techniques even for the major diseases afflicting the people of the third world.

However, the efforts of the Indian government created substantial awareness on the implications of biotechnology for firms. A number of large firms in the pharmaceutical industry began to invest in biotechnology. Several new biotechnology firms were created by public laboratory researchers or industrial scientists. The challenge for the Indian pharmaceutical firms was to expand their knowledge base, which was firmly embedded in organic and synthetic chemistry, to integrate the life sciences based and associated techniques. The latter was new, different, and much more complex to integrate, requiring a multidisciplinary team to create a product (Ramani 2002). Thus, during the 1990s, five types of strategic positioning of Indian biopharmaceutical firms could be distinguished (Ramani and Venkataramani 2001):

1. *Wait and see*: A majority of pharmaceutical firms preferred to adopt the policy of 'wait and see' with respect to biotechnology, as their counterparts had done in the West in the previous decade.

2. *Marketing for Western firms*: Many established pharmaceutical firms marketed biotech diagnostic kits, vaccines, and drugs for Western firms, in order to test the waters.
3. *Producing diagnostics*: A few large integrated pharmaceutical firms entered the market for diagnostics.
4. *Contract research*: Researchers from public laboratories or industrialists with venture capital or foreign capital backing created a handful of dedicated biotechnology firms. Most were into contract research, production of biological products or production of chemicals by rDNA techniques.
5. *Speciality chemicals*: A few enzyme producers got into biotechnology by producing chemicals using rDNA techniques

From the 1990s, the biopharma sector began to emerge in India. Well-established pharmaceutical firms like Wockhardt began to build research capabilities in biotechnology and were joined by other established pharma incumbents like Zydus Cadila, Nicholas Piramal, Dr Reddy's Laboratories, and Ranbaxy Laboratories. Big firms with a non-pharma background such as the petro-chemical giant Reliance through Reliance Life Sciences (RLS) also entered the market. Alongside, enzyme and specialty chemicals biotech firm like Biocon branched into custom research and clinical research through subsidiary companies like Syngene International Pvt. Ltd and Clinigene International Pvt. Ltd. However, the most innovative were a few new entrants like Shantha Biotech, Bharat Biotech, and Lifecare Innovations, which gained international renown through their indigenously developed biotech-based vaccines and drugs, slashing market prices sharply and taking over market shares of branded counterparts produced by Western multinationals. The major biopharma products selling in India today are shown in Table 6.1. All are indigenous products and some like the diagnostics developed by Lifecare Innovations and Cadila Pharmaceutical are the result of cooperation with public research laboratories.

Chaturvedi (2006) notes that while in 2001, the breakdown of players in the industry was 85 in agriculture and 43 in human health, by 2003 this number had shifted to 132 in agriculture and 142 in human health. Furthermore, biotechnology is constantly attracting new entrants. In the 2009 ABLE-BioSpectrum Top 50 Report (ABLE-Biospectrum 2009), 11 of the top 50 companies in terms of sales revenues from 2008 to 2009 were new companies (either new spin-offs from large pharma, or brand-new players). Most of these new companies are investing in human health. The biopharma firms are very outward looking. In the most recent BioSpectrum Industry Review (2010), it was

Table 6.1: Biopharmaceutical products in India

Sector	*Type*	*Product name*	*Application*	*Producer*
Vaccines	Recombinant hepatitis B surface antigen	Shanvac-B	Hepatitis B	Shantha Biotech
	Recombinant hepatitis B surface antigen	Revac-B	Hepatitis B	Bharat Biotech
	Recombinant hepatitis B surface antigen	Gene Vac-B	Hepatitis B	Serum Institute of India
	Purified capsular polysaccharide Vi of *Salmonella typhi*	Typbar Vi	Typhoid	Bharat Biotech
Therapeutics	Recombinant human insulin	Wosulin	Diabetes	Wockhardt (Mumbai)
	Recombinant human erythropoietin α	Epox	Anemia	Wockhardt
	Recombinant human interferon α-2b	Shanferon	Cancer	Shantha Biotech
	Recombinant streptokinase	Shankinase	Cardiovascular	Shantha Biotech
		Indikinase	Cardiovascular	Bharat Biotech
	Liposomal amphotericin B injection	Fungisome	Visceral leishmaniasis	Lifecare Innovations (New Delhi)
	Recombinant human granulocyte colony-stimulating factor	Gramstim	Neutropenia	Dr Reddy's Laboratories
Diagnostics	Immunoblot assays using recombinant HIV-1 antigens gp41 and C-terminus of gp-120 and HIV-2 antigen gp-36	HIV TRI-DOT	HIV-1 and HIV-2	J. Mitra (New Delhi)
	Immunoblot assay using recombinant HIV-1 antigens gp-41 and gp-120, HIV-2 antigen gp-36, and HCV antigens NS-3, NC-4 and NC-5	HIV-HCV Combo	HIV and hepatitis C	Bhat Biotech India (Bangalore)
	Enzyme-linked immunosorbent assay for recombinant HCV core antigens 1b &3g, together with peptides for HCV antigens NS-3, NS-4 1, NS-4 2, and NS-5	HEP-Chex C	Hepatitis C	XCyton Diagnostics (Bangalore)

Contd.

Contd.

Sector	*Type*	*Product name*	*Application*	*Producer*
	Enzyme-linked immunosorbent assay for recombinant version of *Taenia solium* excretory/secretory antigens	Cysti-Chex	Neurocysticercosis	XCyton Diagnostics
	Liposomal amphotericin B	Fungisome IV	Systemic mycosis and visceral leishmaniasis	Lifecare Innovations Pvt. Ltd
	Liposomal dithranol	Psorisome Gel	Mild plaque psoriasis	Lifecare Innovations Pvt. Ltd
	Lecithin-based coal tar	Lipotar S	Mild plaque psoriasis and other skin conditions	Lifecare Innovations Pvt. Ltd
	Lecithin-based coal tar	Lipotar SS	Mild plaque psoriasis and other skin conditions	Lifecare Innovations Pvt. Ltd
	-	NEVA-HIV	Sero-diagnostic test for HIV	Developed by Prof Vijay Chaudhary of Delhi University and commercialized by Cadila Pharmaceutical Ltd

Source: Chaturvedi (2007) and updated by Dr Jitendra Verma of Lifecare innovations

noted that 53 per cent of all revenues of Indian biotech firms were from exports. Indian firms are also seeking more strategic technology alliances. Chaturvedi (2006) estimates that for 2003, in India, the total number of strategic alliances between Indian biotech interests and foreign firms was 129 (35 agriculture, 70 human health, 1 environment, 11 industrial, and 12 others).

To sum up, both for mainstream pharmaceutical and biopharma firms, at the moment, moving to a TRIPS-compliant regime does not seem to have hurt the industry. Indeed, the Indian industry is adapting in a number of ways, and probably would have gone in these directions anyway even in the absence of TRIPS. A study based on interviews just prior to TRIPS enforcement revealed that biopharmaceutical firms were pursuing one or more of three types of innovation strategies (Ramani and Maria 2005). First and foremost, R&D is targeted towards the copy of drugs, vaccines, and diagnostics that are off-patent or are soon to be off-patent, especially in regulated Western markets. Second, Indian firms are vying to participate in the international division of labour for the creation of new drugs by Western MNCs by offering contract research and custom manufacturing services and bioinformatics services for genomics-based drug research, and carrying out clinical trials. Indian companies realize that they cannot match the deep pockets of Western MNCs as far as R&D budgets are concerned but want to avoid exclusion. By providing contract services to Western MNCs in their new drug discovery endeavours, they hope to build new dynamic capabilities either in the pre-competitive upstream research stage or in downstream clinical trials stage. Third, and in a smaller measure, a few firms were investing in the creation of new drugs for global diseases such as diabetes. Besides these internal R&D strategies, in a small measure, Indian firms are initiating strategic alliances (more in USA) and outright acquisitions (more in Europe) for a variety of objectives ranging from access to technology assets, and market penetration to a better understanding of local regulation (Greene 2007). This categorisation of strategies has only slightly evolved and holds true even today for the leading Indian pharmaceutical firms.

Discussion of results

Insights into the catch-up literature

The evolutionary trajectory of the Indian pharmaceutical sector discussed in the preceding section is summarized in Fig. 6.1. It confirms some of the basic tenets of the evolutionary literature on catching-up in terms of technological changes and industrial growth. The national environment and NIS in combination with the responses of the actors concerned, have determined the evolutionary

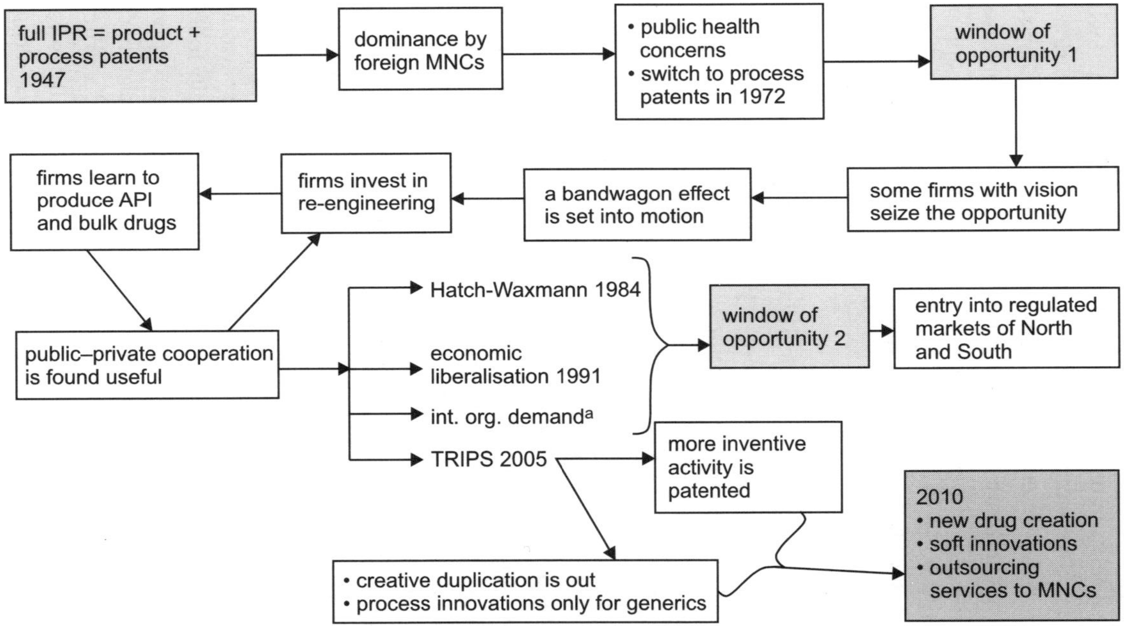

Fig. 6.1: Trajectory of firm strategy in the Indian pharmaceuticals sector

Note: [a] Int. org. demand refers to demand from international public agencies like the WHO

IPR – intellectual property right; API – active pharmaceutical ingredients; TRIPS – trade-related aspects of intellectual property rights

trajectory. Public investment and State policy were at the heart of the catch-up. Beyond this, the case study also yields some new insight on the processes of catching up.

In the existing catch-up literature, 'windows of opportunity' are mostly created by radical shifts in technology paradigms, the exploitation of which by prescient firms leads to sweeping changes in the industrial organisation (Soete 1985). In contrast, the present chapter demonstrates that endogenous changes in State policy can achieve the same effect. In other words, our case study demonstrates that *radical regulatory changes* can open windows of opportunity and generate positive externalities, in a way very similar to radical technological discontinuities. It highlights the role of the State not only as a financier for the development of capabilities, but also as a catalyser of change through initiation of regulatory shifts.

At the same time, the case study clearly demonstrates the limits of State policy and public investment. While these may be formulated and implemented with specific objectives, the final outcome will depend on the perceptions of the stakeholders of the opportunities and threats manifested, their responses to the same, and bandwagon or mimicry effects that are provoked within the system. The catch-up literature is full of examples of windows of opportunity that were effectively seized and this could give rise to an implicit conviction that any window of opportunity will be perceived and acted upon. But this need not be the case. Our case study shows that the beliefs and expectations of firms determine whether or not windows of opportunity are perceived or exploited. Final outcomes are uncertain, unpredictable, endogenous phenomena being guided by the beliefs, and expectations of the different economic actors in the innovation system. In addition, firms may evolve in response to not only regulatory changes within the NSI, but also to those initiated abroad, which cannot be controlled.

When the State adopted the import substitution model of growth after the Second World War in India, it was implicitly assumed that windows of opportunity had been opened and catching-up would automatically follow. But this simply did not happen, because the firms had a different perception and could not identify any windows of opportunity opened by the regulatory change. However, in addition to an import substitution regime, when India switched to a loose patent regime in 1972, suddenly a window of opportunity to increase profits through the development of re-engineering capabilities was recognized.

Finally, it is usually assumed that catching-up is most influenced by policies that impact knowledge or technological learning directly. However, actors may identify distant windows of opportunity, far beyond their domestic market, and act upon them in a desirable way from a national perspective. Definitely, the

US 'Hatch-Waxman Act of 1984,' which was not at all designed with the objective of promoting foreign firms, did exactly that for Indian firms whose astute exploitation of these opportunities contributed to their becoming multinationals.

The challenges ahead

Indian firms face four types of challenges if they are to continue to catch-up, build more dynamic capabilities, and also fulfil their social responsibilities as a provider of essential goods in a country like India with a heavy burden of poverty.

1. *Catching-up in terms of technological knowledge even with minimal financial resources*: In terms of challenges, Indian firms are behind international leaders in recombinant technology and new drug discovery skills, and simply strapped for money to invest in R&D capabilities. For instance, even the sum of the R&D expenditures of the top 11 companies in India in 2005–06 was only US$379 million, while that of Pfizer was almost 20 times more at US$7,440 million (Chaudhuri 2007). Finally, the opportunity cost of resource allocation to R&D is amplified by the fact that Indian firms have lost a lucrative source of profit with TRIPS. They can no longer earn second-innovator rents from re-engineering branded drugs.

   In the context described, can India rise to the new challenge and one day create a blockbuster? As of now, no Indian firm has patented a new chemical entity, and therefore, the innovative achievements of the firms in the future will depend on how their individual R&D initiatives and their participation in the international division of labour with Western and Japanese MNCs on the latter's drug discovery projects bear fruit. The hotspots of the race will be in 're-engineering generics,' 'creating incremental innovations,' and 'learning from international technology alliances'. TRIPS has pushed the focus of innovation investment away from pure re-engineering to re-engineering of drugs that are off-patent or will be off-patent protection in the near future, i.e., generics. Firms are thus making forays into 'soft' innovations around new drugs, in terms of improvements in dosage and drug delivery, new uses for known molecules, and cost reducing process innovations for generics.
2. *Increasing prowess in generic drug manufacturing without losing 'Indian citizenship'*: Indian firms have become so capable that they are becoming attractive to global players. For instance, the Indian industry has been deeply marked by the buying-out of its star performer Ranbaxy, by the Japanese company Daiichi-Sankyo, making the threat of foreign

buy-outs a credible one for all major Indian firms. Daiichi-Sankyo has business operations in 21 countries, while Ranbaxy is present in 56 countries, including emerging and transition economies in which Daiichi has not entered. Ranbaxy is among the international leaders in generics with a renowned low-cost manufacturing infrastructure, but it is struggling to gain expertise in biotechnology and new drug development. Daiichi is weak in generics, but has a good R&D expertise and a solid position in patented drugs. A cash infusion of about US$1 billion brokered under the deal will presumably enable Ranbaxy to retire debt and increase growth. The growth potential for generics, especially in Japan, is high. These complementarities led the founding family of Ranbaxy to accept the offer of Daiichi-Sankyo to buy out their 34.8 per cent stake and make open offers as per Indian regulations for an additional 20 per cent of Ranbaxy's share in June 2008, with the assurance that Ranbaxy will operate as an independent subsidiary of Daiichi under the leadership of its current CEO Malvinder Singh. The acquisition of Ranbaxy, a jewel in India's pharmaceutical crown, by the Japanese firm, struck an emotional blow to the Indian public, even as the business *pundits* pointed out the rationality of the merger (Singh 2008; Basheer 2008).

Following on the heels of the Ranbaxy take-over, in 2009, Shantha Biotech, which was the first to produce an indigenous recombinant product (shanvac B-hepatitis B vaccine), was acquired for €571 million by Sanofi-Aventis of France. One of the key attractors for the investors may well have been the US$340 million UN International Children's Emergency Fund awarded contract for pentavalent vaccines from 2010 to 2012 (Chakma et al. 2010). Investments and partnerships from international or non-traditional sources such as this or another collaboration of Shantha with the National Institute of Health (USA) not only bring in high levels of expertise from research to regulation, but also are of acute interest to investors. Again, the attractiveness of Shantha lay in its manufacturing of bio-generics as well as its lucrative contracts for the supply of vaccines to international agencies like WHO, which proved its capacity to meet international regulatory requirements. In 2010, Abbott acquired Piramal Healthcare for US$3.7 billion to exploit its generics manufacturing capacity as well as to gain the marketing rights to Piramal's branded generics to increase its market share in India. There have also been bids on Cipla, Wockhart, and Dr Reddy's.[9]

[9] As reported in the national dailies.

If firms that are leading in terms of pharmaceutical research are acquired and made to focus on the production of low-cost generics, the future evolution of their innovation capabilities does not look bright. This could revive the pre-1970s' division of labour in India when multinationals imported the raw materials (i.e., bulk drugs) and engaged in formulations in India. If the best Indian performers are acquired and made to focus on manufacturing generics and carrying out clinical trials while strategic R&D is carried on in the mother companies outside India, the future of Indian innovation does not look bright, though in the short run it is clearly a win-win deal for all.

3. *Managing trade-offs between incentives for innovation creation and protection of citizens' welfare*: Under the stronger IPR regime ushered in by TRIPS, there is a growing concern about the conflict between the pursuit of monopoly profit and satisfaction of public interest in terms of access to life-saving drugs. In India, there is an increasing number of patent disputes regarding life-saving drugs between patent owners, generic producers, and the public. There are inherent tensions at the moment in courts between MNCs, which want to protect their branded drugs, and civic associations and NGOs, which demand better access to drugs for the poor through the production of generics. The MNCs call for legal clarity on what can be patented, what can be considered ever-greening of patents, and what kind of patents can be bypassed in the interests of the public. For instance, using a pre-grant opposition mechanism introduced in 2005 in the patent law, Indian firms and a civic association challenged Novartis' application for a patent on its anticancer drug Glivec. The patent was rejected on the ground that the API was based on a derivative of a molecule known before 1995, which 'does not result in the enhancement of the known efficacy of the substance' as stated in Section 3(d) of the Indian Patent Act (Srinivasan 2007). Furthermore, these problems are exacerbated by a lack of coordination between governmental bodies. For example, Cipla gained marketing approval for a generic version of a lung cancer drug from the Central Drugs Standard Control Organization (CDSCO), which operates under the aegis of the Ministry of Health and Family Welfare, while the original innovator Roche was granted a patent by the Indian Patent Office at the same time for its branded drug Tarseva (Gehl-Sampath 2008). Again, in the interests of the public, the Indian court sided with the generic producer. Only the future will tell how the interests of the public will continue to be upheld under the TRIPS regime, given the pressure from the MNC pharmaceutical lobbies. In

turn, such cases are also often cited by MNCs to justify their lack of enthusiasm to introduce new products in India.

4. *Creating pro-poor innovations for neglected diseases*: The R&D efforts of leading Indian firms are focused on 'global diseases' such as cancer, diabetes, cardiovascular diseases, diseases related to central nervous system, etc., which also have an international market. While communicable bacterial diseases have been really brought down due to low-priced antibiotics accessible to most, and there is evidence of R&D on medicines for malaria and tuberculosis, neglected diseases like visceral leishmaniasis, dengue fever, chikungunya, and leprosy beg for better medicines and more research. Innovative public–private partnerships are being proposed as a possible solution, as the not-for-profit scientific effort of the public sector laboratories combined with the technological expertise of the private sector might serve as the springboard for the generation of medicines for targeted neglected diseases.

Conclusions

Today, given TRIPS, in any hi-tech sector, developing country firms, including Indian ones, can move along a combination of three paths to catch-up and reach the market heaven. First, they can invest in the creation of original innovations through internal R&D and cooperation with public laboratories; second, they can focus on gaining first mover advantages as generics producers, keeping a close watch on the products whose patents are close to expiry; and third, they can attempt to learn through the initiation of strategic technology alliances with more technologically competent firms, typically foreign ones, where learning is a key outcome of the process. Which of these three roads is most likely to lead to the best short-term profit and sustainable long-term market share through innovation creation? While the answer to this question is not evident at the present moment, with respect to further catching-up in pharmaceuticals, the case studies yield the following recommendations for Indian State policies and firm strategies.

- *Make policy design more rational:* The experience of India proves that even accumulation of organisational, institutional, and scientific and technological capabilities need not be sufficient for catching-up. In addition, the NSI must induce appropriate endogenous responses from the concerned stakeholders. In other words, if policy is to succeed, the incentives generated must be perceived and exploited appropriately. More than ever, policy-makers and other stakeholders in developing countries have to interact to design policies that match the expectations of the different stakeholders. Only with more dialogue and explicit bargaining can there be fewer surprises and more coordinated development. This implies that the different stakeholders, in particular policy-makers, have to get out of their ivory towers to interact more with one another and contribute to a policy formulation that induces the desired responses to the maximum extent possible. Here civic associations need to be particularly active to ensure that the catching-up in the pharmaceuticals is also increasingly inclusive.

- *Revive public research and improve its contribution to catching-up*: The case study highlights that investment in universities and public research is necessary to ensure pools of qualified labour and technology transfer to private firms. In countries such as India, there is a real decline in the role of universities and public sector laboratories in the catching-up process in pharmaceuticals. Given the challenges of biotechnology (and now nanotechnology), the contribution of public institutions is more important than ever.
- *Build regulation-handling capabilities*: Catching-up in terms of regulatory bureaucracy could also impinge on the accumulation of technological capabilities in the future. For instance, at present in India, a lack of effective regulatory bureaucracy in patenting of new chemical entities or drugs is definitely an institutional problem. Therefore, Indian firms tend to seek EPO and USPTO patents, which are far more costly and risky propositions for developing country firms.
- *Explore options for international cooperation*: A myriad of possibilities can open up with a strategic exploration of South–South and North–South cooperation to develop common R&D programmes, a common patent granting body, and cross-licensing agreements or sharing of patent pools.

Coming now to firm strategies, we propose two main recommendations.

1. *Continue to reinforce comparative advantage in generics and soft innovations*: Many developing countries have more qualified personnel or/and personnel who can be hired at cheaper rates. Therefore, possible secondary windows of opportunity may also open up while investing to create new generics or incremental innovations in terms of drug delivery, dosage, and software to complement an original innovation. Much will also depend on how flexibilities in TRIPS are exploited, for instance, whether developing country firms can innovate *around* a known molecule, exploiting provisions that allow for patenting, if the efficiency of the drug is significantly improved.

2. *Watch out for new opportunities*: New or previously unexploited windows of opportunity may contribute to further accumulation of technological capabilities. Some possibilities are new uses of old technology, traditional knowledge, and traditional medicines. Another possible window of opportunity through which we cannot see through at present, is associated with new forms of collaboration that are emerging between Western MNCs and leading developing country firms.

To conclude, the future of the Indian pharmaceutical industry will depend on how the triple challenge of catching up in terms of new discovery capabilities, surviving in a global market, and being socially responsible in a poverty stricken-country, is played out. By identifying and catering to new windows of opportunity opened by regulatory or other paradigm shifts in India or abroad, Indian pharmaceutical firms can successfully maintain their domestic market domination, continue to grow in terms of export, and even expand their activities outside of India.

References

ABLE (Association of Biotechnology Led Enterprises)-BioSpectrum. 2009. The Seventh Survey of Indian Biotech Industry (Volume 7). ABLE, India. Accessed: 3 September 2013; http://www.ableindia.in/pdf/seventh_biospectrum_survey_june_09.pdf.

Achilladelis, B. and N. Antonakis. 2001. 'The Dynamics of Technological Innovation: The Case of the Pharmaceutical Industry.' *Research Policy* 30 (4): 535–88.

Ahmad, H. 1988. *Technological Development in Drugs and Pharmaceutical Industry in India*. New Delhi: Navrang.

Athreye, S., D. Kale, and S. V. Ramani. 2009. 'Experimentation with Strategy and the Evolution of Dynamic Capability in the Indian Pharmaceutical Sector.' *Industrial and Corporate Change* 18 (4): 729–59.

Basheer, S. 2008. 'Ranbaxy-Daiichi Merger: An Emerging Ardhnarishwar Model?' *Sify Finance*, 10 July. Accessed: 3 September 2013; http://sify.com/finance/fullstory.php?id=14712790.

Burton, H. J. 1998. 'A Reconsideration of Import Substitution.' *Journal of Economic Literature* 36: 903–36.

CCI (Corporate Catalyst India). 2007. *A Brief Report on Pharmaceutical Industry in India*. New Delhi: CCI. Accessed: 3 September 2013; http://www.cci.in/pdfs/surveys-reports/Pharmaceutical-Industry-in-India.pdf.

Chakma, J. H., K. Masum, J. Perampaladas, J. Heys, and P. A. Singer. 2010. 'India's Billion Dollar Biotech.' *Nature Biotechnology* 28 (8): 783.

Chaturvedi, S. 2007. 'Exploring Inter linkages between National and Sectoral Innovation Systems for Rapid Technological Catch-Up: Case of Indian Biopharmaceutical Industry.' *Technology Analysis and Strategic Management* 19 (5): 643–59.

———. 2005, 2006. 'Dynamics of Biotechnology Research and Industry in India.' OECD Science, Technology and Industry Working Papers 2005/6. Paris: OECD Publishing. Accessed: 3 September 2013; http://www.oecd.org/science/sci-tech/34947073.pdf.

Chaudhuri, S. 2007. 'Is Product Patent Protection Necessary in Developing Countries for Innovation? R & D by Indian Pharmaceutical Companies after TRIPS.' Indian Institute of Management Working Paper Series: WPS No. 614/September 2007. Calcutta: Indian Institute of Management.

———. 2005. *The WTO and India's Pharmaceuticals Industry*. New Delhi: Oxford University Press.

Edquist, C. 1997. *Systems of Innovation: Technologies, Institutions and Organizations*. London: Pinter.

Freeman, C. 1995. 'The National System of Innovation in Historical Perspective.' *Cambridge Journal of Economics* 19 (1): 5–24.

Gehl-Sampath, P. 2008. *India's Pharmaceutical Sector in 2008: Emerging Strategies and Global and Local Implications for Access to Medicines*. London: Department for International Development (DFID).

Greene, W. 2007. 'The Emergence of India's Pharmaceutical Industry and Implications for the U.S. Generic Drug Market.' Office of Economics Working Paper. Washington, DC: US International Trade Commission.

Guennif, S. and S. Ramani. 2012. 'Explaining Divergence in Catching-Up in Pharma between India and Brazil using the NSI Framework.' *Research Policy* 41 (2): 430–41.

Kale, D. and S. Little. 2007. 'From Imitation to Innovation: The Evolution of R&D Capabilities and Learning Processes in the Indian Pharmaceutical Industry.' *Technology Analysis and Strategic Management* 19 (5): 589–611.

Kaushal, L. M. 2007. 'Drug Price Control Order and Recent Developments.' New Delhi: NPPA. Accessed: September, 19 2010; http://pharmexcil.com/data/uploads/5.DrugsPriceControlOrderandRecentDevelopments.pdf.

Lall, S. 1974a. 'International Pharmaceutical Industry and Less-Developed Countries: II: Costs and Alternatives.' *Economic and Political Weekly* 9 (48): 1990-96.

———. 1974b. 'The International Pharmaceutical Industry and Less-Developed Countries, with Special Reference to India.' *Oxford Bulletin of Economics and Statistics* 36 (3): 143–72.

Lundvall, B. A. E. 1992. *National Innovation Systems: Towards a Theory of Innovation and Interactive Learning*. London: Pinter.

Malerba, F. 2002. 'Sectoral Systems of Innovation and Production.' *Research Policy* 31 (2): 247–64.

Malerba, F. and L. Orsenigo. 2002. 'Innovation and Market Structure in the Dynamics of the Pharmaceutical Industry and Biotechnology: Towards a History-Friendly Model.' *Industrial and Corporate Change* 11 (4): 667–703.

McKelvey, M., L. Orsenigo, and F. Pammolli. 2004. 'Pharmaceuticals Analyzed through the Lens of a Sectoral Innovation System.' In *Sectoral Systems of Innovation: Concepts, Issues and Analyses of Six Major Sectors in Europe*, edited by F. Malerba, 73–120. Cambridge, UK: Cambridge University Press.

Nelson, R. 1993. *National Innovation Systems: A Comparative Analysis*. Oxford: Oxford University Press.

OPPI. 2001, 2004. Organisation of Pharmaceutical Producers of India: Pharmaceutical Compendium. Mumbai: OPPI.

Ramani, S. V. and A. Maria. 2005. 'TRIPS and its Possible Impact on the Biotech Based Segment of the Indian Pharmaceutical Industry.' *Economic and Political Weekly* 40 (7): 675–83.

Ramani, S. V. 2002. 'Who's Interested in Biotechnology: R&D Strategies, Knowledge Base and Market Sales of Indian Biopharmaceutical Firms.' *Research Policy* 31 (3): 381–98.

Ramani, S. V. and M. S. Venkataramani. 2001. 'Rising to the Technological Challenge: Integration of Biotechnology in the Indian Pharmaceutical Industry.' *International Journal of Biotechnology* 3 (1/2): 95–115.

Redwood, H. 1994. *New Horizons in India*. Suffolk. England: Oldwicks Press Limited.

Reid, S. and S. Ramani. 2010. *The Harnessing of Biotechnology in India: What Are the Roads to Market Heaven?* Online Publishing: Mimeo.

Silverman, E. 2011. 'India may Cap Foreign Direct Investment in Pharma.' *Pharmalot*. Accessed: June 2012; http://www.pharmalive.com/india-may-cap-foreign-direct-investment-pharma .

Singh, N. 2008. 'A Win Win Deal.' *Biospectrum (India)*. Accessed: 19 September 2010; http://biospectrumindia.ciol.com/content/BioBusiness/10807111.asp.

Singh, S. 1985. *Multinational Corporations and Indian Drug Industry*. New Delhi: Criterion Publications.

Soete, L. 1985. 'International Diffusion of Technology, Industrial Development and Technological Leap frogging.' *World Development* 13 (3): 409–22.

Srinivasan, S. 2007. 'Battling Patent Laws: The Glivec Case.' *Economic and Political Weekly* 42 (37): 3686–91.

Tribune des droits de l'homme. 2007. L'industrie Indienne N'aide Pas Les Indiens (The Indian Industry does not help Indians). Accessed: 3 September 2013; http://www.infosud.org/La-pharma-indienne-n-aide-pas-les,2152.

Waning, B., E. Diedrichsen, and S. Moon. 2010. 'A Lifeline to Treatment: The Role of Indian Generic Manufacturers in Supplying Antiretroviral Medicines to Developing Countries.' *Journal of the International AIDS Society* 13 (35): 1–10.

WHO. 2010. *W.H.O. List of Prequalified Medicinal Products*. Accessed: October 2010; http://who.int/prequal/.

CHAPTER 7

Nanoscience and Nanotechnology

The New Sunrise Industry?[1]

Manish Anand

Introduction

The desire to harness cutting-edge science and technology for enabling development has prompted global interest in emerging technologies such as IT, biotechnology, and of late, nanotechnologies. The latter refers to a set of technology-platforms whose common aim is to exploit the unique properties of matter at the nanometre scale (one billionth of a metre), which are not displayed at a larger scale, for scientific research and commercial innovation creation. The term 'nanotechnology' represents a variety of technologies with potentially different applications that nevertheless share the common features of either incorporating nano-scale materials or having evolved from processes at the nano-scale. At this scale, which is one billionth of a metre, materials display properties and functions that are different from their larger analogues, and it is these unique properties that nanotechnologies aim to utilise in order to create novel applications or enhance functions of existing products or processes. While definitions vary, a broad consensus has evolved that nano-scale refers to dimensions in the ranges of 1–100 nm. The scientific foundation for nanotechnology stems from the field of nanoscience, which involves the study of the 'nano' phenomenon, the manipulation of materials at nano-scales, and the investigation of the properties of various nanomaterials. Nanotechnology is defined as the design, production, and application of structures, systems, and

---

[1] Portions of this article are based upon: TERI (The Energy and Resources Institute) 2009. *Nanotechnology Developments in India: A Status Report*. New Delhi: TERI [Project Report No. 2006ST21], Research Deliverable No. D5.

devices through control at nanometre scales (The Royal Society and The Royal Academy of Engineering 2004). However, the term 'nanotechnology' is by and large used as a reference for both nanoscience and nanotechnology (NST from now on), especially in the public domain. In fact, in many cases, the term 'nanotechnologies' is used in place of its singular counterpart, as in effect the field denotes numerous technology prospects in various sectors rather than a single kind of technology intervention.

Nanotechnologies are hailed as being transformative, with speculations that they might even trigger a new industrial revolution. They have clear points of convergence and intersection with biotechnologies and IT. It is, therefore, not surprising that NST has been accorded national R&D priority in both developed and developing economies. Indeed, the USA formally marked the start of an international technology race in NST, with the launching of its 'National Nanotechnology Initiative' in January 2000, consecrating the phenomenal amount of US$270 million to this promising field. The rest of the world has followed suite with minimal delay – presumably to avoid the catch-up lag. The EU Framework Program for Research and Technological Development in Europe made it one of its themes in the 5th Framework Programme (1998–2002), with an overall budget of €14.96 billion. Developing countries and transition economies of Brazil, Russia, India, China, and South Africa (BRICS) have also given nanotechnology a high priority, and all countries have defined a national plan for its development. Brazil, India, and China already account for more than 1 per cent each of the world's total investment in nanotechnology, which is expected to grow further in the coming years (Aguirre-Bastos 2010).

India's engagement with the nanosciences, especially in the area of material science, had already commenced during the 1980s. Its active scientific community was aware of the international developments in the area. Thus, the launching of Nano Science and Technology Initiative (NSTI) by the Department of Science and Technology (DST) in 2001 was simply the culmination in the form of a formal and visible engagement. This programme is credited as being instrumental in propelling several developments in the nanotechnology R&D landscape in India.

Using the achievements under the NSTI as a foundation, the nation embarked on a mission mode programme in 2007, namely the Nano Science and Technology Mission (NSTM). About Rs 1000 crore (US$254 million) has been allocated for the NSTM programme, which was to run from 2007 to 2012.[2]

[2] http://dst.gov.in.

Besides promoting research in the area of nanoscience and technology, these umbrella programmes were initiated as avenues to facilitate capacity building to undertake R&D and innovation generation. The Indian government supports and guides capacity building in NST through a myriad of public agencies.

In the above context, it is pertinent to examine, as done in this chapter, the ways in which India is participating in the nanotechnology revolution. How can this participation be guided to maximise returns in terms of economic growth and inclusive development? The remainder of this chapter is organised as follows. First, the main actors in the system are presented. Second, the nature and forms of State investment in NST capacity creation are analysed. Third, the outcomes in terms of construction of capabilities are examined. Finally, the main findings are discussed and recommendations are proposed.

The main actors in the nanoscience and nanotechnology innovation system

Since nanotechnologies can contribute to various sectors and also span several disciplines, a wide variety of state agencies are involved in supporting NST (see Table 7.1). Under the Ministry of Science and Technology, the primary agencies involved with nanotechnology are DST, Department of Biotechnology (DBT), and Council for Scientific and Industrial Research (CSIR).

The DST is the nodal department for coordinating the activities of NST in India. The DBT, on the other hand, is primarily involved in promoting the field of biotechnology in India, and therefore, has been involved in the sphere of nano-biotechnology. The CSIR, which is constituted by a network of 39 laboratories that undertake research in the areas of scientific and industrial importance, also supports R&D in the area of nanoscience and nanotechnology in its laboratories. Other public agencies involved in NST capacity creation are indicated in Table 7.1.

Table 7.1: Main Indian agencies involved in nanoscience and nanotechnology capacity creation

Agency	*Ministry*	*Focus*
Department of Information Technology (DIT)	Ministry of Communication and Information Technology	ICT and Electronics
Indian Council of Medical Research (ICMR)	Ministry of Health and Family Welfare	Health

Contd.

Contd.

Agency	*Ministry*	*Focus*
	Ministry of New and Renewable Energy	Energy production and storage
Department of Atomic Energy (DAE)	Ministry of Defence	General applications
Defence Research and Development Organisation (DRDO)	Ministry of Defence	General applications
Indian Council of Agricultural Research (ICAR)	Ministry of Agriculture	Agriculture
Department of Industrial Policy and Promotion (DIPP)	Ministry of Commerce and Industry	Industry participation
Council for Scientific and Industrial Research (CSIR)	Ministry of Science and Technology	Biological, Chemical, Engineering, Physical, and Information sciences
Department of Biotechnology (DBT)	Ministry of Science and Technology	Healthcare, Agriculture, Animal sciences, Environment, and Industry
Indian Space Research Organisation (ISRO)	Department of Space	Space technology and its applications

Besides the State, the other actors in the innovation system are as follows:

- Scientists in universities and public laboratories, who are the main recipients of State funding.
- A handful of established firms that have started exploring the incorporation of NST in their production systems.
- Some start-ups either from public research or as new divisions of established firms.
- A larger group of distributors of nano-products produced abroad.
- Some industry associations, whose purpose is to link up firms to the other actors in the innovation system.
- Private research institutes, some of which have also received funding.
- Not-for-profit foundations.

The last two types of actors are distinct and specific to NST in the sense that they have been accompanying the emergence of the new sector of NST right from the start, which was not the case with the biotechnologies or the IT. Hence, I briefly touch upon them, before examining the strategy of the State.

Prominent among the private research institutes involved in NST are the Amrita Institute of Medical Sciences, which has a faculty devoted to nanosciences and is a designated Centre of Excellence (CoE) in NST. Other active institutes include the Sankara Nethralaya, Birla Institute for Technology and Science, Amity University, and SASTRA University. Several of these institutes have displayed a keen focus in the areas of nanomaterial synthesis, nanobiotechnology, and the area of nanosciences and health.

A few not-for-profit organisations (listed below) are associated with nanotechnology developments in India; however, their role is mostly in the capacity of linking academia with industry.

- *The Nanotechnology Research and Innovation Foundation*: This is a non-profit organisation supported by academic and industry experts such as National Chemical Laboratory, Girvan Institute of Technology, and The Centre for Materials for Electronics Technology, and is funded by IndiaCo, a private equity investment holding company that invests in hi-tech companies. Its aim is to provide a platform to stakeholders for R&D collaboration in advanced technologies, including nanotechnology. In January 2006, a strategic partnership between the Nanotechnology Research and Innovation Foundation and the Regional Research Laboratory, Trivandrum (RRL-T), was forged to leverage their combined expertise to respective expertise to facilitate the commercialization of the latter's technologies with industries in India and across the globe.
- *The Nano Science and Technology Consortium*: This non-governmental, industry-managed and promoted organisation seeks to create a platform conducive for the growth, promotion, and collaboration among industries, academics, and government in the nano domain through consultative, advisory, and educative processes.
- *The Energy and Resources Institute*: This is a not-for-profit research institute, which undertakes research in the area of synthesis of nanosize materials using microbial strains. TERI, as it is called, is also involved in policy-related research in the area of nanotechnology. The various stakeholder dialogues organised by the institute provide a platform to discuss the aspects of R&D innovation and capacity building, benefits, risks, and regulatory and socio-ethical issues in the context of nanotechnology and its implications for developing countries. Moreover, these platforms have successfully brought together scientists, social scientists, technology developers, policy-makers, lawyers, NGOs,

and the media to discuss and debate the aforementioned issues. Through these stakeholder engagement and dissemination initiatives, the institute has donned the role of a civil society organisation that seeks to contribute to the sustainable development of nanotechnology in India.

The role of the State

Building scientific capabilities

The strategy for the creation of scientific capabilities involved a four-pronged approach: (i) creation of quality manpower through the establishment of CoE; (ii) support of R&D projects in the existing and the new academic institutes; (iii) promotion of international collaborative programmes; and (iv) strengthening of industry–university links through public–private partnerships (PPP). Examples of each of these types of actions are given below.

Establishment of Centres of Excellence

Under the NSTI, the DST established 19 CoE in various institutes (see Fig. 7.1).

They are of two types. First, there are 'units' that have the mandate of pursuing basic research in the various areas of nanoscience and nano-scale systems. Second, there are 'centres' whose mission is to engage in time-bound R&D in niche areas for the development of specific applications. Furthermore, CoE has been designed to ensure a good diversity and spread of organisations throughout India. All except one of the centres have been lodged in public-supported institutions.

The Amrita Institute of Medical Sciences and Research Centre, Cochin, established in 1998, is the only private research institute, which has received funding from India's Nano Mission for the establishment of CoE in Nanotechnology. It undertakes a range of activities, including patient care, developmental projects, sponsored research projects, societal missions, and in-house research projects. The institute's engagement in the domain of nanoscience started with the establishment of India's first nano-bio centre, the Amrita Centre for Nano Sciences and Molecular Medicine (ACNSMM), in 2006. The Centre has excelled with its pioneering research work in tissue engineering and stem cells under the leadership of Dr Shanti Kumar Nair, regarded as India's foremost scientist, who received the prestigious National Research Award from the Ministry of Science and Technology in 2012. The

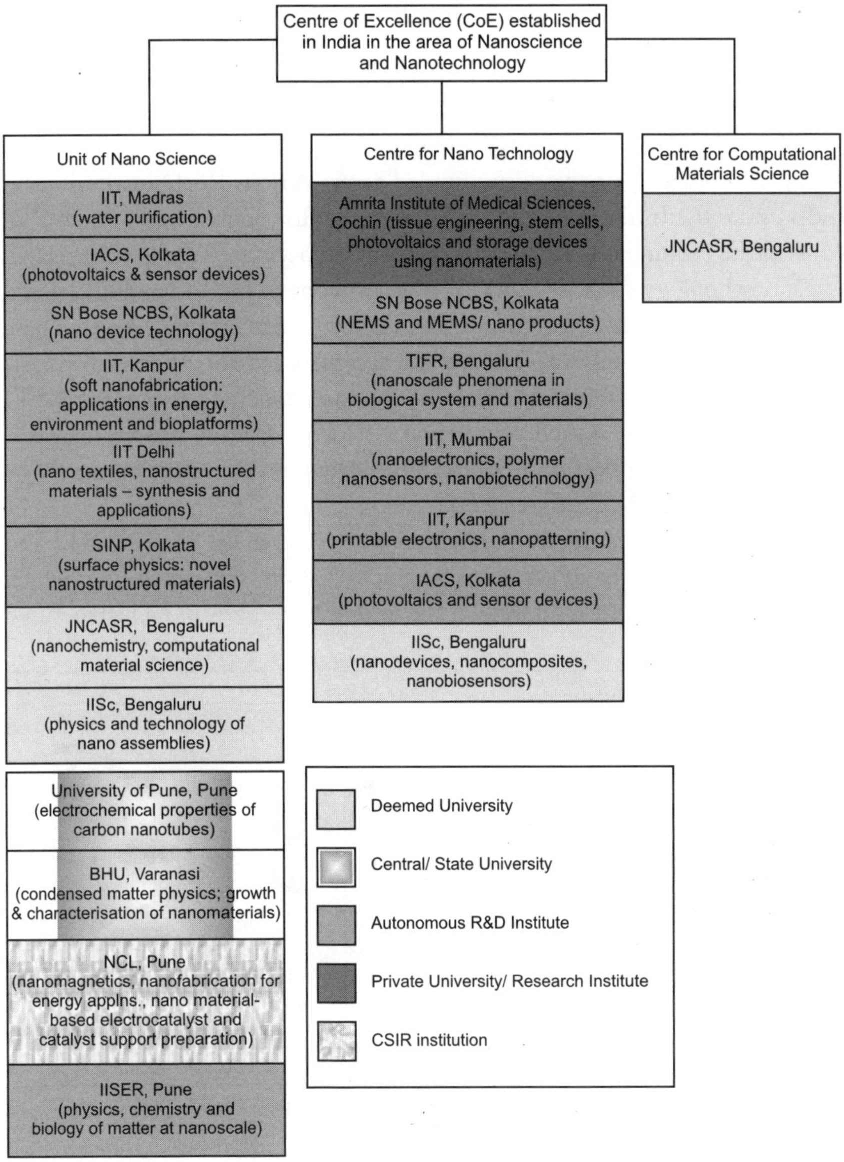

Fig. 7.1: Architecture of centres of excellence in nanoscience and nanotechnology in India

Centre also hosts a Nano Solar CoE which focuses on innovative product development in photovoltaics and storage devices using nanomaterials. Two products, viz., the world's first integrated solar power storage tile named Amrita Smart and nanophotomedicine to be soon used in clinical trials, were launched in the NanoBio Conference 2012, organised by the Amrita Centre for Nanosciences.

Support of R&D projects

About 100 research projects were funded across various R&D institutions in India under the Initiative of the DST under the first nano programme NSTI. Most of the funding was for the procurement of specialized equipment related to nanotechnology, such as atomic force microscopy (AFM), scanning electron microscopes (SEM), and transmission electron microscopy (TEM). These are extremely costly instruments that are necessary for undertaking any basic research in the nanosciences. A difference between the first programme, NSTI (2001–05/06) and the second programme, NSTM (2007–12), was the increased emphasis laid on application-oriented research and product development. The research thrusts of NSTM are given in Fig.s 7.2 and 7.3.

As Fig. 7.2 reveals, the bulk of the projects supported by the NSTI and NSTM are in the area of synthesis, characterization, and processing, which help to develop an understanding of the unique properties of nanostructured

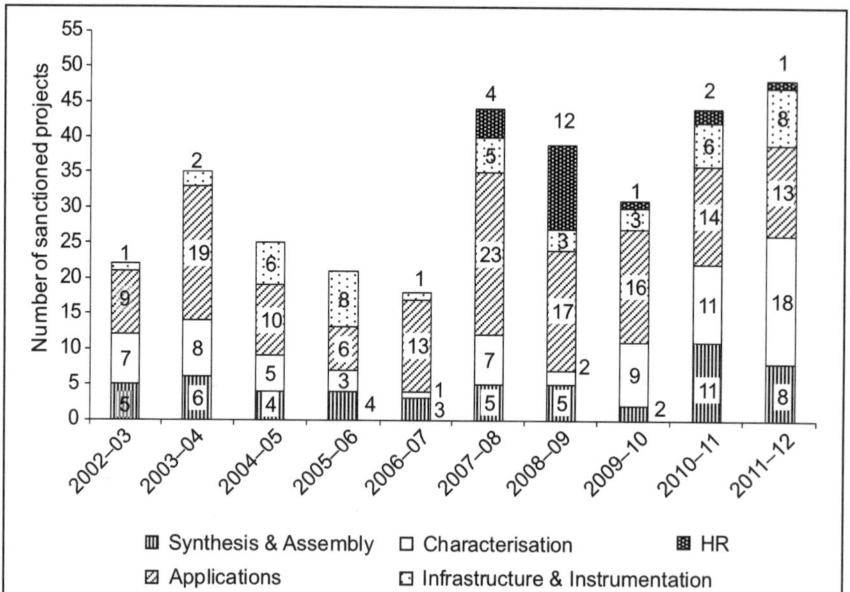

Fig. 7.2: Distribution of nanotechnology projects supported by the Department of Science and Technology across basic and applied research/product development and other areas

Source: http:/nanomission.gov.in and http://dst.gov.in/scientific-programme/ser-serc.htm

materials. Research on a variety of nanomaterials such as metallic, metal oxide, semiconductor, magnetic nanoparticles, quantum dots carbon nanotubes and other nanocomposites, polymeric nanomaterials, nanofilms, nanowires, nanoalloys, and nanoporous solids has been undertaken and a wide range of routes for the synthesis of nanomaterials, including sol-gel, salvo-thermal, spray pyrolysis, chemical vapour synthesis, etc., have been explored. Also, emphasis has been placed on exploring the industrial applications of these materials, besides focusing on the development of nanolithography and nanofabrication techniques. Furthermore, a significant number of projects have been supported in basic and applied research in the area of physical and chemical sciences. For example, in physical sciences, nanotechnology research in India covers the areas of condensed matter physics and materials science, plasma, high energy nuclear physics, astrophysics, and nonlinear dynamics as well as laser, optics, atomic, and molecular physics.

As Fig. 7.3 shows, initially the projects supported by the NSTI and NSTM had no particular sectoral orientation, but increasingly now, they are being focused on the health and environment sectors. For example, in the area of health, there has been considerable research focus on targeted drug and gene

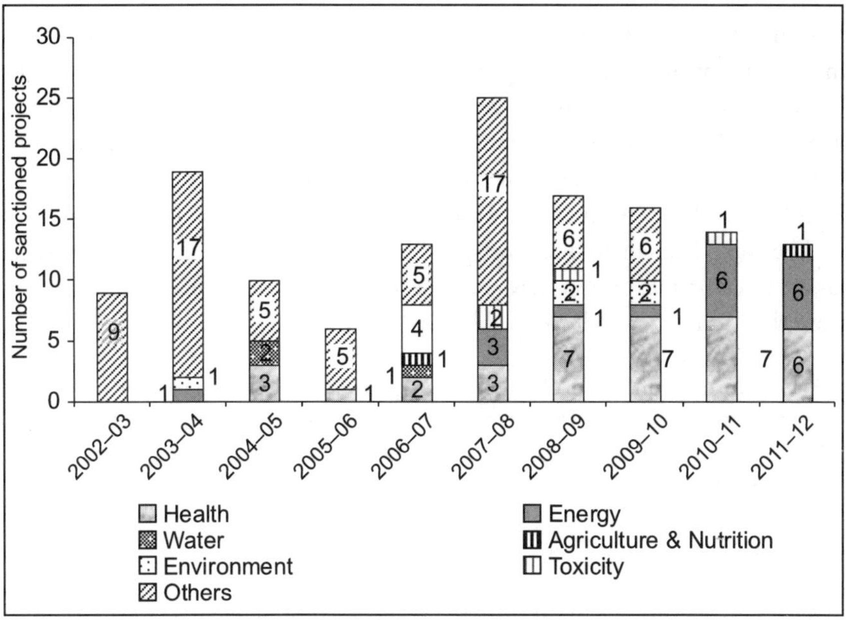

Fig. 7.3: Distribution of application-oriented R&D projects, by sector in nanotechnology supported by the Department of Science and Technology

Source: http:/nanomission.gov.in and http://dst.gov.in/scientific-programme/ser-serc.htm

delivery (therapeutics), diagnostic systems, and biosensors as well as the use of biomaterials for various applications, including regenerative medicine. Significant technological achievements have been achieved in this application area, e.g., encapsulation of water-soluble drugs using hydrogen nanoparticles developed through reverse micelles-based process.

With respect to environment, the thrust is on the application of nanomaterials, particularly nano-scale silver and carbon nanotubes in water purification. Various technologies have been developed in this area, such as incorporation of nano-silver in traditional candle filters for disinfection, nano-silver-based carbon blocks for pesticide removal, nano-silver-coated alumina catalyst for controlling microorganisms in water, carbon nanotube filters for bacteria removal, and nano iron oxide/mixed oxide for arsenic removal.

In the energy sector, the basic aim is to improve the solar photovoltaic process efficiency, reduce the cost of solar cells, and search for alternate semiconductors in lieu of silicon shortage through the development of new materials, processes, and device structures.

Public–private partnerships

Recently, there has been an increasing trend in the nanotechnology industry in India pointing towards the private sector beginning to invest in the R&D laboratories at university and government institutions. Innovative mechanisms like incubators are supported by the State funding agencies and nanotechnology innovation parks are being planned to encourage the industry–academia–public R&D networking (for instance, IISc MEMS lab).

The Nano Mission also supports joint institution–industry linked projects. So far, six collaborative projects have been developed with some financial assistance from the industry as well (see Table 7.2).

Table 7.2: Institute–industry collaborations supported under the Nano Mission

Institute	*Industry*
Nano Functional Materials Centre, IIT Madras	Murugappa Chettiar & Orchid Pharma
Nano Technology Centre, University of Hyderabad	Dr Reddy's Labs
Centre for Interactive and Smart Textiles, IIT Delhi	International Advanced Research Centre for Powder Metallurgy and New Materials (ARCI), Hyderabad and Textile Industry

Contd.

Contd.

Institute	*Industry*
Centre for Pharmaceutical Nanotechnology, National Institute of Pharmaceutical Education and Research (NIPER), Chandigarh	Pharma industry
Rubber Nanocomposites, MG University, Kottayam	Apollo Tyres
Nanophosphor Application Centre, University of Allahabad	Nanotech Corp., USA

Source: http://nanomission.gov.in/

Other incentives provided to scientists to create useful technology

For application-oriented technology development, setting up of 'Nano Applications and Technological Development Centres' and 'Nanotechnology Business Incubators' has also been envisaged under the mission along with a Research Industry Collaboration Hub (RICH) to promote Mergers & Acquisitions (M&A) deals, collaborations, and investments in nanotechnology.

Recently, the government announced new policy measures to encourage the 'development and commercialization of inventions and innovations'.[3] These include allowing researchers working in public academic and research institutions to also hold equity in spin-off companies and enterprises and also facilitating mobility of researchers between academic institutions and industry. The former initiative would enable new avenues for the translation of innovative lab research into commercial applications and technology. This new initiative would encourage scientific entrepreneurship and allow scientists with promising research and expertise in the nano domain to create start-ups and aim for applications development. As nanotechnology is a cost-intensive endeavour, scientists would be able to leverage the facilities, infrastructure, and manpower already developed at their centres to undertake technology development ventures in nanotechnology. On the other hand, the exchange of personnel amongst academic institutions and industry will help the transfer of expertise and skills to bring forth new ideas and knowledge for application development. The understanding of the concerns and perspectives on both sides would enable successful PPPs and technology development, both of which are proving to be a major challenge in the nanotechnology arena.

3 http://dst.gov.in/.

Performance of the Indian nanoscience and nanotechnology system

Scientific publications

A few studies have been carried out on the publications coming out from Indian institutions (see Table 7.3). But it is difficult to compare them because they have used different databases and different research equations, and have covered different years.

Karpagam et al. (2011) use the Scopus database to identify NST-related publications over the period 1990–2009. Bhattarcharya and Bhati (2012) explore the Science Citation Index Expanded (SCI-E) to search for articles from India on NST during 2000–09, and Ramani et al. (2010) use the same database, but with a different research equation to identify NST articles over the period 1998–2008.

Table 7.3: Characteristics of Indian nanoscience and nanotechnology publications produced

	Bhattacharya and Bhati (2012)	*Karpagam et al. (2011)*	*Ramani et al. (2010)*
Years covered	2000–09	1990–2009	1978–2008
Total number of publications with at least one author affiliated to an institution in India	13,092 (3.4% of world total)	22,765 (3% of world total)	2,477
Trend	Growing steadily but much behind China	Growing	Growing steadily but much behind China
Percentage of total co-authored papers (more than 1 author)	–	96%	96.4%
Percentage of international collaborations (with at least one author from outside India)	–	20.84%	21.78%
Leading institutes in terms of number of publications	IIT network, NCL, IISc		
Focus areas	Materials sciences, applied physics, physical chemistry, polymer science	Materials science, chemistry, physics, engineering	Nano-materials, nanoparticles and nanomicroscopy, nanobio

It is clear from Table 7.3 that different research equations and databases can yield different results. But they all show that the publications are growing in number and most are collaborative papers between Indian institutions, of which the top three laboratories noted by all the authors are the same. Moreover, Bhattacharya and Bhati (2012) observe that in 2009, India accounted for 5 per cent of total world production in NST publications, ranking seventh globally. Its highly cited papers are often those in which the collaborators are from different institutions. Having at least one of the authors from a high-income country seems particularly helpful. This condition seems to be useful also for getting articles published in the top journals. Only one Indian firm, Reliance Life Science Pvt Ltd, seemed to have published with a public lab. They also note cluster effects and relate it to the fact that undertaking research in NST requires access to costly equipment, so researchers form clusters around laboratories that house the costly machines. This becomes evident, as they note, even in Chinese publications. In recent times, the SEM and the TEM appear as central nodes among the keywords used to mark the contribution of the publication. They note that unlike China, India does not produce these instruments, which may be a factor holding back the production of publications.

Patenting activity

In terms of patent analysis also, there have been different studies applying different measures to identify patent applications issuing from India. Bhattacharya and Bhati (2012) examine the patents issuing from India in the US Patent Office (USPTO) under the class '977' signifying affiliation to nanotechnology and identify 12 filings between 2000 and 2009, out of which 7 have been granted. Ramani et al. (2010) also confirm 35 patent applications in the European Patent Office (EPO) under the class 'Y01N', which is the tag applied to nanotechnology. On the other hand, Gupta (2009) uses a string of indicative keywords to identify a total of 167 patent applications related to NST issuing from India between 2001 and 2007 deposited in patent offices in the USA, Japan, Germany, Europe (EPO), World Intellectual Property Organization (WIPO), and International Patent Documentation Centre (INPADOC). All these studies show that in the international arena, universities and public laboratories are the most active patentees.

Using the keyword 'nano' yields a total of 166 patents granted by the Indian Patent Office (IPO) during the period 1997–2009. Of these, a total of 91 patents had been assigned to Indian entities. Of the total 166 patents, 75 patents (45 per cent of the total patents) are owned by the firms in the industry, 40 patents (24 per cent of the total) by the academic institutions, and 35 patents (21 per cent of the total)are owned by the government institutions. There are 15 patents (9 per cent of the total) that are owned by individual inventors and 1 patent is jointly held by the government institutions and the firms from industry. Table 7.4 shows the distribution of patent applications in nanotechnology at the IPO across institutional sectors and nationalities.

The scientific research institutions owning majority of the patents include laboratories of the CSIR. Among the academic institutions, the IITs have been the major contributors of nano patents at the IPO. In the industry sector, foreign companies are the major patent holders. Among Indian firms, nano patents are being held by firms like Ranbaxy Laboratories Limited (Indian company till 2007), Dabur India Limited, Nano Cutting Edge Technology Private Limited, and Lifecare Innovations Private Limited.

The major technology areas of patenting in nanotechnology include applications related to drugs and pharmaceuticals, chemical science, and technologies, e.g., catalysis, colloid chemistry, separation of non-metallic elements and materials, analysing materials by determining their chemical or physical properties, and methods/filters implantable into blood vessels.

Table 7.4: Analysis of 166 patent applications at Indian Patent Office (IPO) between 1997 and 2009: Type of patent assignees and sectors

Category	*Indian*	*Foreign*	*Total*
Type of patent assignees			
• Individual inventors	13	2	15
• Academic institutions	33	7	40
• Government institutions	34	1	35
• Industry	10	65	75
• Joint patents	1	–	1
Sectors			
• Nanomaterials	–	–	17
• Process for making nanomaterials	–	–	139
• Devices	–	–	10

Source: IPO

The nascent Indian nano sector

Manufacturing – four main fields of activities: Incumbent firms as well as start-ups

Nanotechnology industry comprises firms involved in bringing nanotechnology-based delivery platforms, processes, materials, tools, and devices into the market.

Fig. 7.4 maps these sets of industrial players in nanotechnology in India along with the fields in which they specialized.

As Fig. 7.4 shows, on the one hand, there are start-ups, which focus mainly on nanotechnology. On the other hand, there are certain established companies that have tried to engage with nanotechnology either to improve their existing product line or to explore businesses for the next generation. 'Materials and surface engineering,' 'biotechnology,' 'electronics,' and 'instrumentation and equipment' are the four nanotechnology fields to which Indian companies have drawn to.

'Nanomaterials' is the major focus of most Indian companies in conformity with global trends. Both established companies as well as start-ups share this niche in materials and surface engineering. Some start-ups, which are becoming renowned are: Monad Nanotech Pvt. Ltd, Innovations Unified Technologies, INNANO Technologies, NANOSHEL, United Nanotechnologies Pvt. Ltd,

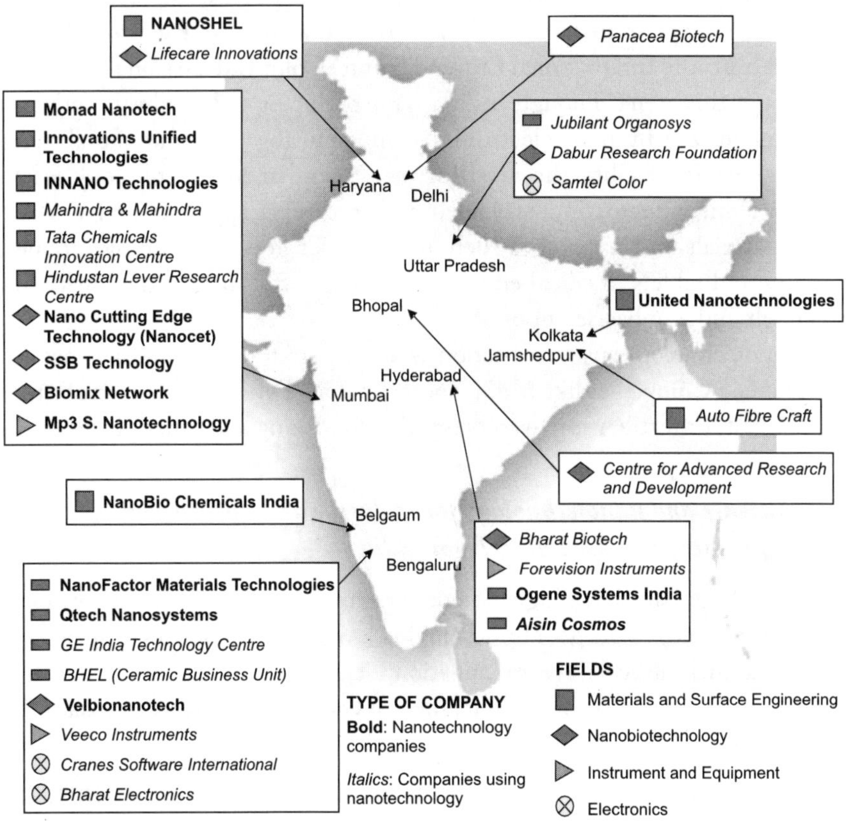

Fig. 7.4: Nanotechnology industry in India

NanBio Chemicals India Pvt. Ltd, NanoFactor Materials Technologies, Qtech Nanosystems, Ogene Systems India (P) Ltd, and Aisin Cosmos R&D Co. Ltd. Among established companies, Mahindra & Mahindra Ltd, Tata Chemicals Innovation Centre, Hindustan Lever Research Centre, Jubilant Organosys Ltd, Auto Fibre Craft, BHEL (Ceramic Business Unit), and GE India Technology Centre are the major players.

'Nano-bio' is the second most popular sub-field of NST, which is the prime focus of some companies. These are essentially biotechnology companies with a particular expertise in nano-bio. Here, among start-ups, some successful firms are Velbionanotech, Nano Cutting Edge Technology Pvt. Ltd (Nanocet), SSB Technology, and Biomix Network Ltd. Among established biotechnology companies that have ventured in the area of nano-bio interface, some of the major players are Panacea Biotech, Dabur Research Foundation, Bharat Biotech, and Lifecare Innovations Pvt Ltd.

'Nanoelectronics' is another important application area in the nanotechnology industrial landscape in India. Bharat Electronics Ltd, Cranes Software International Ltd, and Samtel Color Ltd are some important players in this arena. Though Indian semiconductor and hardware industry is characterized by poor electronics manufacturing base and the absence of chip-making facility, the need for investment in nanotechnology R&D in the semiconductor industry to innovate diverse applications based on nanomaterials and nanotubes such as consumer goods, including mobiles, laptops and tablets; medical equipment; energy efficiency; and security has been felt and emphasised upon.[4]

'Nanotechnology instrumentation' is an important area requiring massive investment. Companies like Mp3Nanotechnology Pvt. Ltd, thus, manufacture equipment for textile waste water recycling for dyeing and other processes.

Nanoscience and nanotechnology for inclusive and sustainable development? Surprise stories from India

NST has been hailed as a high-potential technological source for countries to address the needs of the poor by developing low-cost technologies in areas such as sanitation, healthcare, energy, nutrition, elementary education, and housing. Surprisingly, despite the heavy State investment in public research, to date, the

[4] Indian semicon industry told to invest in nanotechnology, 21 February 2011, The Times of India.

NST-based innovation that has had the maximum impact on the lives of the poor has been created by a private firm – without being involved in any NST project directly.

Given the scarcity that marks the Indian context, Indian firms are renowned for their creativity and innovation generation using frugal resources.[5] A typical example is the 'Tata Swach' created by the Tata Research, Development and Design Centre (TRDDC) and Tata Chemicals with contributions from other Tata group companies. In India, access to potable water is a major challenge for most of the poor. The 'Tata Swach' is a radically new water purifier system that is being sold at a price that most can afford. The water purifier, 'Swach', was developed in 2009 using rice-husk ash and silver nano-particles to filter out the bacteria and other germs and deliver safe drinking water at a minimal cost of Rs 749 and Rs 999 (less than US$25).[6] Swach, an example of nano-based innovations for bottom of the income pyramid communities, does not require electricity. Instead it runs on a replaceable cartridge with a shelf-life of 3,000 litres that is expected to serve a family of five for approximately 200 days. Further, the technology has been adapted to suit geography-specific challenges, such as providing safe drinking water in regions where groundwater contains high levels of arsenic and fluoride impurities.

In many countries, civil society groups have expressed concerns about the risks of NST for the environment. The Samsung washing machine 'Silver Nano' was introduced in India in 2010. A range of semi-automatic washing machines were developed with an investment of over US$10 million on R&D and priced between Rs 6,590 and Rs 9,990. The 'Silver Nano' is being promoted as the first technology that combines the disinfectant and antibiotic properties of electrolytic silver nano-particles in washing machines to remove harmful bacteria. However, when the washing machines were released in the USA market, the US Environmental Protection Agency temporarily banned the machine on account of claims made by environmental groups in the USA that the silver-laced waste water might damage the earth.[7] Many environmental activist groups all over the world are advocating for its ban. There was no controversy over the sales in India and the regulatory agencies did not subject it to further scrutiny.

---

[5] See http://en.wikipedia.org/wiki/Frugal_innovation.

[6] See, http://www.business-standard.com/india/news/nano-filters-down/380243/.

[7] See, http://www.hindu.com/thehindu/holnus/006200710042122.htm; http://www.livemint.com/2010/02/22234526/Nanotech-devices-must-be-regul.html.

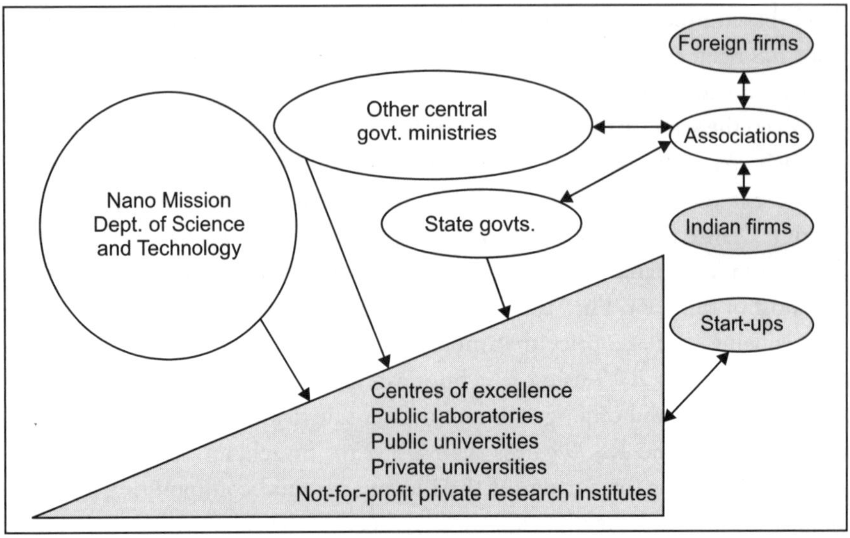

Fig. 7.5: Innovation system of nanoscience and nanotechnology in India

An illustrative case study: The application of nanoscience and nanotechnology to health

From the information provided in the previous sections, the Indian system of innovation with respect to NST can be represented as in Fig. 7.5. Further illustration of this system with a brief outline of the processes by which NST is being integrated in the health sector is provided below.

At the outset, it must be noted that the health sector is an ideal recipient for NST outputs because of India's already acquired capabilities. India has a strong pharmaceutical industry, with firms that are renowned for their technological prowess and are supported by public laboratories with strong scientific capabilities.

State-funded research and products that are coming from public laboratories

The perceived societal benefits that R&D in nanomedicine can provide in the health sector together with the economic opportunity it provides to the pharmaceutical sector to leverage its strength in generics has led to both state-aided funding as well as industry participation in this domain. Several of the state agencies including DST, DBT, Indian Council of Medical Research (ICMR),

CSIR, and Defence Research and Development Organisation (DRDO), are involved in funding R&D in the area of nano-biomedical sciences. The present research activity in nanomedicine in Indian scientific establishments could be divided into the following broad areas:

- *Nanomaterial-based diagnostics, nanosensors, and nanodevices*: The use of nanomaterials in the field of diagnostics and the development of nanosensors and devices is actively progressing in India, as especially in the latter case, nanomaterials provide great opportunities to improve the functionality and service of devices and adjust their properties for better biological compatibility. Micro diagnostics are intended for targeted drug delivery and diagnostics for major diseases such as cancer and tuberculosis. Nanosensors are components that are integrated into diagnostics to detect diseases. These are especially important in micro-devices for cardiac use.
- *Understanding the interaction of nanomaterials with human systems and toxicological studies*: The main purpose of this research is to understand the toxicological impacts of specific nanomaterials and provide a safety/toxicity assessment of engineered nanomaterials (ENMs) and nanodevices. The prominent areas of research are the effects of nanomaterials on cells and tissues and on ecosystems.
- *Drug delivery and design*: Drug delivery products/processes refer to applications that can facilitate sustained release of drugs and their targeted delivery to the diseased parts of the body and also improve the bioavailability of drugs. Typical target diseases are cancer (some types), Alzheimer's disease, diabetes, tuberculosis, visceral leishmaniasis, and bone diseases.
- *Post-curative regeneration medicine via tissue engineering*: Here, the main targets are cardiovascular grafts and artificial skin, bone tissue engineering, and grafts for wound healing.
- *Development of nanomaterials for therapeutic applications and integration into biomaterials*: The typical expected outcomes are materials for bone grafting, bone and dental implants as well as wound healing.

Table 7.5 shows how State support of research is creating scientific capabilities in NST in various scientific establishments all over India. Some of the public research has been transformed into innovations that are currently in the pipeline. Some are particularly noteworthy.

Table 7.5: Academic institutions involved (not a comprehensive list)

Nanomaterial-based diagnostics, nanosensors, and nanodevices

- IIT Mumbai, Indian Institute of Science (IISc), Central Scientific Instruments Organization (CSIO), Centre for Cellular and Molecular Biology (CCMB)

Understanding the interaction of nanomaterials with human systems and toxicological studies

- Amrita Centre for Nanosciences, Indian Institute of Toxicology Research (IITR), Jawaharlal Nehru Centre for Advanced Scientific Research (JNCASR)

Drug delivery and design

- Jawaharlal Nehru University (JNU), CSIO, All India Institute of Medical Science (AIIMS), IIT-Mumbai, University of Calcutta, Aligarh Muslim University, National Institute of Pharmaceutical Education and Research (NIPER), H.S.G. Vishwavidhyalaya Sagar Institute, Jamia Hamdard University, Patel Pharmaceutical Education and Research Development, Amrita Institute for Nanosciences, Rangaswami College of Technology, Rajiv Gandhi Center for Biotechnology, Bharathidasan University, SASTRA University

Tissue engineering and regenerative medicine

- National Metallurgical Laboratory, SASTRA University, IIT Guwahati and National Centre for Cell Science, Amrita Centre for Nanosciences, Nichi-In Center for Regenerative Medicine (NCRM) – Indo-Japan joint venture institute

Development of nanomaterials for therapeutic applications and other biomaterials

- Anna University, IIT Kanpur, National Chemical Laboratory (NCL), Indian Institute of Chemical Technology (IICT)

Source: Author's own compilation

Significantly, scientists at IIT-Mumbai have developed the biosensor 'iSens,' a point of care assay system that detects molecular markers for myocardial infractions that typically are indicators for heart attacks. They have also developed a silicon locket as a part of a complete suite for cardiac monitoring and diagnosis. The Central Scientific Instruments Organisation (CSIO) has developed a nanotechnology-based diagnostic kit for tuberculosis that is claimed to be portable, inexpensive, and efficient in use. Similarly, labs under DRDO have fabricated a nanosensor-based typhoid diagnostic kit. Finally, the University of Delhi has developed a process of entrapping genetic materials in nanoparticles of inorganic compounds to form non-viral carriers for drug delivery.

Nanoscience and nanotechnology products/services of firms and modes of innovation creation in the health sector

Most of the firms in this sector are either biotech companies or pharma companies. A few have created new nanotech-based drug delivery systems and diagnostics. Many are successful contract research companies – with the development of new technologies to their credit. These form the Indian 'production of invisible innovations' that support the global pharmaceutical industry.[8] The commercialization of new products can either stem from own R&D or an external entity. Examples of each of these are now presented:

- *From the in-house R&D of Indian firms*: Leishmaniasis, also known as Kala-Azar in India, debilitates and kills thousands in India every year. The most effective existing drug known as Amphotericin B also has deadly side effects. Lifecare Innovations Pvt. Ltd of Gurgaon, a biotech company, developed a safe wrap for the Amphotericin B such that the 'Liposomal Amphotericin B' delivers the drug to the desired site safely while amplifying its curative properties. The Dabur Research Foundation in Ghaziabad (Uttar Pradesh) – which is the research arm of the manufacturer of traditional medicines, Dabur – has developed the world's first polymeric nanoparticle-based drug delivery system in the world called Nanoxel for cancer.
- *Invisible innovation from India via contract research or technology development*: Many Indian companies have become famous through the contract house research. They provide the invisible innovation that supports the innovation commercialization process of large global multinationals. Velbionanotech, a nano-bio company based in Bengaluru, has been ranked among Asia's top 100 nano-biotech by Red Herring, 2005. It has developed nanotech-based treatments for atherosclerosis (arterial plaque), nephrolithiasis (stone in the urinary tract), and diabetes. Bilcare Ltd based in Pune has developed anti-counterfeit technology for drugs using nano-fingerprinting technology.
- *Introducing new products through collaboration or technology transfer*: In the health sector, there has been an increasing trend of collaboration, both among private organisations and between public and private entities. Table 7.6 indicates some of the important collaborations in nanotechnology in the health sector.

[8] See India's invisible innovators by Nirmalaya Kumar; http://www.ted.com/talks/nirmalya_kumar_india_s_invisible_entrepreneurs.html.

Table 7.6: Collaborations in nanotechnology in the health sector in India

Supplier/Manufacturer	*Activity/Technology*	*Source of technology*
Private-Public collaboration		
• Dabur Pharma, New Delhi	Cancer drug Nanoxel (with nano polymer base) and Docetaxel using nanotechnology drug delivery	Delhi University and Dabur Research Foundation
• Panacea Biotech, New Delhi	Pharmaceutical formulation of non-steroidal anti-inflammatory drugs into polymeric micelles nanoparticles for ocular delivery in Nimesulide	University of Delhi
• Shantha Biotech, Hyderabad	Process for preparing pharmaceutical formulation of Amphotericin B or other polyene antibiotics entrapped into nanoparticles of copolymeric micelles	University of Delhi
• Transgene Biotek Ltd, Hyderabad	Drug delivery system for oral insulin (using biodegradable polymeric nanoparticles loaded with human insulin or hepatitis B surface antigen as a carrier)	IICT Hyderabad and Transgene Biotek Ltd
• Cadila Pharma, Ahmedabad	Nano sensor-based typhoid detection kit (recombinant DNA technology and immunological technique for rapid detection)	IISc Bengaluru (nano sensor), DRDO (typhoid detection kit)
• Panacea Biotech, Chandigarh	Ophthalmic delivery using polymeric micelles nanoparticles	Delhi University
Private-Private collaboration		
• Bharat Biotech, Hyderabad	Topical emulsion for oestrogen therapy (micellar nanoparticles technology) for drug delivery	Novavax, USA
• Biocon, Bengaluru	Breast cancer nanodrug Abraxane (paclitaxel protein and albumin bound medicine).	Abraxis Biosciences, USA
• Nano Cutting Edge Technology (Nanocet), Mumbai	Biostabilised nanoparticles like Ag, Au, Fe, Pd, etc. in the areas of cancer hyperthermia, targeted drug delivery, diagnostics, antimicrobial agents, and mitigation of pollutants	Agharkar Research Institute, Pune
• Natco Pharma, Hyderabad	Nanotechnology drug, Albupax, which is used in the treatment of breast cancer. Albupax is the first generic version of the international brand Abraxane and consists of Paclitaxel in an albumin-bound nanoparticle form	Abraxis Biosciences, USA

Note: IICT – Indian Institute of Chemical Technology; DRDO – Defence Research and Development Organisation
Source: Nanotech Insights (2010)[9]

[9] Nanotech Insights, Volume 1, Issue 1, January 2010, Centre for Knowledge Management of Nanoscience & Technology, Secunderabad.

- *Manufacturing of innovations developed elsewhere*: Indian biotech firms carry out contract research for foreign firms and in return commercialize some of their other products. For instance, Bharat Biotech based in Hyderabad has a strategic alliance with Novavax (USA) to develop an influenza vaccine. Novavax has also developed a virus-like particle that can deliver drugs used in estrogen therapy against negative menopause symptoms called Estrasorb, which will be produced and marketed by Bharath Biotech in India.
- *Marketing foreign brands*: SSB Technology (Mumbai) markets, among other chemicals, nanoparticles, some of which are imported from abroad, for use in the healthcare industry.
- *Presence of foreign companies*: Various foreign companies are present in India, simply to sell their NST products, and/or to engage in strategic alliances for manufacture outsourcing and/or R&D. An interesting example of the latter is Veeco Instruments Inc. (USA), a leading supplier of instrumentation to the nanoscience community, which established the Veeco-India Nanotechnology Laboratory to be jointly operated with the Jawaharlal Nehru Center for Advanced Scientific Research (JNCASR) in 2005. This strategic alliance will permit access to very costly instruments such as Veeco's latest AFM and scanning tunneling microscope (STM) products.

Concluding remarks and challenges ahead

The essence of the NST capacity building process is captured in Fig. 7.6. The motor of NST capacity building in India was the Indian government. Under the two major government programmes NSTI and NSTM, public investment in the scientific community and infrastructure in universities and public laboratories has resulted in a substantial expansion in the knowledge base on NST.

This expansion in the knowledge base in universities and public laboratories has given rise to a few start-ups and some technology transfer to established firms. The distances between the academic units and firms have also decreased due to collaboration and technology activism by industry associations. In addition, Indian firms, which have entered into the NST arena have acquired *savoir-faire* from in-house R&D and foreign firms. A variety of foreign firms are active in the Indian market too, and both Indian and foreign firms have created new markets for NST-based products and services. Though it is difficult to measure the size of the NST sector, or the income and employment generated by it, given its multidisciplinary and general purpose nature, clearly,

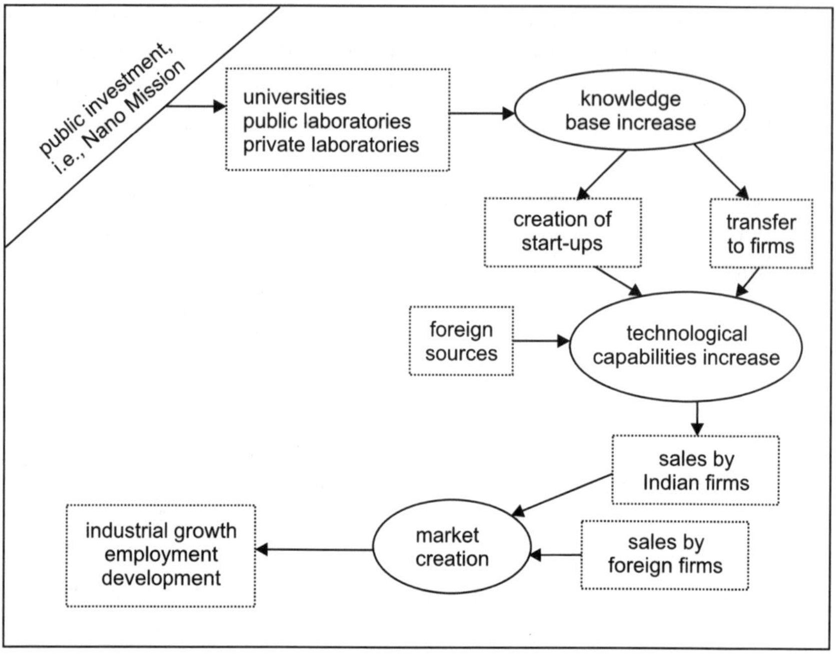

Fig. 7.6: Trajectory for capacity building in India in nanoscience and nanotechnology

it has been positive. The promised fruits of NST to satisfy the pressing needs of the poor are not yet in sight – though there are silver linings, like the Tata Swach water purifier.

Despite the relatively smaller investment of the Indian State in nanotechnology as compared to China and other developed economies, India's performance in NST is noteworthy. It has fared well in terms of the creation of scientific capabilities as demonstrated by the sizeable generation of scientific publications and collaborations. Public investment in new knowledge absorption has had a high impact, only because the Indian scientific establishments already possessed the required knowledge necessary to learn about new emerging sciences. That said, even a cursory glance at the health-related research taking place in the different academic institutes reveals 'clouds of efforts'. While it is not a good idea to curb research liberty with too much of planning, perhaps more R&D collaboration can be encouraged to prevent duplication of effort. Moreover, the accumulation of scientific capabilities does not guarantee the automatic build-up of technological capabilities. This is proved by the few start-ups and the handful of large firms, which have incorporated NST-based processes in their production systems, despite the hype about NST.

The in-house nanoproducts, which have been developed by the leading local Indian companies are basically the low-hanging fruits such as nanocomponents in diagnostics, nanosensors, and nanodevices. This is an excellent start – but the technological challenges are going to increase as the low-hanging fruit gets picked, and there is a need to reach higher. Again, there will be a need for upscaling the production of the low-hanging fruit and reducing costs, because neighbouring China will be the biggest competitor in local and international markets. Another unpleasant surprise is that while India has established technological competencies in ICT, somehow the surge to integrate NST has been minimal in ICT sectors.

Like new drug discovery, technology development in the nano domain is extremely costly with long gestation periods. Reid (2014) identifies the following eight features as being crucial to the evolution of successful nanotech start-ups/ firms in the USA:

1. Patent ownership.
2. Visionary leadership often from a scientist coming from academia.
3. Strong revenue and science base/deep pockets with good government contacts and an ability to develop radical technologies quickly (whether internally because of extant expertise or through acquisition) if an incumbent firm moves into new territory.
4. Alliances with both universities and leading companies, particularly important for newcomers.
5. Ability to attract capital and not run out of money for newcomers.
6. Strong distribution and promotion of known brand to reach end customers.
7. Economies of scale in production (ability to achieve scale up).
8. A large size market with disposable income will be useful for final commercialization of new products.

She continues, 'In other words, firms that are likely to make inroads into NST must start with the classic triangle: "established technological capabilities", "managerial vision" and "deep pockets". Thereafter, connections with both knowledge creators – such as universities and money lenders – help to maintain the tempo of research. However, it cannot appropriate innovation rent unless it has manufacturing capabilities and penetration to ensure brand loyalty.'

Clearly, visionary leadership, networks among firms and universities, manufacturing capabilities, and brand reputation are not lacking in India, but funds are. There is a real shortage of funds for the State to invest in NST, the venture capital market is very sluggish, and firms do not have deep pockets.

Another major problem is the shortage in the skill-set required to take ideas to market. The creation of nano-based product and process innovations demands emphasis on not only basic and applied research, but also on technology development, efforts towards commercialization, and developing inter-linkages and partnerships amongst the actors in these various domains. There are a number of challenges to transform new science into new technology in India as mentioned below:[10]

- Lack of sufficient engineering expertise to design a prototype from an idea.
- Inadequate access to equipment to test the prototype in near-endless repetitive trials.
- Lack of business angels to support the costs of design and market testing.
- Problems of synergy creation between researchers of different academic disciplines.
- Academic reputation for those who publish rather than those who create new technology.
- Technology transfer units not staffed by those who have made a mark in both science and business.
- The mindset of academics who prefer to work with established foreign academic institutions on technology creation projects rather than engage in a PPP with Indian firms.
- Lack of enough role models and myths to inspire dreams. Barring a few instances of small and medium enterprises (SMEs) and start-ups there are not enough success stories in Indian nano.

For the regulation of NST in India and its channeling to promote economic growth with inclusive development, there are three main challenges. The Indian science and technology system has become very complex, like that of any developed country, being characterized by diverse capabilities and a wide variance in governance structure made up of multiple ministries, local governments, regulators at various levels (central, state), and federal law-making institutions. They have to deal with a large number of public, private, and public–private sector entities engaged in national and international collaboration. So the first challenge is developing a clear delineation of roles and responsibilities at multiple tiers including at the level of local governments for knowledge

[10] Dr Vivek Srivastava, Would public research investment create innovative Indian nanotech companies, http://www.nanotech-now.com/columns/?article=193

generation, its utilisation, and adaptation. Otherwise, the system is beleaguered by repeated problems of coordination.

This could be illustrated with an example. Nanotechnology development is coordinated by the DST in India, but its mandate does not include risk governance. Management of environmental risk comes under the mandate of the Ministry of Environment and Forests (MoEF), but none of its legislations has identified nanoparticles as a potential hazard, and therefore, it has no jurisdiction to impose any precautionary measures (Jayanthi Beumer, and Bhattacharya 2012). Moreover, the MoEF shares its regulatory mandate with the Ministry of Health and Family Welfare (MoHFW), which has made it difficult to provide a comprehensive and coherent regulatory cover on the issue of environmental health. In fact, environmental health as a policy discipline is underdeveloped in the Indian context.[11] Thus, the fragmentation of mandates further exacerbates regulatory fissures. By setting up individual State departments with the sole objective of technology promotion and facilitation, the Indian government indirectly undermines regulatory overtures by privileging technology within the development agenda (Anand Srivastava, and Sarma 2012).

A smooth flow of information is necessary for enabling any governance framework and more so if it is multi-level. One of the main concerns around nanotechnology is its potential environmental and health risks, governance of which necessitates availability of information. Moreover, the information amongst agencies with different mandates, such as DST, Department of Scientific and Industrial Research (DSIR), MoHFW, Ministry of Consumer Affairs, Food and Public Distribution, and the research institutes has to be both shared and channelised in a way that all of the these institutions perform their functions and further their mandate in an informed manner. Therefore, the second challenge is to ensure that all public agencies have similar/same information base through efficient information flows for risk governance.

Debates on nanotechnology, going beyond regulation, are struggling to keep pace with technological advancements, leaving room for the possibility of abuse by malicious and unstable political elements. At present, there is a paucity of data and risk assessment studies to guide the risk governance process.[12] Therefore, concerns take multiple forms and sometimes get exaggerated about

[11] See for similar conclusions, World Bank (2001) Environmental Health in India: Priorities in Andhra Pradesh, Environmental and Social Development Unit, South Asia Region, 4-6, New Delhi.

[12] The Energy and Resources Institute (TERI) 2010. Summary of Proceedings of Workshop on Issues of risks in regulation, 8 January 2010, New Delhi.

the risks associated with nanotechnology on human health, environment, and society. However, risk and toxicity studies are specialized disciplines, which only a few institutes are equipped to carry out and are beyond the capabilities of regulatory institutions. Therefore, the third challenge is to generate information, which will be comprehensive enough to avoid polarisation of the debate on nanotechnology's benefits and dangers. There is a danger of the debate going the biotechnology way in which the supporters and the critics were arrayed at extreme positions on the issue.

From the above analysis for the future of the NST sector in India, two policy recommendations are apparent:

1. *Understand sector specificities*: The desire to utilise nanotechnology as a tool to facilitate socio-economic development necessitates the development of products and applications and their commercialization. Currently, the presence of a small number of suppliers, manufacturers, and user firms across various sectors in which nanotechnology has found application potential has led us to assume that there exists an emerging nanotechnology innovation system in India. Therefore, although nanotechnology is at an emergent stage, it should be emphasised herein that the growth and trajectory of innovation system of NST in India are sector dependent and would thus require sector-specific policy interventions.
2. *Create an active triple helix with State, academic, and industry cooperation*: An examination of the Nano Mission reveals the composition of committee members to include only one industry representative – all others are either scientists or policy-makers from selected R&D agencies like DST, DRDO, and Department of Atomic Energy (DAE). Therefore, industry representatives, industry associations, and other ministries must be included in the deliberation. For example, since nanotechnology is a cost-intensive development, professionals might help identify priority areas that nanotechnology might address, or risks for which professionals might help develop strategies for management, etc. Policy-makers from other agencies like the Planning Commission and Ministry of Commerce as well as industry associations like Confederation of Indian Industry (CII) and Federation of Indian Chambers of Commerce and Industry (FICCI) that are also leading industry engagement in nanotechnology in India might also be involved. This approach would help identify and explore the benefits, challenges, and risks this technology poses to India and strategise to channel its potential in a sustainable manner, for national benefit.

To conclude, India's good performance in the creation of scientific and regulatory capabilities in NST is largely due to the remarkably supportive role played by the State to create scientific and technological capabilities in NST. Given these initial optimistic results, the future evolution of the NST sector in India promises to boost economic growth. However, there are some self-evident queries that emerge. First, what are the prospects for the application of the present results to promote inclusive development significantly? Second, nanotechnology has application potential in several sectors, with each operating over divergent timescales and exposed to different market dynamics. Therefore, how should priorities be established to legitimise efforts? Third, if concrete results are to be obtained, would it be better if at least a part of India's investment in NST is focused along a 'mission mode' on alleviating current environmental problems and agricultural productivity with focused time-bound targets? These are the issues that all NST stakeholders will have to address sooner or later.

Acronyms

NSTI	Nano Science and Technology Initiative
DST	Department of Science and Technology
NSTM	Nano Science and Technology Mission
CSIR	Council for Scientific and Industrial Research
DBT	Department of Biotechnology
PPP	Public–Private Partnerships
BRICS	Brazil, Russia, India, China, and South Africa
DIT	Department of Information Technology
ICMR	Indian Council of Medical Research
DAE	Department of Atomic Energy
DRDO	Defence Research and Development Organisation
ICAR	Indian Council of Agricultural Research
DIPP	Department of Industrial Policy and Promotion
ISRO	Indian Space Research Organisation
CoE	Centre of Excellence
AFM	Atomic Force Microscopy
SEM	Scanning Electron Microscopes
TEM	Transmission Electron Microscopy
RICH	Research Industry Collaboration Hub
USPTO	United States Patent Office
EPO	European Patent Office
WIPO	World Intellectual Property Organization

INPADOC	International Patent Documentation Centre
IPO	Indian Patent Office
TRDDC	Tata Research, Development and Design Centre
ENMs	Engineered Nano Materials
MoEF	Ministry of Environment and Forests
MoHFW	Ministry of Health and Family Welfare
CII	Confederation of Indian Industry
FICCI	Federation of Indian Chambers of Commerce and Industry
ACNSMM	Amrita Center for Nano Sciences and Molecular Medicine
NIPER	National Institute of Pharmaceutical Education and Research
IIT	Indian Institute of Technology
IISc	Indian Institute of Science
CSIO	Central Scientific Instruments Organization
CCMB	Center for Cellular and Molecular Biology
IITR	Indian Institute of Toxicology Research
JNCASR	Jawaharlal Nehru Centre for Advanced Scientific Research
AIIMS	All India Institute of Medical Science
NICRM	Nichi-In Center for Regenerative Medicine
NCL	National Chemical Laboratory
IICT	Indian Institute of Chemical Technology
JNU	Jawaharlal Nehru University
IACS	Indian Association for the Cultivation of Science
SN Bose NCBS	S.N. Bose National Centre for Basic Sciences
SINP	Saha Institute of Nuclear Physics
BHU	Banaras Hindu University
TIFR	Tata Institute of Fundamental Research

References

Aguirre-Bastos, C. 2010. 'Capacity Building for Innovation and Sector Systems of Innovation: The Case of Renewable Energy Technologies and Nanotechnology in Brazil, India, China, South Africa and Mexico (BICSAM).' A report at Austrian Institute of Technology Foresight & Policy development Department (AIT-F&PD) Volume 17, Austria.

Anand, M., N. Srivastava, and S. D. Sarma. 2012. 'Governance and Nanotechnology Developments: A Focus on Health Sector in India.' *SCRIPTed – A Journal of Law, Technology and Society* 9 (1): 7–24.

Bhattacharya, S. Shilpa and M. Bhati. 2012. 'China and India: The Two New Players in the Nanotechnology Race.' *Scientometrics* 93 (1): 59–87.

Gupta, V. K. 2009. 'Indian Patents Output in Nanotechnology.' *Journal of Intellectual Property Rights* 14 (2): 164–65

Jayanthi, A. P., K. Beumer, and S. Bhattacharya. 2012. 'Nanotechnology "Risk Governance" in India.' *Economic and Political Weekly* 47 (4): 34–40.

Karpagam, R., S. Gopalakrishnan, M. Natarajan, and B. Ramesh Babu. 2011. 'Mapping of Nanoscience and Nanotechnology Research in India: A Scientometric Analysis, 1990–2009.' *Scientometrics* 89 (2): 501–22.

Ramani, Shyama V., N. Chowdhury, R. Coronini, and S. Reid. 2010. 'On India's Plunge into Nanotechnology: What are Good Ways to Catch-up?' Globelics Conference Proceedings, October, Kuala Lumpur, Malaysia.

Reid, S. E. 2014. 'On Nanoscience, Nanotechnology and Nanoproducts: Why Everyone Wants to Join This Game.' In *Nanotechnology and Development: What's in it for Emerging Countries?,* edited by S. V. Ramani. New Delhi: Cambridge University Press.

The Royal Society and The Royal Academy of Engineering. 2004. *Nanoscience and Nanotechnologies: Opportunities and Uncertainties*. London: The Royal Society.

CHAPTER 8

Energy Innovation (sub)Systems in India

Ambuj Sagar and Ankur Chaudhary

Introduction

Energy plays a central role in the human, social, and economic development of any country. Households rely on energy services for their cooking, heating/cooling, lighting, and other needs; modern buildings – commercial or residential – depend on electricity for lighting as well as space conditioning; economic activities are underpinned by energy, whether in modern knowledge-centred industries that rely on electricity to run computers and communication networks or traditional industries that also utilise other sources of energy such as natural gas, oil, or coal; and petroleum products are the basis of most transportation systems. There is a wide range of technologies that are involved in the extraction/capture, conversion, and utilisation of energy across these various disparate application areas. The development and commercialization of these technologies often involve very different innovation processes and systems, yet there is very little examination of these, despite the undisputed importance of energy innovation for any country.

The innovation systems (IS) approach has been widely used to understand the innovation processes by looking at not just the individual firms or research institutes but the broader societal structure into which these firms and institutes are embedded. In the innovation studies literature, several IS approaches have been discussed and employed. These approaches, although conceptually similar, differ mainly in terms of the differing 'system boundary' definition. Among these approaches, the national system of innovation (NIS) approach considers the nation-state to be the unit of analysis, thereby evaluating the innovative performance of a nation (Lundvall 1992). However, given the number of actors and institutions involved while looking at an NIS, any analytical exercise becomes quite complex. A more recent, albeit similar, IS approach has been to look at a geographical region within a country as the unit of analysis, thus trying to assess the innovative performance of a 'regional system of innovation' (RIS) (Cooke et al. 1997; Saxenian 1994). However, a limitation of geographically focused IS approaches is their inability to capture the differences in the innovative activities across various sectors. To address this issue, the 'sectoral innovation system (SIS)' approach focuses at the innovative capability of specific sectors, analysing the technology, knowledge, practices, and linkages across a specific sector (Breschi and Malerba 1997).

Innovation ecosystems are characterized by a set of actors with inter-relationships embedded in an institutional context that together determine the performance of the system (Fagerberg 2004). In the case of developing countries, all three elements – actors, their inter-relationships, and institutions – are generally weaker than those in industrialised countries that possess well-functioning IS. Although the IS approaches discussed earlier have been extensively employed in policy contexts by regional, national, as well as international agencies in a developed country context, a sector as diverse and complex as the Indian energy sector is resistant to the straightforward application of any of the particular frameworks mentioned above and would need a closer look into the complexities of the energy innovation process and the additional challenges posed by the context of a developing country (both in terms of dealing with a wider set of energy challenges and weaker capabilities and institutions).

This chapter examines four case studies in diverse areas – wind power, coal-power generation, automobiles, and improved cookstoves. The first two are energy-generation technologies, while the last two are end-use technologies. Within the former category, the first case is that of a decentralised electricity-generation technology, which in recent years has begun to emerge as a significant renewable source, while the second case dwells on a more traditional generation option. In the latter category, automobiles are a high-cost end-use technology targeted at the upper income class within the country, while improved cookstoves are intended to be used by the rural and urban poor. Together, these sub-sectors of the energy IS in India represent diverse IS with differing roles of the actors and institutions as well as fundamentally different networks within the IS, and as a result, for a better understanding, we need to look at these sub-sectors more closely in terms of the evolution of the specific IS. Further, while looking at the role of the respective IS in inducing innovation processes, i.e., contributing to the development, diffusion, and use of innovations, we analyse the actors, institutions, and networks in the IS for their roles in performing the IS functions.

Towards this end, the 'functional approach' to studying the IS remedies the lack of system-level explanatory factors for the IS performance and looks at the IS actors, institutions, and networks as performers of IS functions that enable it to induce innovation in the system. Further, an analysis in terms of system functions allows for a focus on the IS dynamics and evolution over time, an important aid in trying to draw retrospective insights from IS studies. A definition of functions in an IS is provided by Bergek (2002): 'All activities that contribute to the development, diffusion, and use of innovations are considered system functions.' Though there is little consensus in literature on the exact IS functions, we analyse our case studies employing an adapted version of the IS functions-events framework presented in Suurs (2009). We consider the following functions to structure our discussion of the energy innovation system (EIS) in India: *knowledge creation/sourcing, knowledge diffusion, guidance of the search, market formation, and resource mobilisation*. While the original version, based on the work of Bergek and Jacobsson, considers *entrepreneurial activities* and *support from advocacy coalitions*, we have omitted these functions owing to their heavy dependence on the *guidance of the search* function as well as for the maturity of the considered sub-sectors.

The *knowledge creation/sourcing* involves learning activities in terms of technology as well as markets, users, networks, etc., through activities such as learning-by-searching and learning-by-doing and is typically fulfilled by universities, research institutes, and industry labs. As emphasised by Lundvall, interactive learning is an integral part of any IS and is fulfilled through the *knowledge diffusion* activities among the variety of actors and networks within the IS. The primary function of a network is to facilitate knowledge sharing and transfer among the actors involved. In an IS, the *guidance of the search* function is encoded in the promises and expectations of the various constituent actors as well as the stated institutional frameworks such as policy targets. This function provides a selection process for the various options that emerge through the *knowledge creation/sourcing* and *knowledge diffusion* phases and can be fulfilled through industries or governments, either within or outside the context of a market.

In order to enable emerging technologies to compete with the incumbent ones, markets for the newer technologies need to be created to foster innovation within the IS through a *market formation* activity. The actor undertaking this function differs, depending on the country and technology context as well as the targeted application areas for the technology. Further, any innovation needs a support mechanism in terms of financial, material, and human resources to support its deployment. This *resource mobilisation* function requires the establishment of educational, support, and R&D infrastructure to enable the widespread adoption and sustenance of innovations fostered in the system and could be shared across the various actors of the IS as well as the government.

To emphasise, in an IS, these functions are not linear in nature but are expected to interact and influence each other with the result that the IS evolves over time. Over the next few pages, we analyse the Indian EISs through a 'functional approach' as outlined earlier while looking at the evolution of the actors, institutions, and networks constituting the IS, and seek to provide the context within which these EIS evolved.

Wind energy: An Indian 'success' story

The roots of modern wind energy technology can be found in the R&D that began during the 1970s in some industrialised countries, especially Denmark, the Netherlands, Germany, and the USA. These early players had the first-mover advantage and continued to dominate the industry for more than 25 years (Kamp, Smits, and Andriesse 2004). In the 1990s when India started to develop its wind energy industry, there were companies in the global market with more than 20 years of experience in manufacturing wind technology equipment. With limited technological capabilities, it was difficult for India to foster indigenous companies that could compete at the global scale. The technical know-how, manufacturing capability as well as the support infrastructure for the sector were largely missing from the picture.

To overcome this lack of critical IS constituents, the Government of India (GOI) started active promotion of wind power in the 1990s. With the 'National Guidelines for Clearance of Wind Power Projects' in 1996, the government mandated that all state electricity boards ensure grid compatibility with the planned wind power projects, thus providing the initial *guidance of the search* for the power sector towards wind power's direction. Financial incentives with 100 per cent depreciation of wind equipment allowed in the first year of project installation, a five-year tax holiday, and concessional land allotment was also offered to the wind power projects being established (Rajsekhar, Hulle, and Jansen 1999), with the aim of encouraging the setting up of new wind farms in the country. Later, these benefits were supplemented by generation-based incentives (GBI) in the form of feed-in-tariffs[1] and duty concessions that were

[1] Feed-in-tariffs are policy instruments designed to encourage the adoption of renewable energy by offering long-term contracts to renewable energy producers, generally involving a payment for electricity produced through renewable means, thereby making renewable energy competitive with conventional methods.

offered to entities involved in wind power generation. These policy measures resulted in the creation of a niche power market that offered incentives for capacity building as well as power generation through wind farms. The inclusion of GBIs into the incentive scheme further signalled a shift of focus on generation rather than just capacity building. Generally, GBIs did not favour those who set up capacities for captive consumption,[2] third-party sale,[3] merchant plants, etc.[4] Quite recently, in April 2012, the accelerated depreciation benefits were discontinued, with an accelerated depreciation at 15 per cent available for the wind farms, as against 80 per cent available earlier (WPM 2012).

A significant institutional support for the wind power sector came in the form of the 'Electricity Act' of 2003. This act required all state energy regulatory commissions to encourage fixing minimum levels of electricity to be procured by the electricity distributors from the renewable energy resources. Moreover, recent guidelines and practices by the state electricity boards reflect these changes in policy. As a result, feed-in-tariffs, which have been shown by studies to be effective in encouraging wind power development, have been implemented by several Indian states. The average feed-in-tariff for wind-power-generated electricity in India is around Rs 3.50 per kilowatt hour (kWh) with additional Rs 0.5 being announced by the Ministry of New and Renewable Energy (MNRE), GOI, for projects smaller than 49 megawatts (MW) in 2008, which makes wind-generated electricity quite competitive with the coal-generated electricity (IWEO 2009).

As a result of these *market creation* activities, growth of wind power in India has been quite strong. Over the past decade, the installation base of wind-based power has increased ten-fold: from 1,077 MW in 1999 to 10,200 MW in 2009, before galloping to over 14,000 MW by 2011. The installation pace in the sector outstripped the GOI targets by over 50 per cent during the Tenth Five-year Plan. Given the environmental advantages and low running costs for wind-based power generation, it makes sense for India to proliferate wind power in the country; consequently, the GOI has set a target of adding 2 giga watts (GW)/year of installations for the Eleventh Five-year Plan (IWEO 2009). With an estimated generation potential of over 100 GW[5] from wind power in India,

---

[2] Generation and consumption within a factory.

[3] Selling of electricity to other businesses/firms.

[4] A recent announcement by the MNRE has allowed for GBI benefits to be availed by captive power plants (www.mnre.nic.in).

[5] A recent assessment by the Lawrence Berkeley National Laboratory (LBNL) has pegged India's wind energy resource base at a significantly higher number of 2,000–3000 GW (http://newscenter.lbl.gov/feature-stories/2012/03/21/wind-energy-in-india/).

this energy source is expected to play an important role among the renewable technologies as India grows its power sector [the International Energy Agency (IEA) predicts that India will require 327 GW of power generation capacity by 2020, connoting a capacity addition of around 16 GW per year, with some contribution to the generation mix coming from the renewable resources]. Against the backdrop of this high growth, the constituents of the wind power IS have been mobilised to foster innovation in the sector. However, more than any other actor, the evolution of the wind power IS in India is closely tied to the story of a firm that made significant contributions in shaping the Indian wind power sector.

A star firm

Suzlon started its operations in 1995 when wind power in India represented a meagre 1 per cent of the total power generated in India, and institutional support from the GoI for wind power was still nascent. By 2009, within 14 years of its foundation, the company became the third largest wind turbine manufacturer in the world, with over 9 per cent of the global market share, and a domestic market leader, with over 50 per cent market share, as can be seen in Fig. 8.1).[6]

Suzlon's high growth has been fuelled by the unique strategy that it adopted after its foray into the wind technology market. The first key step in Suzlon's technology journey was taken in 1995 with a technology collaboration agreement with Südwind GmbH Windkrafttanlagen, a wind turbine generator (WTG) manufacturer. When Südwind went bankrupt in 1997, Suzlon acquired it. This acquisition resulted in Suzlon gaining technical know-how for manufacturing of several wind turbine models. In the same year, it entered into agreements with Aerpac B.V. and Enron Wind Rotor Production B.V. to obtain technology and equipment for the manufacture of wind turbine equipment. While acquiring technology from firms in Germany and USA, Suzlon kept building upon its manufacturing facilities in India. Manufacturing in India made sense because of low labour costs, easy access to capital, and an emerging stable domestic market.

In 2004, two of the world's largest wind technology firms in the world, Vestas and NEG Micron, merged. This merger resulted in the laying-off of several skilled workers. Again, Suzlon was able to capitalise on this opportunity by hiring several of these skilled workers when it established its international headquarters in Denmark during 2004 (Lewis and Ryan 2007). Aided by the ease of capital access and with an aim to build up its expertise in the gearbox

[6] See www.suzlon.com.

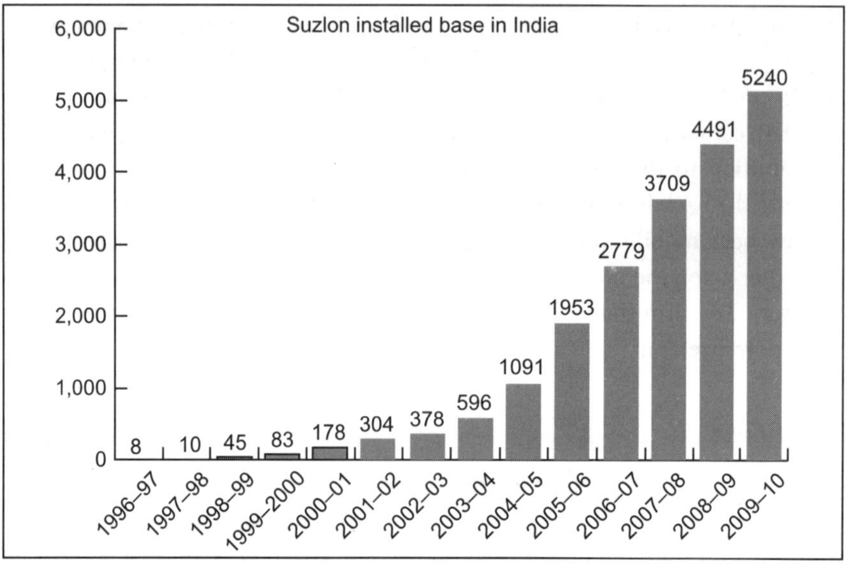

Fig. 8.1: Growth of Suzlon's installations in India

Source: Suzlon Annual Report (2010–11); www.suzlon.com/pdf/indiasnapshot.pdf

manufacturing, Suzlon acquired Hansen in 2005, which was at that time the second largest manufacturer of gearboxes in the world. Hence, Suzlon's own R&D centres and overseas subsidiaries focused on developing advanced wind turbines and components and were simultaneously able to leverage the recurrent inflow of technology in the firm. In essence, Suzlon was able to create effective firm-wide yet international *knowledge sourcing, creation* and *diffusion networks* that leveraged its geographically distributed expertise.

Table 8.1 shows the various international activities of Suzlon that have helped it obviate the need of a developed domestic IS for the wind sector. Although similar strategies have been followed by other global players in the wind technology domain, Suzlon has complemented its own R&D centres with several international partnerships in the form of licensing agreements and joint ventures (JVs) with other companies worldwide. As a result, its expertise/knowledge development exercise has not been limited by its organisational boundaries. The presence of a leading firm at the helm of global technology centres has meant that the wind power product offerings in India have been at par with the global best available technologies in terms of cost-competitiveness as well as technological performance.[7]

[7] For example, Suzlon has been offering wind turbine sizes comparable to the best in the world in the Indian market.

Table 8.1: Suzlon's international activities

India	group headquarters; manufacturing of turbines, rotor blades, generators, control systems, towers
Germany	technology development (gearboxes), cooperation with REpower
The Netherlands	global management centre, technology development (rotor blades)
China	manufacturing of turbines, rotor blades, control systems, generators, and hubs
Denmark	international headquarters
USA	manufacturing of rotor blades
Belgium	manufacturing of gearboxes (Hansen)
Austria	joint venture to manufacture generators
Australia, Portugal, Brazil	sales

Source: www.suzlon.com

One could almost say that Suzlon's success is mostly a result of a firm strategy rather than a coordinated innovation policy, although the market-stimulating policies of the government fulfilled the key function of market creation. Effectively, the core strategy for Suzlon has been to establish R&D centres in the middle of the regional learning networks for the respective technologies. Therefore, its R&D centre in the Netherlands utilises the Dutch expertise in aerodynamic rotor blade design and the R&D centre in Germany leverages the German wind turbine engineering skills (Lewis and Ryan 2007). These R&D centres, coupled with licensing agreements and JVs have enabled Suzlon to develop an extensive R&D knowledge base. With technology development taking place globally, Suzlon has leveraged its local manufacturing capabilities, backed by GoI's support for local manufacturing, to cater to the domestic as well as global market.

To address the *resource mobilisation* needs of the wind IS in India, the government also established the 'Centre for Wind Energy Technology (C-WET)', to provide for public sector R&D and a testing and certification agency. The C-WET, which functions as an autonomous R&D institution under the MNRE, has also assessed over 629 (2008) sites in India to determine their wind power generation potential, thus reducing the commercial uncertainty for the companies operating in this sector and ameliorating the investor skepticism. To further mitigate the testing and certification related uncertainties (and their adverse impact on the technology adoption rates), C-WET has developed facilities for testing and for standards certification of wind turbines in India.[8]

[8] www.cwet.tn.nic.in, 2009

However, this capability is limited to 1.25 MW or lower-class turbines and may not suffice with higher-class turbines already in the Indian market (1.5 MW and 2.1 MW).[9]

Apart from the C-WET, the needed *resource mobilisation* for supporting the wind sector has been lacking, with the role of the universities and other research labs [Council of Scientific and Industrial Research (CSIR), National Aerospace Laboratories (NAL)] being quite limited. Unlike the OECD countries such as the USA and Norway, where universities offer degrees in wind energy technology, the Indian universities do not have any significant specialization programmes in this field, which hampers the competence building as well as R&D in the sector; in fact, human resource training is necessary to cater to the technical competence requirements of the wind technology industry and is thus an essential part of the IS. Perhaps the only major competence building undertaking at present is the Information, Training and Commercial Services Unit at C-WET, which is yet to demonstrate any major impact. Beyond this, there is little evidence of any major technical competence building measures undertaken by the GoI for the wind sector.

A national/sectoral IS generally plays an important generative as well as a catalytic role in the development of any new technology in a country. The extent to which a transferred technology would be successfully adopted is dependent to a great extent on the actors of a domestic IS and the interactions among them. However, in the absence of an active role by the other actors in the IS, an alternative form of adopting and developing on a new technology has been demonstrated by transnational companies (Archibugi and Miche 1995). Suzlon represents a good example of how firms can successfully complete their 'technology leapfrogging' phase, using acquired, licensed, or shared technology to grow and become competitive in the global market (Box 1).

Box 1

Suzlon vis-à-vis Goldwind

An interesting parallel can be seen in Goldwind, a Chinese wind technology company, which started in the mid-1990s and despite more aggressive local content requirement policies of the Chinese government has lagged behind Suzlon in growth. Goldwind concentrated its R&D efforts in its domestic learning networks and did not establish any substantial R&D centres in

[9] See www.suzlon.com.

the global regional centres of innovation in wind technology (Lewis 2007). Till 2008–09, although growing rapidly, Goldwind was able to capture only about 4 per cent market (2008) (BTM 2009) compared to nearly 12.33 per cent (2009) market share of Suzlon. A noticeable difference between these two companies was their global presence. Suzlon, owing to its established subsidiaries in the overseas market, was able to sell its equipment in Europe, North America, and Australia, whereas Goldwind has been limited to the Chinese market. However, more recently, due to the aggressive support policies of the Chinese government and rapid expansion of the Chinese market, Goldwind has surpassed Suzlon in terms of global market share (BTM 2011).

Coal-based power: A 'hand-me-down' story

The pre-Independence coal-based power generation sector in India was dominated by private sector companies with a majority of them being foreign owned. During this period, the technology that was used in the power plants as well as the equipment were imported. With the power plant units of this time being characterized by small plant sizes (1 MW to 15 MW) and use of 'stoker-fired' boilers, the technology was inferior even at that time (Shah 1949). The more advanced 'pulverised coal' technology was not introduced in India until the early 1950s despite its development during the 1920s in the USA. The earliest power plants based on pulverised technology were installed in 1952–53 at Bokaro for the Damodar Valley Corporation and used boilers and turbine generators imported from foreign firms (e.g. International General Electric). Over the next two decades, India bought a large number of turbine generators and boilers from American and European companies as well as the former USSR (Subramanian 1997). This import of technology and equipment provided for the immediate needs of a power-starved country, albeit leading to several associated problems. The primary reason for these problems was the difference in the quality of the coal found in India and the coal for which these equipment were designed, leading to maintenance problems and requiring subsequent modifications for such plants. With very limited domestic technology capability, little knowledge creation or diffusion was possible during these early years. Furthermore, with a paramount focus on generation capacity building, there was little policy guidance or push for innovation or technology advancement in the sector.

The GoI, through the Electricity (Supply) Act of 1948, placed the power generation sector in the public domain. However, by late 1950s, the heavy electrical machinery industry (including the power generation equipment manufacturing industry) was opened up to the private sector, although with limited success at ramping up manufacturing. Despite the presence of some small private companies that had collaborations with foreign manufacturers for manufacturing power plant equipment in India, it was only with the advent of public sector manufacturing units in the 1960s that domestic manufacturing commenced on a large scale. Many of these public sector units were managed under Bharat Electricals Limited (BEL), which had significant technology licensing collaborations with foreign manufacturing companies. These technology design licenses from foreign companies helped BEL obviate the immediate need for huge investments in R&D both in terms of time and finance. Meanwhile the 'oil shock' of the 1970s and the delays in the commissioning of hydel power projects further accelerated the pace of coal-based power plant installation in India (Chikkatur and Sagar 2007). Despite the opportunity provided by this large-scale deployment to move towards newer technologies and prop up the missing elements of the coal power IS, the lack of any *guidance of the search* impeded any developments on this front. This lack of policy direction was critical for the public sector units that dominated the sector.

Bharat Heavy Electricals Limited (BHEL) was set up to consolidate coal power plant manufacturing in the country, which began supplying indigenously manufactured equipment by 1970. BHEL's mandate was to coordinate the manufacturing of power plant equipment, adding new manufacturing plants as well as introducing new technology in the older plants. In the absence of any significant private firm in the sector, BHEL was all set to dominate the power plant supply scene during the next couple of decades with the introduction of larger size units (100 MW+) and formal corporate R&D beginning during the 1970s. However, all the plants manufactured by BHEL during this period continued to rely on the dated 'sub-critical pulverised' technology, while the developed countries had begun moving towards the more advanced and efficient 'supercritical pulverised' technology. This period of market dominance by BHEL was characterized by imports of technology and designs from foreign companies, with little indigenous contribution in terms of innovation. Thus, for introducing a new technology, the route taken was that of licensed technologies under foreign collaborations rather than indigenous R&D (Govil 1998). The foreign collaborators, whose investments in the country were restricted, had little motivation to contribute towards the adaptation of their technology to the Indian context. This trend of technology import by BHEL through foreign collaborations has continued with imports in gasification, combustion, and boiler technologies. As a result, despite having a sizeable market for technologies, the incomplete knowledge sourcing and R&D activities coupled with the lack of incentives for improved technology adoption and adaptation led to the build-up of a dysfunctional coal power IS during the 1970s and 1980s.

In 1991, the GoI passed an amendment to the Act of 1948 to allow private players to own power plants and generate electricity as independent power producers (IPPs). While this step was expected to increase competition and boost efficiency in the coal power sector, due to a variety of reasons, this scheme to attract huge investments into the growing Indian power sector failed to take off, with several of the planned projects failing to materialise (Chikkatur and Sagar 2007). With over twice as much capacity addition happening in the public sector as in the private sector even in the late 1990s, the coal power sector continued to rely on public sector players for its growth till early years of the last decade (Tongia 2003). Throughout this journey of over five decades, a major inefficiency plaguing the Indian power sector for decades had been the poor management of the state electricity boards. As a result, with the Electricity Act of 2003, the GoI separated generation from transmission and distribution with the hope of subjecting the coal power generation sector to intensified competition. The new market-driven system and its inherent competition were expected to drive power generation companies into adoption and better adaptation of newer technologies. However, the continued dominance of a limited number of public sector companies in equipment manufacturing and generation has moderated the process to some extent.

Between 1973 and 2009, BHEL manufactured boilers that accounted for over 65 per cent of the total installed thermal power capacity in India, thus representing over 80 GW in installed capacity. Moreover, BHEL-manufactured 60 per cent of the plants were set up during the 1970s, and almost all the plants were installed during the 1980s and early 1990s (Chikkatur and Sagar 2007). These installations, mostly done using the sub-critical-technology-based boilers, represent a large chunk of the present installed coal-based power capacity in India. While BHEL has a substantial R&D spending (Rs 3 billion), this R&D effort is directed towards a broad category of BHEL's production lines, and the actual contribution towards developing coal power technologies is a fraction of this amount (BHEL Annual report 2008–09). Even then, BHEL's R&D spending to sales ratio at 2.36 per cent is lower than that of other equivalent large foreign companies in the sector: Areva (4.2 per cent), Alstom (3.3 per cent), etc. This limited investment coupled with a lack of incentives for efficiency improvement has meant that BHEL has failed to develop any major indigenous technology relating to coal-based power generation over the past several decades of its existence. These limited knowledge creation efforts by the market leader have hampered the development and evolution of a strong IS in the coal power sector.

The coal-based power generation sector in India is similar to the electric power technology manufacturing sector in terms of dominance by GoI-owned entities. The 'National Thermal Power Corporation (NTPC)' has emerged as the leader in this sector, with its contribution at over 30 per cent of the total coal-based power generation capacity in the country[10] and has been responsible for the first deployment of supercritical pulverised coal technologies in India. While NTPC has developed some knowledge linkages with the 'Center for Power Efficiency and Environment Protection [CenPEEP, an NTPC–USAID (United States Agency for International Development) collaboration programme aimed at improving the efficiencies and lowering emissions from the power plants through the acquisition, demonstration, and dissemination of relevant technologies], these linkages are quite recent and their impact on the IS is yet to be seen (Chikkatur and Sagar 2009a). Further, there are state established energy and fuel research centres in the country, viz. Central Fuel Research Institute (CFRI), Central Power Research Institute (CPRI), Central Mining Research Institute (CMRI), etc., meant to provide effective resource mobilisation for the coal power IS. Even though these institutions have formally existed for long, in reality, these centres have acted more like testing houses rather than undertaking R&D work in the latest power generation technologies, thus being just fringe actors in the overall IS. With little evidence of any major research in the academia during the last 60 years in the field of coal-based power generation technologies, not much technology development efforts have been undertaken indigenously by the industry–academia nexus. Overall, a juxtaposition of the Indian IS (for coal-based power) with that in developed countries reveals the presence of weak players with a lack of synergy among them, thus highlighting a dysfunctional knowledge creation and diffusion within the IS. Moreover, given the lack of innovative activity in the IS, the human resource development and mobilisation aspect has remained undeveloped.

Coal-based generation has powered Indian energy needs for the past six decades and is the keystone for a few more decades to come. Despite the usual vexations associated with fossil fuels, it seems likely that the Indian economic growth for several years to come would depend on coal for its energy needs, given the substantial domestic coal reserves and the complexity, non-availability, and cost issues associated with the other alternative sources (Chikkatur and Sagar 2007). Historically, the total installed capacity of coal-powered plants has risen from 1 GW in 1950 (CWPC 1951) to 116 GW in June 2012, representing over 56.7 per cent of the total installed electricity generation capacity in India

[10] www.ntpc.com, 2010.

(CEA 2012). According to the 'Central Electricity Authority' estimates, over 80 GW of new coal-based power capacity would be added by 2017, which would represent about 60 per cent of the total electricity capacity addition during the same period (CEA 2007), thus providing opportunities for market formation of newer technologies and more effective knowledge sourcing via foreign firms as well as resource mobilisation.

Despite the singular importance that the coal-based power generation sector has for India's present and future energy security, its performance has not been at par with that of the world. The Indian coal-based power generation sector has shown the lowest generating end lower heating value (LHV) efficiency among the major countries for the past two decades. Compare the Indian figure of less than 30 per cent with the LHV efficiency of Japan, which has consistently been above 40 per cent (IEA 2004). The efficiencies of older plants (less than 200 MW) are particularly poor with net efficiencies about 22–23 per cent. This difference represents huge economic and environmental costs at the scale of a country like India. In the face of the huge interests that India has in improving the efficiency of its coal-based power generation apparatus, the shift from outdated technologies to the recent improved ones has been slow and incomplete to a great extent. Although new installations in the coal-based power generation sector are shifting towards newer supercritical steam-based generation, the change is slow and a large majority of the pre-existing power plants are still based on the now outdated sub-critical pulverised coal technique.

The system of innovation that characterizes coal-based power generation in India has been focused on developing manufacturing capabilities through a 'hand-me-down' approach, with BHEL being the primary player. The power generating entities, the state electricity boards SEBs, and coal supplying companies have played a fringe role with most of the innovation activity centred on developing manufacturing capability for coal-based power plants. The public sector companies (such as BHEL) have had limited indigenous R&D capability and have relied solely on either fully imported plants or imported technology for manufacturing (Chikkatur and Sagar 2007). The Western model for technology generation through a nexus of academia, research institutions, and the private sector labs or the alternative model of adopting and tweaking with technologies and indeed making them better than the parent was completely missed in India. The actual realization of the IS with a strong technological capacity that the Indian coal-based power sector could have achieved is limited due to a number of reasons. Over a large part of the past four decades, there remained a conspicuous lack of guidance of the search from the government in terms of moving to newer technologies. With a lack of financial incentives for investing in technology development, adaption, and adoption, the R&D efforts at BHEL, the major player, remained lacking. Further, even the limited R&D efforts at BHEL do not seem to have integrated well with the other actors (or potential actors) in the field of coal power generation. The IS, already plagued by the absence of key players such as the academia, has further suffered due to the 'need-based' approach towards technology acquisition – often leading to the adoption of dated technology. This has adversely impacted the efficiency and operational performance of the coal-powered plants in India while crippling the further evolution of the IS.

Box 2

Recent developments

The Government of India has signalled a policy shift towards leveraging newer technologies for future installations with supercritical technology becoming the installation norm for players like National Thermal Power Corporation (NTPC) (Chikkatur and Sagar 2009b). A positive trend in this new development has been the increased participation of private players, who are keen on sourcing and supplying better technologies in the Indian market. Companies like L&T, JSW, and Mitsubishi Heavy Industries have recently forayed into the Indian power sector and have started bagging contracts for coal-based power plants in the country, and despite their late entry, have been successful in competing with Bharat Heavy Electricals Ltd (BHEL) for the award of newer supercritical technology-based plants (Hindu Business Line, Nov. 2009). So far, the poor coal quality in India has proved to be a deterrent to the introduction of supercritical and ultra-supercritical technologies in the country; however, techniques such as coal washing promise to improve the usability of Indian coal with these technologies. That said, as the R&D efforts in academia as well as at government research labs have remained dormant, the private sector is yet to invest significantly, with the recent design centres set up by companies like L&T being exceptions rather than norms (http://www.larsentoubro.com, 2010).

Improved cookstoves for India

Rural India, home to 70 per cent of Indian population and over 25 per cent of the world's poor, depends to a large degree on biomass fuels for its energy needs. Over 70 per cent of the total households – and 90 per cent of the rural households – in India depend on biomass fuel such as fuelwood, cow dung, and agricultural residue for their primary energy needs (IIPS 2007). The traditional cookstoves are plagued with poor combustion, which results in very low efficiencies (10–15 per cent) and high indoor air pollution levels, which pose health hazards for the exposed populations (i.e., women and children) – in fact, it is estimated that every year, indoor air pollution causes more than 450,000 deaths in India. Disturbingly, the absolute number of biomass users in the country is projected to remain unvarying or even increase over the next decade, excluding major interventions for fuel switching to liquefied petroleum gas (LPG) (Venkataraman et al. 2010). Given the large primary energy usage contribution from cookstoves, especially in rural India, and the health and climatic repercussions of non-intervention, development and wide-scale dissemination of improved cookstoves have been a relevant energy issue for India over the past few decades (Simon 2009). Even though the first improved cookstove, 'Magan', debuted in 1947, it was only during the 1970s and 1980s that scientific research in improved cookstove design proliferated.

The GoI initiated the 'National Programme on Improved *Chulha*' (NPIC) as a demonstration programme in 1983 through the Ministry of Non-Conventional Energy Sources (MNES, now renamed MNRE and later as a dissemination programme in 1986 (ESMAP and World Bank 2001). Although the primary objective of the programme was to reduce the demand for fuelwood and hence check deforestation, it was also aimed at reducing the drudgery of rural women and creating employment opportunities in rural areas (Singh 1997). The NPIC encompassed 23 states and 5 union territories and introduced over 35 million biomass-based improved stoves in rural areas during 1984–2002. The NPIC was primarily involved in two main functions: a target-oriented welfare activity (with heavy subsidies provided by the government for each improved cookstove) and carrying out R&D in the 'improved cook stoves'. The programme was structured along the same lines, having an R&D and a target fulfillment components (Hanbar and Karve 2002). Thus, the NPIC, through its design, oversaw the *knowledge creation, diffusion, market creation, as well as the resource mobilisation* for the cookstoves IS in India.

Even though the first improved cookstove was developed way back in the 1940s, the need for R&D in *chulha* design was critical for the *chulhas* in use in the early 1980s, which were quite inefficient[11] and unclean. To tackle this problem, NPIC created 14 technical back-up units (TBUs) to provide R&D

[11] In terms of the heating energy provided per unit fuel burnt.

and training inputs for the programme, the rationale being that cooking (and fuels) and, therefore, stove design needs were different in different parts of the country and each TBU could better respond to local needs. As a result, the knowledge creation as well as resource mobilisation functions were delegated to location-specific TBUs. These TBUs were assigned to specific departments in universities, engineering colleges, research institutes, etc., across the country. As the prime actor responsible for providing the learning needs of the IS, the TBUs designed and tested new models for the improved cookstoves while also training stove makers, trainers, users, NGO representatives, etc. (Hanbar and Karve 2002). To address the design requirements, either the standard stove designs were adapted to the respective area of operation or radically different designs were developed to cater to the local need. Consistent with the objectives of the programme, the key criteria while designing the stoves were improved fuel-use efficiency and reduced drudgery for users (by making the stoves 'smokeless'). However, the stove models were to be designed within certain constraints. One, the overall cost of the stove had to be affordable for the end-users, who in NPIC's case were the rural poor in India. Moreover, as a result of this user base, some of the early stove models suffered due to their complex design, which made the construction and operation of the stove difficult in the rural areas.

The TBUs were further responsible for facilitating technology transfer to the stove part manufacturers working with NGOs for testing and collecting regular information about the working status of the cookstoves. The design and technology of the stoves were to be improved by the TBUs with inputs from the users, self-employed workers (SEWs), and manufacturers. In essence, the TBUs were designed as the nodal centres with linkages to the other IS actors while being the primary centres for knowledge creation.

The target fulfillment component of the NPIC was delegated to the state government machinery and was administered through various implementing agencies at different levels. The village *panchayat* was responsible for the programme operation at the lowest level through SEWs and the cookstove manufacturers. The dissemination function was also performed by the NGOs, who motivated the local villagers to adopt the improved cookstoves. A state nodal agency was responsible for providing R&D support to the TBU units, selecting cookstove parts manufacturers, and administering the subsidies. Each year, the nodal agency received targets for stove installations in the state, which it allocated at the district and further at the block and village *panchayat* levels, thus resulting in an uncertain and controlled market creation for the improved cookstoves, much unlike the market-driven commercialization model (ESMAP and World Bank 2001).

The GoI spent over Rs 1.6 billion on the NPIC from 1985 to 2002 in the form of capital subsidies and other forms of support; however, a 2003 National Council of Applied Economic Research (NCAER) report termed the NPIC programme a failure. The same report also raised questions on the usefulness of these improved *chulhas*. Reports from the World Bank raised doubts over the usefulness of the subsidised, target-based dissemination model, instead suggesting a more commercial approach to distribution. The lack of synergy among the various actors in the IS in general and the NPIC in particular is apparent from the findings of Chandrasekar and Kundpal (2007), which reveal the clear differences in the perceived importance of activities of the NPIC by the different constituent actors. Such diverging priorities within the actors themselves point towards a dysfunctional *guidance of the search* in the IS and would have made the success of the NPIC quite difficult.

Several reasons have been attributed to the failure of the NPIC to initiate and sustain the dissemination of improved *chulhas* in rural households. The NPIC was designed and implemented, from its inception till its sudden termination, as a target-oriented welfare activity (Hanbar and Karve 2002). The poor interaction interfaces among the technology producers (TBUs), propagators (NGOs, SEWs, etc.), manufacturers, and the end-users further led to poorly optimised technologies and products. Overemphasis on the capital subsidies and a lack of commercialization efforts encouraged promotion of many substandard designs that were frequently discarded or malfunctioned within a year, eventually resulting in poor repurchase patterns and maintenance rates, with the target-based funding getting priority over the long-term research funding for the R&D actors (Kishore and Ramana 2002). As a result of inadequate *guidance of the search* and market creation activities, the cookstoves IS remained chronically dependent on government subsidies.

The TBUs also could not train the users effectively to help them understand the implications of the stove design on quality and efficiency. This resulted in a sub-optimal competence building in the IS, further resulting in a lack of resource mobilisation in the IS (Hanbar and Karve 2002). Moreover, the uncongenial environment for entrepreneurship in India (Transparency International 2006) coupled with inadequate support for capacity building and high levels of subsidy building routed through bureaucratic channels, kept the entrepreneurs from venturing into the improved *chulha* market in significant numbers. This led to the absence of an important and dynamic actor in the system of innovation for the improved *chulhas*.

The NPIC programme lasted for around 18 years and introduced over 35 million improved cookstoves in India. Effectively, the 'IS' for the improved cookstoves could be thought of as a decentralised (but coordinated) system. Thus, the 'nodes' (so to speak) of the system were built around TBUs, which themselves were academic (or quasi-academic) laboratories. While the technology R&D was almost fully located in the TBUs, the state machinery also played a major role in the dissemination of the cookstoves, which were target driven; this dissemination often also involved partnership with NGOs and other grassroots groups, which failed to materialise significantly. Even the R&D efforts that were undertaken at the TBUs suffered due to little external networking of the TBUs with other such institutions, a result of the dysfunctional knowledge diffusion initiatives.

Although the NPIC was designed to facilitate design and technology improvements based on the inputs from users, SEWs, and manufacturers, there was an absence of a uniform policy to approve and incorporate such changes. As a result, despite the presence of TBU-driven R&D efforts, the performance of the distributed stoves (efficiency range: 18–28 per cent) was not significantly improved. In fact, the programme itself had limited success compared to the Chinese programme that was put in place during almost the same period.[12]

[12] The National Improved Stove Programme (NISP) was a Chinese initiative to provide rural households with improved cookstoves, aimed at improving efficiency and reducing indoor pollution levels. Initiated in 1982, it progressed from providing household subsidies till the early 1990s to providing support for manufacturers and energy services companies till the mid-1990s. Thereafter, the support was limited to provision of extension services and certification systems for standardisation. During the 1982–98 existential period, the NISP provided improved cookstoves to over 185 million households with just 15 per cent of the total programme cost borne by the state. Contrast this with the 50–75 per cent direct cash subsidy for the National Programme on Improved *Cholhas* (NPIC) (Kishore and Ramana 2002). Under the NISP, the households shared the cost of ownership and maintenance of the improved cookstoves, which was lacking under the NPIC riddled with poor maintenance and repurchase records. The Chinese programme surpassed the NPIC in terms of penetration, better cookstove efficiency (up to 30 per cent), and actual usage of improved cookstoves, at a much lower cost (subsidy) for each cookstove introduced. Better planning, consideration for local conditions, user involvement, and timely commercialization in the NISP, with focus on support rather than subsidy, have been the distinct factors for the success of NISP (Sinton et al 2004).

Box 3

The 'next-generation' programme?

Since the termination of the National Programme on Improved *Chulhas* (NPIC) in 2002, the cookstove dissemination activity in the country has been carried forward by NGOs through donor support (Bailis et al. 2009), albeit with limited penetration. More recently, in 2009, the Government of India has initiated the new National Cookstoves Initiative (NCI) with a stated goal 'to achieve the quality of energy services from cookstoves comparable to that from other clean energy sources such as liquefied petroleum gas (LPG)' (MNRE 2009). From a technology perspective, the new programme aims to increase the efficiency of cookstoves, thus mitigating pollutants in the first place rather than just transferring them outside the kitchen using a chimney. While the details of this programme are still being finalised, the intention is to take a different approach to the development and deployment of improved cookstoves. Drawing upon a deeper and more nuanced understanding of the cookstove problem and from the lesson from NPIC and other dissemination programmes as well from the field of technology innovation more generally, this programme intends to define the problem as one of cleaner and efficient combustion (rather than more efficient combustion and avoidance of exposure to smoke in indoor environments) and set as target group all the households in the country that do not have access to modern cooking energy services. Accordingly, this suggests the need for participation of different actors in different parts of the innovation chain, i.e., scientific and technical researchers for stove design, engineering firms for large-scale manufacture, large industrial conglomerates (with existing supply chains, or the potential to build new supply chains) to assist with the dissemination, and partnership with local actors (and the development of new entrepreneurs) to facilitate the local delivery, using new and innovative delivery mechanisms rather than a pure subsidy model.

The Indian automobile sector

The genesis of the automobile industry in India was in the late 1920s with the establishment of plants by Ford Motors and General Motors for assembling of cars and trucks. Following that, the first partially manufactured cars in India rolled off the HM assembly line in 1949, with several firms importing and assembling cars around the same time. Over the next several years, the industry remained in a state of technological stagnation due to the tight government controls and heavy duties levied on cars and other automobiles. Further, as a result of the licensing controls and high duties, the production levels remained low and the existing companies could not achieve meaningful economies of scale. In terms of the market size, government licensing ensured that the production figures remained well below consumer demand and the limited number of manufacturers enjoyed a captive market with little incentives for the introduction of new technologies. The limited production coupled with lack of incentives for technology R&D meant that the Indian car industry's IS remained virtually non-existent. Circa 1981: the GoI established Maruti Udyog Limited (MUL), which signed a JV with the Suzuki Motor Company of Japan to produce a small "people's" car, with significant capacity addition occurring as a result of the concomitant production. The JV resulted in the introduction of modern technology in the Indian market with more efficient engines, thus stimulating the existing players to start modernising their product offerings significantly. Soon afterwards, the economic liberalisation policies of the GoI in 1991 resulted in the delicensing and deregulation of the industrial sectors of the Indian economy, removal of licensing restriction, reduced duties, and greater countenance for foreign capital in 1993, thus providing a much-needed boost to the Indian automobile sector via *market creation*. As a result, the 1990s marked the beginning of intense competition in the Indian automobile industry with the entry of global automobile companies and introduction of emissions standards by the government. The sectoral growth began to pick up towards the end of the 1990s and with this increased market access and demand, the IS too began to evolve further.

The Indian automobile industry produced almost 3 million passenger vehicles (excluding two- and three-wheelers) in 2010–11, a 26.7 per cent growth over 2009–10, with all the other segments in the automobile sector registering impressive growth figures too. With this high growth sustaining over the past decade, the Indian automobile industry, which represented just 1.91 per cent of the global production figures in 2002–03, has grown to 4.27 per cent (including cars and commercial vehicles) of the global production figures in 2009.[13] Technologically, the sector that was plagued by outdated technology in the pre-liberalisation years has now caught up with the global industry and uses designs and technologies comparable with the prevailing world standards (see Fig. 8.2). The Indian automobile sector IS, with its new-found dynamism, has been successful in transforming the technology landscape with enabling institutions and actions of the willing actors.

[13] www.siamindia.com.

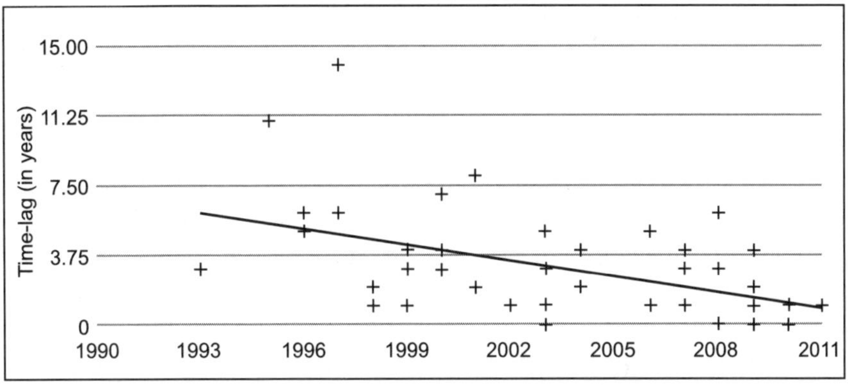

Fig. 8.2: The time-lag between the global and Indian launches of a particular car model over the years, based on engine version

Source: Company annual reports, press releases, and product information pages

Almost all major global players have launched their products and set up manufacturing in the country, transforming the Indian market into one of the most fiercely competed automobile markets in the world. With several of the Indian players having multinational company (MNC) linkages, the role of global firms has been substantial in terms of transfer of technology as well as tacit knowledge, thus enabling the *knowledge sourcing* for the IS. This technology transfer has facilitated the catch-up process of the Indian automobile industry, with the intense competition compelling the global players to introduce newer products and associated technologies in the Indian market. More recently, India has emerged as a hub for small cars, with Indian manufacturers leveraging their local engineering base and low-cost manufacturing experience to capture export markets, while the global automotive majors are looking at India both as a large market as well as a small car manufacturing base for catering to other emerging markets.

The contrasting transformation that the Indian automobile market has witnessed, from being a captive space for a handful of players to a highly competitive one now, is marked with different strategic considerations and decisions by the now numerous players in the sector. The Indian players like MUL, Tata, and Mahindra have engaged in capacity building while trying to tap into international technological know-how through their international linkages (Suzuki for MUL) and outward foreign direct investment (FDI) flows. Coupling this knowledge scouting with their domestic manufacturing capabilities, these players are vying for the international markets (all the three above-mentioned companies export their products) in addition to the growing domestic market.

The R&D expenditure of the Indian automobile majors, though historically much lower than the global norms, has begun to pick up in the recent years to supplement and/or develop upon the knowledge gained from abroad with their domestic experiential learning, resulting in domestic *knowledge creation*. Moreover, the domestic companies have been quite successful in sourcing knowledge from abroad, for example, Tata Motors hired the IDEA Institute of Turin for assisting in the design effort of its small car, and it was also assisted by Le Moteur Moderne of France in configuring the gasoline engine. Further, it took the help of Japanese firms for assistance with instrumentation and electronics. Thus, Tata Motors was able to leverage technology know-how from across the globe to introduce a product into the Indian market (Sagar and Chandra 2004). More recently, the much-touted Tata Nano largely utilised the domestic R&D capability that Tata Motors has developed over the past decade, a departure from the crucial foreign dependency involved in Indica's design process in the 1990s. Similarly, Mahindra has been able to leverage a strategic mix of in-house R&D, technology licensing, manufacturing JVs, and firm acquisitions to build up a strong vehicle development capability over the past decade. Interestingly, despite being the top manufacturer for over two decades, MUL is still dependent to a great extent on its Japanese partner, Suzuki Motors, for technology R&D as well as product development, thus limiting its contribution to the IS.

Among the multinationals, the Hyundai Motor Company (headquartered in South Korea), the second largest car manufacturer in India, entered the Indian market in the late 1990s and has established itself as a major player in the market while building its Indian manufacturing capabilities (along with R&D capabilities) with an export perspective in mind. For such companies, their substantial operations in the international (including developed economies) markets and strong manufacturing capabilities in India have provided for a strong firm-wide innovation network. Among the other major players in the domestic market are large multinational firms like Ford Motors, General motors, Volkswagen, Renault, etc., who have access to cutting-edge technology as a result of their global apex stature but still have limited manufacturing capabilities in India. That said, recent investments as well as future plans for these firms signal their intent to establish strong manufacturing bases in India with the domestic market as well as export perspective in mind. Moreover, the intense competition in the domestic market has stimulated these multinationals into introducing technologically recent and updated products in the Indian market while investing in R&D to adapt them to Indian conditions and tastes (read: bad Indian roads and fuel efficiency consciousness of the Indian customer). As a result, over the past decade, the IS has evolved much in terms of *knowledge creation* and *sourcing*.

Interestingly, as shown in the graphics in Fig. 8.3, the R&D intensity of Indian firms has been increasing in the recent years. On the other hand, the multinational firms have shown an increasing propensity to import technical know-how from the parent companies, whereas domestic firms like Tata, Mahindra, etc., have shown an inclination towards in-house R&D.

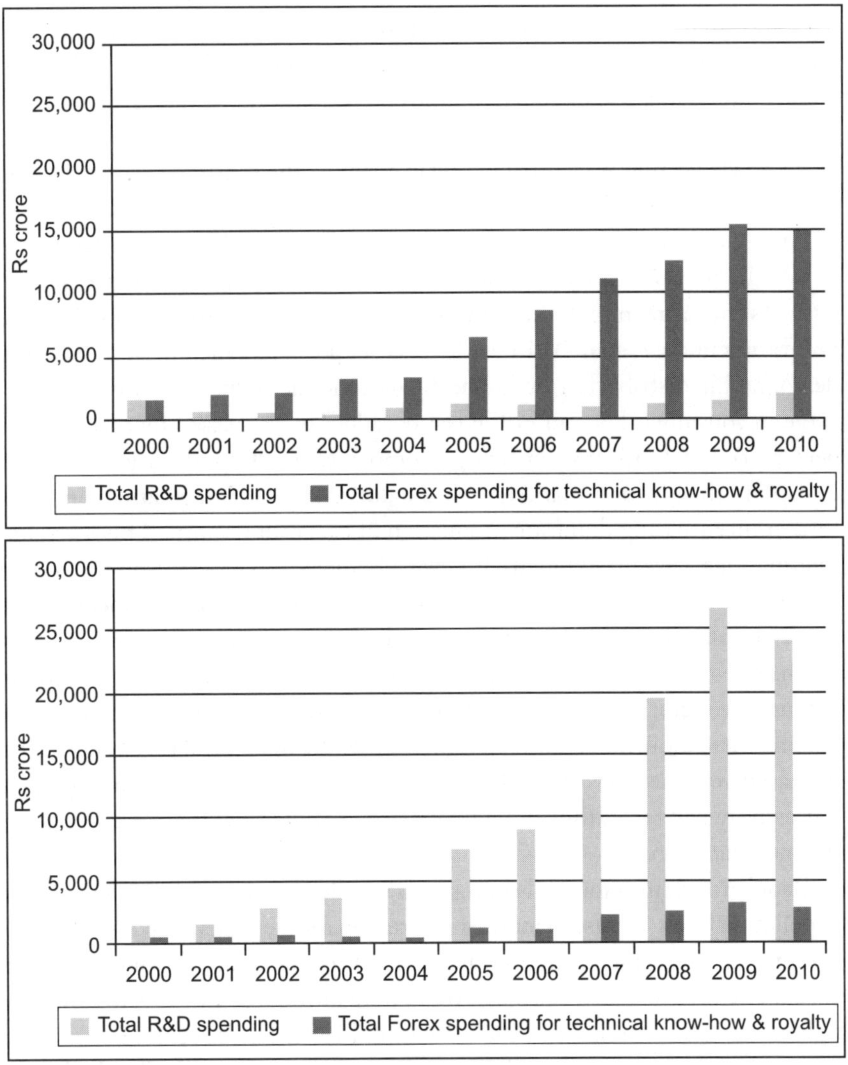

Fig. 8.3: Spending on R&D and technical know-how import by multinational (top) and domestic (bottom) firms in the Indian automotive market

Source: CMIE Prowess Database

Several factors have contributed to changing the landscape (and ensuring the dynamism) of the Indian automobile industry, with various IS actors and institutions playing significant roles in the process. However, a unique advantage for the Indian automobile IS emerged in terms of the Indian-origin human resource base that played a critical role during the 1990s as the domestic players were beginning their vehicle development projects. For example, both Tata and Mahindra recruited Indian-origin industry veterans working for the global car-makers (for instance, V. Sumantaran came on board Tata Motors, while Pawan Goenka joined Mahindra, both expatriates with long experiences at General Motors) for developing their first in-house vehicles. This top-level tacit knowledge flow was instrumental in enabling the Indian players to develop their in-house capabilities.

An important determinant of the performance of any IS is the environment in which the actors of the system are embedded. In the present case, the business environment in the automobile industry has had a major impact on the innovation in the system. With an increase in per capita income and changing demographic distribution in India, the automobile industry has witnessed a surge in consumer demand in the last decade. This increased market size has been a primary motivator for foreign companies to set up operations in India to capture the growing market, with the duties and tariff regulations in India further incentivising domestic production of cars over imports. This arrival of multinationals has brought with it new technology and processes in the Indian automobile industry whilst making the competition intense for the incumbents.

A peripheral yet crucial issue is the increasing consumer awareness about automobiles. The advent of websites, TV shows, and magazines dedicated to the automobile sector has resulted in an increased awareness amongst consumers, making them more demanding in terms of their technological preferences. The increased awareness has also helped the consumers understand technology and successively pay for it, thus providing a strong *guidance of the search* signal for the manufacturers.

Rapid growth in automobile numbers is usually accompanied with an increase in pollution levels, a cause of increasing vexation for the policy-makers. The GoI has introduced several norms over the years to keep the vehicular pollution under check. The first vehicular norms in India were introduced in 1989, followed by tightening of the standards in 1991, 1996, and 1998. These control norms were followed by the introduction of 'Bharat Stage (BS)' norms in India; the BS-II, BS-III norms have been recently tightened to BS-IV norms, mandating the introduction of cleaner and more efficient technologies, an important *guidance of the search* directive. Meeting these norms has entailed the introduction of technologies such as fuel injection, multi-valve engines, catalytic converters, on-board integrated circuits (ICs), and fixed exhaust gas recirculation (EGR). Globally, the introduction of emissions control norms has resulted in stimulating technology development in the automobile industry; similar changes have also taken place in India with tighter controls, resulting in the introduction of more efficient engines.

Apart from the car-makers, several other industry players are involved in the automobile sector in the form of component suppliers (also known as vendors). A high level of component outsourcing by the original equipment manufacturers (OEMs), i.e., car-makers such as Geneal Motors, Ford, MUL, Tata, etc., necessitates a close vendor–firm relationship for the automobile industry. The presence of a large automotive component manufacturing industry (US$19 billion in 2009) in the country has aided the growth of the domestic automobile industry, with high degrees of domestic component sourcing helping OEMs achieve better cost and process efficiencies, while at the same time encouraging *knowledge sharing* across the IS. OEMs have been focusing on integrating closely with the vendors to implement more efficient manufacturing processes and tighter quality norms (For instance, the 'Vendor Development' programme of MUL). This process has also seen firms buying shares into their suppliers and setting up JVs to facilitate technology development at a common supplier. These measures have helped ensure the active participation of the automotive component manufacturers in the IS. That said, even through a period of rapid growth and capacity building in the Indian auto-component sector, the share of R&D expenditure as a fraction of the turnover has been low, in the range 0–1.5 per cent. While the tier 1 suppliers are achieving global standards, the tier 2 suppliers (those who supply to tier 1 suppliers) are yet to catch-up to the global levels (Narayanan and Vashisht 2008).

As the global OEMs set up operations in the country, the global automotive suppliers like Bosch, Cummins, Denso, Eaton, etc., followed by setting up manufacturing capabilities in India. This influx has resulted in significant knowledge and technology flow into the IS at the auto-component level. Interestingly, some of the automotive component suppliers to the Indian automobile industry have become leading suppliers even in the international markets, catering to the global giants such as General Motors and Ford. Companies like Bharat Forge and Sundaram Fasteners, which started as auto-component suppliers, have in some cases moved beyond the automotive sector with products for other sectors while expanding internationally through manufacturing facilities abroad (Pradhan and Singh 2008). Such outward FDI flows have enabled the automotive component manufacturers to leverage the local knowledge base of the foreign country and integrate it into their IS. Moreover, due to the centrality of cost-efficiency to retaining/gaining market share in the competitive global automotive industry, the exports from Indian auto-component manufacturers have been increasing over the years, affording ample learning-by-doing opportunities.

Among the public sector actors, the presence of automotive research institutes such as 'Automotive Research Association of India (ARAI)' and 'Vehicle Research and Development Establishment (VRDE)', which provide for product design and development, evaluation of automotive equipment and ancillaries, and other testing and certification facilities, has helped firms collaborate for common research projects, resulting in effective *knowledge sharing* as well as increased *resource mobilisation* in the IS. With a primary focus on developing testing facilities in the country for automobile testing as a part of the 'Automotive Mission Plan 2006–16', the number of testing centres in the country is set to increase in the near future (see Box 4). Further, automotive industry associations such as the 'Society of Indian Automobile Manufacturers (SIAM)' and 'Automotive Components Manufacturers Association (ACMA) have played a role in helping the policy-makers and manufacturers better assess the market, technologies, certification, and other regulations for reducing the uncertainty in the sector, a key IS enabler.

Despite the strong growth in the automobile sector, the Indian automobile industry is marred with a sub-par design capability, and weak non-proprietary R&D and design capacity. For example, in terms of vehicle design, most of the global OEMs still rely on either their parent companies (such as Ford, General Motors, Toyota, Honda), with little efforts to establish domestic design capabilities. Academia, which could have been an active player in this regard, has thus far been absent from the IS characterizing the Indian automobile industry. To address the issue of human resource capability building, something which has been contemplated as a possible stumbling block for the high industry growth, the 'National Automotive Plan of the Ministry of Heavy Industry'[14] aims at establishing a national auto institute as well as encouraging collaborative research with Indian Institutes of Technology (IITs), Council of Scientific and Industrial Research (CSIR), etc., to attain enabling manpower capacity. This policy measure is intended to ensure that *resource mobilisation* in terms of a trained workforce is not a limiting reagent in the future evolution of the IS.

[14] www.dhi.nic.in/Final_AMP_Report.pdf.

Box 4

NATRiP

The National Automotive Testing and R&D Infrastructure Project (NATRiP) is a major initiative of the Ministry of Heavy Industry, aimed at establishing world-class testing and R&D infrastructure in the country. The project envisages stronger manufacturing capabilities, increased localised R&D, higher exports, and a convergence of IT, electronics, and automotive engineering fortes to helps India become a global leader in the automotive sector. In its conception, the project is geared to bridge the gaps in knowledge creation and resource mobilisation aspects of the Indian automotive innovation system (IS). As a part of NATRiP, nine state-of-the-art R&D centres, and testing, certification, and homologation facilities are to be developed at four Indian cities by the end of 2011. The centres would specialize in different aspects of automotive engineering ranging from noise, vibration and harshness (NVH) to crash testing. Further, NATRiP is also tasked with coordinating activities for road data collection and making it available for research centres across the country.

Conclusion

From the case studies discussed earlier, following conclusions emerge:

- Energy innovation is a very complex arena, reflecting the diverse nature of the energy challenges and technologies involved. The particularities of the Indian historical context and post-Independence evolution add a further dimension to this complexity.
- India has had a mixed scorecard in energy innovation. On the one hand, global players have emerged in some technology areas like wind power and automobiles, catering to the Indian markets as well as seizing the opportunities offered by globalisation through long-reaching firm strategies. On the other hand, in the area of energy applications like improved cookstoves for low-income communities, India has not been able to leverage technology to meet a major challenge. Similarly, India has scored poorly in satisfying the infrastructural demands. For instance, in the coal-power generation arena, BHEL has been able to build some capabilities for low-cost manufacturing, which have contributed to the Indian power sector, yet it has not been able to deepen these capabilities to develop cleaner and more efficient technologies. The reasons for the relative success or failure in the various energy sectors have been the unique functioning and interactions among the actors in the different IS.

- The role of public policy is clear in all these cases. In the cases that involve commercial technologies, the main role of the government has been to create demand for these technologies (through market-pull policies). In the case of cookstoves, the low income of the 'consumers' required the direct participation of the government on both the technology push and dissemination elements, through the establishment of a publicly funded development and deployment programme. But the cases also highlight the need for getting the policies 'right' so as to promote innovation without creating perverse incentives for the relevant firms/actors.

  The State clearly can function better to facilitate innovation if the costs to 'creation' or 'switching' to a more efficient technology paradigm are low. For instance, in the case of wind power, the GoI took proactive measures to promote the market for wind power through regulation, public investment, and public procurement. However, in the case of power generation, given the existing network of public sector monoliths and the absence of interest from the private sector, there was a 'lock-in' with passive acceptance of 'hand-me-down' from the international market. Also, the impact of the State occurs not only via industry-specific acts but also through general macroeconomic policies as illustrated by the automobiles sector story. The adoption of economic reform during the early 1990s promoted this sector more than any specific policy.

  The clear failure of the State functioning seems to be in the realm of human capital creation and in the design of delivery platforms of innovations to low-income communities. As we move towards becoming more knowledge-intensive economies, there is a need to build qualified personnel with very specific expertise – in addition to engineers and scientists in different fields. Thus, it is only now that the Indian State is taking steps to generate personnel who can perhaps dream up innovations in the wind power, automobiles, or power generation sectors.

  In terms of delivery design, it is overlooked that effective adoption by low-income communities occurs only if there is effective demand. Focusing only on the supply side, assuming that demand exists, can lead to failure of public programmes. For example, in the NPIC, emphasis of the programme was more on target fulfillment rather than technology innovation, thus reducing the incentive for better products in the IS. NGOs had limited involvement in terms of awareness generation and providing maintenance support in some areas. The stove manufacturers and SEWs operated at small scales with insufficient room for expansion and scale generation. Such small scales and local manufacturing also contributed to poor maintenance, durability, and performance records for the distributed cookstoves. The design of this programme seemed to have drawn from the appropriate technology movement, where the emphasis was on local design and production, drawing on local inputs (materials and skills) to meet local needs, with an additional goal being to provide employment in the area. Thus, there was almost no linkage to industrial production organisation or structures. The rationale behind establishing several TBUs for developing stove designs consorting with the local needs was positive. However, the absence of proper interfaces of the TBUs with the users, NGOs, and manufacturers deprived the design process of effective feedbacks. Failure of the government to commercialize the cookstoves meant that cash subsidies were the lifeline of the programme with little participation from the end-users and other players in the IS. While there was an attempt to promote local entrepreneurship, this was not through formal entrepreneurship support programmes. Thus, despite the presence of actors as well as institutions to support the IS, innovation in terms of technology, design, and dissemination suffered due to the impaired implementation and interaction mechanisms as well as a misdirected financial incentivisation.

- As one examines the reasons for the sectoral performances, the criticality of the presence of functional and interacting elements within the IS is quite clear. The absence or presence of certain actors or institutions (while not an if-and-only-if requisite as demonstrated by the Suzlon case) has a strong bearing on the evolution of the IS, and as pointed earlier, a sector-specific design of the IS makes a strong case. Suzlon represents a good example of how firms can successfully complete their 'technology leapfrogging' phase, using acquired, licensed, or shared technology to grow and become competitive in the global market even in the absence of an active role by other players in the IS.

Moreover, the unique position of the global technology paradigm and local capabilities in an area needs to be taken into account while looking to develop an innovation system. For example, in hindsight, a possible alternative to developing and supporting indigenous firms would have been providing the global players with incentives to establish local manufacturing capacities. This route, although able to provide access to local means of adopting wind power, would not have resulted in any significant capacity development for the country in moving towards self-reliance in terms of technology competence. A purely local R&D effort to develop indigenous technology would have been a long drawn effort and in the presence of already highly developed wind power technology in other countries, would have made little sense. International technology transfers could have been a viable intermediate solution but global MNCs are unlikely to engage in such transfers with companies that could become future competitors. Furthermore, companies from developing countries could always build upon the transferred technology and make cheaper products owing to low labour costs in their domestic markets. Interestingly, with the Indian market providing strong manufacturing capabilities (and a somewhat evolved IS) in addition to the growing domestic market, international majors like Vestas, Enercon, and Gamesa have established manufacturing capacities as well as R&D centres (Vestas, Enercon) in India and are active actors in the new wave of wind power growth in India.

On the other hand, despite the presence of potential players in the IS, a weak *guidance for search* and *knowledge sourcing* in the innovation system can lead to sub-optimal performance. For example, in the case of coal-based power generation in India, a 'hand-me-down' approach to technology sourcing coupled with a weak R&D focus of the major IS player (BHEL) and an even weaker role of power generating entities – the state electricity boards and coal suppliers – has led to an inefficient coal-based power generation system in India. Further, with little incentive to deploy more efficient technologies, and a clear lack of *guidance of the search* by the state, the IS has often adopted dated technology.

- Despite a low innovation index rank[15] (as a result of poor literacy record, labour market issues, uncongenial business environment, excessive regulation, etc.) (Augusto López-Claros 2009), some of the sectors have managed to match global standards in terms of competitiveness.

The motor of accumulation of innovation capabilities seems to be only localised in the private sector. Clearly, the managers of leading firms in knowledge-intensive sectors take a global rather than a local view of technology acquisition and integration. The success stories in the wind power and automobiles clearly illustrate the benefits of such vision. However, the public sector firms in the power generation

[15] India was ranked 85th in the Innovation Capacity Index rankings reported by Augusto López-Claros, 2009 below developing countries like Ghana, Namibia, Azerbaijan, etc.

sector passively accepted such knowledge and technology, without transforming it and innovating upon it. Being a passive buyer without developing adsorptive capacity to build upon it led to 'lock-in' with inefficient technologies because of the high-costs of switching over to cleaner and more efficient technologies. Therefore, international acquisition without further transformation and innovation is likely to lead to sub-optimal outcomes. In other words, taking a 'global' view is not sufficient; the vision must be 'glo-cal' – global search for appropriate and efficient technology to be adapted to local conditions.

While unique reasons exist for the success across different sectors, the unique position enjoyed by India in terms of a rich Indian-origin human resource base dispersed globally has been quite crucial. For example, both in the wind and automobiles sectors, the presence of Indian-origin human resource as already being a part of the global industry, has aided a better search for adaptation and deployment of technologies in the Indian context. Furthermore, the size of the Indian market (in sectors where privatisation has been allowed/encouraged) has been a major factor in gathering the elements needed for an IS to evolve. As a result, it is non-trivial for a developing country to emulate the Indian IS model for these specific sectors. Nevertheless, crucial lessons in terms of the unique specifications of IS are quite evident.

References

Archibugi, D. and J. Michie. 1995. 'The Globalization of Technology: A New Taxonomy.' *Cambridge Journal of Economics* 19 (1): 121–40.

Bailis, R., A. Cowan, V. Berreueta, and O. Masera. 2009 'Arresting the Killer in the Kitchen: The Promises and Pitfalls of Commercializing Improved Cookstoves.' *World Development* 37 (10): 1694–705.

Bergek, A. 2002. 'Shaping and Exploiting Technological Opportunities: The Case of Renewable Energy Technology in Sweden.' Thesis, Chalmers University of Technology, Göteborg, Sweden.

BHEL. *Annual report 2008–09*. Accessed: June 2012; http://www.bhel.com/images/pdf/annual_report_2008-09.pdf.

Breschi, S. and F. Malerba. 1997. 'Sectoral Innovation Systems: Technological Regimes, Schumpeterian Dynamics, and Spatial Boundaries.' In *Systems of Innovation—Technologies, Institutions and Organizations,* edited by C. Edquist, 130–56. London: Pinter.

BTM. 2011. *International Wind Energy Development, World Market Update 2011, March.* Copenhagen: BTM Consult ApS.

———. 2009. *International Wind Energy Development, World Market Update 2009, March.* Copenhagen: BTM Consult ApS.

CEA. 2012. *Power Sector at a Glance.* New Delhi: CEA, Government of India. Accessed: June 2012; http://www.powermin.nic.in/JSP_SERVLETS/internal.jsp

———. 2007. *Report of the Working Group on Power for 11th Plan.* New Delhi: CEA, Government of India. Accessed: June 2012; http://planningcommission.nic.in/aboutus/committee/wrkgrp11/wg11_power.pdf

Chandrasekar, B. and Tara C. Kandpal. 2007. 'An Opinion Survey Based Assessment of Renewable Energy Technology Development in India.' *Renewable and Sustainable Energy Reviews* 11 (4): 688–701.

Chikkatur, A. P. and A. D. Sagar. 2007. *Cleaner Power in India: Towards a Clean-Coal-Technology Roadmap.* Cambridge, MA: Kennedy School of Government, Harvard University. Accessed: June 2012; http://cleanairinitiative.org/portal/system/files/72569_resource_1.pdf.

———. 2009a. 'Positioning the Indian Coal-Power Sector for Carbon Mitigation: Key Policy Options.' Coal initiative Reports, Pew Center on Global Climate Change.

———. 2009b. 'Carbon Mitigation in the Indian Coal-Power Sector: Options and Recommendations.' *Energy Procedia* 1 (1): 3901–07.

Cooke, P., Uranga M. Gomez, and G. Etxebarria. 1997. 'Regional Innovation Systems: Institutional and Organisational Dimensions.' *Research Policy* 26 (4–5): 475–91.

CWPC (Central Water and Power Commission). 1951. *Public Electricity Supply: All India Statistics, 1950.* New Delhi: CWPC (Power Wing), Ministry of Natural Resources and Scientific Research, Government of India.

ESMAP, and World Bank. 2001. 'India's National Program of Improved Cookstoves: A Growing Challenge.' *Indoor Air Pollution,* 5: 1–8. Delhi: The World Bank.

Fagerberg, J. 2004. 'Innovation: A Guide to the Literature.' In *Oxford Handbook of Innovation*, edited by J. Fagerberg, D. C. Mowery, and R. R. Nelson, 1–21. Oxford: Oxford University Press.

Govil, K. K. 1998. *Electricity Generation: Policy, Technology and Economy.* New Delhi: Venus Publishing House.

Hanbar R. D. and P. Karve. 2002. 'National Program on Improved Chulha (NPIC) of the Government of India: An Overview.' *Energy for Sustainable Development* 6 (2): 49–55.

Hindu Business Line. 2009. 'L&T Bags 6900 Crore Order.' *The Hindu,* 3 November. Accessed: June 2012; http://www.thehindubusinessline.in/bline/2009/11/04/stories/2009110452350100.htm.

IEA (International Energy Agency). 2004. 'Energy Balances of OECD and Non-OECD Countries.' Paris, France.

IIPS. 2007. *National Family Health Survey (NFHS-3), 2005–06.* Mumbai: International Institute for Population Sciences.

IWEO. 2009. *Indian Wind Energy Outlook 2009.* Global Wind Energy Council.

Kamp, Linda M., Ruud E. H. M. Smits, and Cornelis D. Andriesse. 2004. 'Notions on Learning Applied to Wind Turbine Development in the Netherlands and Denmark.' *Energy Policy* 32 (14): 1625–37.

Kishore V. V. N. and P. V. Ramana. 2002. 'Improved Cookstoves in Rural India: How Improved are They? A Critique of the Perceived Benefits from the National Programme on Improved Chulhas (NPIC).' *Energy* 27 (1): 47–63.

Lewis, Joanna I. and Ryan H. Wiser. 2007. 'Fostering a Renewable Energy Technology Industry: An International Comparison of Wind Industry Policy Support Mechanisms.' *Energy Policy* 35 (3): 1844–57.

Lewis, Joanna I. 2007. 'Technology Acquisition and Innovation in the Developing World: Wind Turbine Development in China and India.' *Studies in Comparative International Development* 42 (3–4): 208–32.

Lundvall, B. A. 1992. *National Systems of Innovation: Towards a Theory of Innovation and Interactive Learning*. London: Pinter.

MNRE (Ministry of New and Renewable Energy). 2009. 'National Biomass Cookstove Initiative.' Press release at the launch of the National Biomass Cookstoves Programme (NBCP), 2 December 2009. New Delhi: MNRE. Accessed : June 2012; http://www.mnre.gov.in/schemes/decentralized-systems/national-biomass-cookstoves-initiative/.

Narayanan, B.G. and Pankaj Vashisht. 2008. 'Determinants of Competitiveness of the Indian Auto Industry.' ICRIER Working Paper, No. 201. New Delhi: Indian Council For Research On International Economic Relations (ICRIER). http://www.icrier.org/pdf/working%20paper%20201_final.pdf.

Pradhan, J. P. and Neelam Singh. 2008. 'Outward FDI and Knowledge Flows: A Study of the Indian Automotive Sector.' ISID Working Paper No. 2008/2010. New Delhi: ICID.

Rajsekhar, B., F. Van Hulle, and J. C. Jansen. 1999. 'Indian Wind Energy Programme: Performance and Future Directions.' *Energy Policy* 27 (11): 669–78.

Sagar, Ambuj, D. and Pankaj Chandra. 2004. *Technological Change in the Indian Passenger Car Industry.* Harvard University: BCSIA Discussion Paper 2004–05, Energy Technology Innovation Project, Kennedy School of Government.

Saxenian, A. L. 1994. *Regional Advantage: Culture and Competition in Silicon Valley and Route 128*. Cambridge, MA: Harvard University Press.

Shah, K. T. 1949. *Power and Fuel: Report of Sub-Committee.* Bombay: Vora and Co. Publishers Ltd.

Simon, Gregory. L. 2009 'Geographies of Mediation: Market Development and the Rural Broker in Maharashtra, India.' *Political Geography* 28 (3): 197–207.

Singh N. P. 1997. 'Biomass Programme in India: An Overview.' *Proceedings of the International Conference on Biomass Energy Systems*, edited by P. Venkataramana and S. N. Srinivas, 65–72. New Delhi: Tata Energy Research Institute.

Sinton, J. E., K. R. Smith, J. W. Peabody, L. Yaping, Z. Xiliang, R. Edwards, and G. Quan. 2004. 'An Assessment of Programs to Promote Improved Household Stoves in China.' *Energy for Sustainable Development* 8 (3): 33–52.

Suurs, R. A. A. 2009. *Motors of Sustainable Innovation: Towards a Theory on the Dynamics of Technological Innovation Systems*. Utrecht, Netherlands: Utrecht University.

Subramanian, S. A. 1997. *Monograph on a Number of Aspects Relating to Thermal Power Plant.* New Delhi: Central Board of Irrigation and Power.

Tongia, Rahul. 2003. 'The Political Economy of Indian Power Sector Reforms.' Program on Energy and Sustainable Development Working Paper 4. Stanford: Stanford University.

Transparency International. 2006. *Corruption Perceptions Index: 2006.* Accessed: June 2012; http://www.transparency.org/policy_research/surveys_indices/cpi/2006.

Venkataraman, C., A. D. Sagar, G. Habib, N. Lam, and K. R. Smith. 2010. 'The Indian National Initiative for Advanced Biomass Cookstoves: The Benefits of Clean Combustion.' *Energy for Sustainable Development* 14 (2): 63–72.

WPM. 2012. *Wind Power Monthly*. April 2012. Accessed: June, 2012; http://www.windpowermonthly.com/article/1125335/ india-cuts-wind-tax-incentive-15

III
Innovations for inclusive development

CHAPTER 9

Evolution of Traditional Medical Systems Sector in India

What Has Innovation Got to Do with It?

*Arijita Dutta**

Introduction

The World Health Organization (WHO) defines traditional medicines (TM) as 'diverse health practices, approaches, knowledge, and beliefs incorporating plant, animal, spiritual therapies, manual technique, singularly or in combination, to maintain well being, as well as to treat, diagnose and prevent illness' (WHO 2001). By this definition, TM have been in existence worldwide, representing the knowledge accumulated over centuries – transmitted from one generation to another through various mediums. Some of the systems of TM are *codified* in writing and describe sophisticated medical methods supported by theories and rich experience. Such systems are widely diffused within and beyond national boundaries. Traditional Chinese medicine, *Ayurveda*, *Unani*, Mongolian, and Thai medicines are the main traditional medicine systems (TMS) of Asia. On the other hand, there are also *non-codified* orally transmitted systems, which are practised by village physicians or faith healers within a limited circle of people, and these are termed as 'folk medicine' or 'indigenous medicine' (Koning 1998). In India, the 'folk medicine' operates at two levels: one is 'rural village-based', where the knowledge is transmitted through generations via the housewives

* The author is extremely thankful to Prof Sachin Chaturvedi, Prof Sarmila Banerjee, and Prof Samira Guenif for their comments, suggestions, and constant encouragement. The author is indebted to Prof Shyama Ramani for her excellent comments and suggestions on the first draft.

and mothers and faith healers, while the second one is 'tribe-based', wherein knowledge is preserved via the tribal people inhabiting the forest and semi-forest regions (Pushpangadan 2003).

In developed countries, from the nineteenth century onwards, TMS were increasingly driven out of the market by chemical companies, which emerged as the first pharmaceutical firms. These chemical companies branched into the production of more standardised and effective drugs by exploiting the scientific discoveries and technological innovations of academics. In India, too, given the remarkable growth of the pharmaceutical industry producing modern allopathic medicines since the 1980s, one could have easily supposed that the TM niche in India would have died a natural death. However, this is simply not the case. Even in the new millennium, 65 per cent of the rural population in India uses TM and herbal drugs to meet its primary healthcare needs (WHO 2003). India's *Ayurvedic* market itself expanded at a rate of 20 per cent per annum over 1987–1996, the highest in world TM markets (Verma and Singh 2008). This phenomenon leads to the enquiry: what are the factors that have permitted TMS to flourish in India? Does innovation have anything to do with their survival? These are the questions that this chapter will seek to answer.

Innovation by definition refers to new ideas, new products, new processes, or even new forms of organisation, marking a clear departure from the past. In contrast, TM refers to products and techniques obtained from hundreds of years of practice, through incremental experiments and finally embedded in a rich set of experiences of people (rather than designed clinical trials). At first glance, therefore, innovation and TM might seem to be inherently incompatible. However, any sector in an evolving system is also likely to evolve in response, implying that innovation is also likely to characterize traditional sectors such as TM. So the question asked and answered in this chapter is: how has innovation occurred in TMS sectors in India?

Such a query is important because it has been clearly demonstrated by the proponents of 'endogenous growth theory' that innovation plays a key role in promoting economic growth through increasing productivity and generating new products, processes, and organisational routines (Solow 1957; Romer 1990; Denison 1985; Jorgenson 1990). Moreover, since the signing of the UN Millennium Declaration in September 2000 by 189 Heads of State, including India, there is a worldwide policy commitment to achieve socio-economic equity in terms of food security, health security, and environmental security. All eight central Millennium Development Goals (MDG) can be linked directly or indirectly to health security, and among the 48 indicators proposed by the UN to measure the MDGs, 17 relate directly to health. Innovation is going to be a key factor in promoting economic growth and addressing inclusive development. In this context, TM will play a crucial role in promoting health in developing countries, given its age old ability to cure diseases and accessibility to the poor.

In terms of methodology, we adopt a 'systems perspective' to study innovation in TMS rather than the standard linear model of macroeconomics, because especially in TM, innovation is not just an industrial output, but a social construct reflecting the interplay of different agents. The main actors in any innovation system are the State, public laboratories, universities, firms, financial organisations, consumers, and civil society groups interacting both on the supply and the demand sides of the markets. With respect to TM, this chapter will explore the nature of innovation generation through the interaction among the different actors, their impact, and the possible trajectories that this sector can follow in the future.

The rest of the chapter is organised as follows. The first section sets the background and explains the nature of TMS in India. Then, the next two sections examine the role of the State and firms on the supply side of the market in supporting and generating innovation. This is followed by the section focusing on the drivers of innovation and the impact of innovation in TMS on economic growth and inclusive development. The final section concludes with our propositions on future trends and policy recommendations.

Setting the background: What is traditional medicine and why is it popular?

Historical background

In most developing countries, there are two systems of healthcare: one is 'modern' and 'scientific' in nature and of Western origin, whereas the other is traditional, time-tested, and of local origin. TMS the world over are derived from practices over thousands of years, embodying the rich experiences of practitioners transferred from generation to generation through written texts or rituals and folklore. History of TMS in India is not an exception in this regard. India has a very rich culture of TMS, which are time-tested and use local natural products. Indeed, India has nurtured five major traditional medical systems: *Ayurveda, Unani, Yoga and Homeopathy, and Siddha.*

Of the five systems, *Ayurveda* is the oldest and also by far the most used TMS in India. *Ayurveda*, which means 'science of life,' is also one of best documented medical systems in the world. It was even referred to in the *Vedas* (1500–600 BC), as well in texts such as the *Charak Samhita* (600 BC) and the *Susrut Samhita* (500 BC). *Ayurveda*, like all TMS, starts from the premise that any medical treatment should be holistic and all-encompassing, rather than simply identifying a set of symptoms and treating them in isolation, as is done in allopathy. Therapies in *Ayurveda* are individualised to each person's constitution or profile of health strengths and weaknesses, and include dietary changes, *yoga*, meditation, massage, and medicines. The last are mainly herbal, mineral and ashen products. The practice of *Ayurveda* has received patronage from the ancient princely states and local philanthropists since the twelfth century. Prominent trading communities like Marwaris and Gujratis also supported *Ayurvedic* dispensaries and hospitals (Baru 1996).

Unani was initially developed in Greece in 460–377 BC. This system is documented in texts like *Al Anoon* (AD 980–1037) and *Al Havi* (AD 850–923). *Unani* reached India through the Muslim invaders and is very much a part of Indian systems. *Siddha* is another medical system that flourished in Sri Lanka and South India, and it mainly stresses on the use of magical masters. *Siddha* literature is in Tamil and is practised largely in Tamil speaking part of India and abroad. The *Siddha* medicines that contain mercury, silver, arsenic, lead, and sulphur have been found to be effective in treating certain infectious diseases, including venereal diseases.

Dr Christian Friedrich Samuel Heinemann (1755–1843), a German doctor, developed homeopathy in the late eighteenth century and gave it a scientific basis in the early nineteenth century. This 200-year-old system of medicine is one of the most widespread modes of treatment in the world. Apart from India, it is also widely used in South America, North America, Australia, and Europe. Homoeopathy is being practised in India since about 150 years.[1]

All forms of TMS received a terrible blow with the onset of British colonial rule. European rulers thrust the concept of modern biomedical systems in their colonies as the 'only scientific' option.[2] Thus, the advent of the British rule marked the end of the State patronage of TMS, which were even officially relegated to be 'unscientific' and 'informal' healthcare options.

[1] Department of AYUSH, MoHFW, GoI, 2007; www.indianmedicine.nic.in.

[2] Even the early Communists in China did not accept the traditional medical systems as scientific and modern (Croizier 1965).

Rising demand

Despite the suppression of TM during the previous centuries, after Indian Independence in 1947 (and post World War II in Western countries), the demand for TM has been rising both within India and abroad. Indeed, TMS is attracting people all over the world, developed and developing, due to its focus on preventive healthcare, instead of just curative healthcare.

In developed countries, while more people are coming to recognize the importance of TM, the governments are not willing to integrate it into their national healthcare systems or healthcare reimbursement programmes. In the USA alone, US$17.8 billion was spent on dietary herbal supplement and US$4.2 billion on botanical remedies. The World Bank has identified that trade in medicinal plants, botanical drug products, and raw materials is growing at a rate of 5–15 per cent per annum (WHO 2000). In the decade 1993–2003, the percentage of population that used TM as a part of complementary and alternative medicine therapies was 48 per cent in Australia, 31 per cent in Belgium, 49 per cent in France, and 70 per cent in Canada (WHO 2003). This trend is also observed in developing and even least developed countries.

At the outset itself, at least two factors seem responsible for this trend. First, 'lifestyle diseases' such as diabetes, cardiovascular problems, cancer, psychological disorder, etc., which are normally created by stress and strain of the modern life, pollution, social problems, and dietary habits, are on the rise. In the Third World Countries, too, we find a gradual shift in disease burden already in progress. This movement of disease pattern is outlined in 'epidemiologic transition theory' (Omran 2005). While modern allopathic treatments, based primarily on the philosophy of curative healthcare, have been successful in handling the communicative diseases via the antibiotic revolution, they do not perform well with respect to non-communicable lifestyle diseases. TMS, on the other hand, are rather successful in the latter category of diseases owing to their 'bundle-prescription' involving not only medicines, but also dietary regime, exercise charts, meditation, and above all, a change in lifestyle. In one word, TMS treat the patient more holistically, moving away from just 'point prescription' with drugs.

Second, the allopathic drugs often threaten patients with side-effects or adverse reactions, and in some cases, they can be extremely severe. In contrast, TM is believed to treat the body holistically with little scope of side-effects. This aspect has particularly attracted educated people towards TM treatments. The pluralistic societies automatically created parallel healthcare systems beside the modern biomedical systems (Croizer 1965; Harilal 2009). The shift from *singularity* to *plurality* in healthcare system is taking place because it is becoming increasingly evident that no single healthcare system has the potential to solve all the needs of the society, particularly in Indian society where people are diverse not only with respect to economics and social identities, but also in relation to cultural variance (Working Group on Traditional Health Sciences 2008).

Innovation in traditional medicine system: Similarities and differences with modern allopathic medicine systems

There are a number of similarities and differences between mainstream 'allopathic medical systems (AMS)' and TMS, on both supply and demand sides. Innovation in TMS can take the form of 'new medication' or 'new complementary non-medication' as in AMS. Innovations in medication can take the form of new products, new usage of earlier products, or new processes to produce the drugs. Normally in TMS, the products are herbal, mineral, or animal parts, and most often, they have more than one ingredients. The innovations of non-medication, on the other hand, mean new therapies, in form or usage.

The TM innovation system also comprises three classes of agents as in AMS: (i) distributors and pharmaceutical companies, (ii) research institutes and professional scientists, and (iii) practitioners and patients. But, the participation of the last group is both crucial and unique to innovation generation in TMS. They are the most important inputs in the process of innovation, unlike in AMS.

While in AMS, the output of the innovation system is traditionally measured in terms of patents, in TMS, it has to go beyond patent counts as patentability is low in this sector (discussed later). It must also include improved varieties, new usages, and new methods of marketing within the set of innovation outputs generated by the sector.

Finally, unlike in AMS, wherein the 'new drug discovery' process follows the path of 'from molecule to mice to men', the TMS research proceeds in the conduit 'from men to mice to men' (WHO 2006). In AMS, to create a new drug, the first step is to identify the biochemical molecule in the laboratory, then test it and treat it for formulation and toxicological study, and finally follow it up with clinical trials. On the other hand, in TMS, the lead molecule as well as its formulation are already known from the historic texts (referred to as the sacred texts written by sages) and aboriginal people. Only the final formulation has to be designed to make it ready for consumption. Thus, reverse pharmacology in this sector needs both more clinical research and extended exploratory studies.

Role of the State: Direct investment and indirect capacity building in innovation process

During the first few decades after Independence, in order to improve access to medicines and develop self-sufficiency, the focus of State policy was on catching-up in terms of production capabilities of allopathic medicines. However, since 1980s, given the market demand for TM and the knowledge and production capabilities of India in TM, the State began to take a real interest in the evolution of this sector. This has taken a variety of forms common to State investment for capacity building in any industrial sector, as discussed here.

Ministerial coordination

The Government support of TMS was concretised by a formal ministry, named 'Department of Indian Systems of Medicine and Homoeopathy (ISM&H)' established under the Ministry of Health and Family Welfare in March 1995. It was renamed as the 'Department of Ayurveda, Yoga and Naturopathy, Unani, Siddha and Homoeopathy (AYUSH)' in November 2003. This department has continually worked for the standardisation of drugs, quality control, and improvement in upstream educational facilities and standard. It also aims at building awareness within and outside of India about the efficacy of the TMS and herbs. Finally, the AYUSH department plays a pivotal role in strengthening the research facilities for TMS by identifying time-bound specific tasks of innovations on pre-identified diseases.

Investment in upstream research

In 1969, an apex research council for all AYUSH systems was established under the name of 'Central Council for Research in Indian Medicine and Homeopathy.' In 1978, the integrated council was dissolved to form four individual 'research councils' to cater to the specific need of a specific TMS. These councils are the apex bodies for research in their specific niche (i.e., *Ayurveda*, *yoga* and naturopathy, *Unani*, *Siddha*, or homoeopathy) and are fully financed by the Government of India. Public investment in TMS research, which was a mere 1 per cent of total outlay on healthcare during the First Five-year Plan (1951–56) had moved up to only 1.3 per cent by the Seventh Plan period (1985–90). However, by 2004, the TMS sector received 2.5 per cent of the health budget (Dalal 2005). Within this small budget, the research councils received the highest allocation (see Table 9.1 and Fig. 9.1).

Table 9.1: Research councils and the network

Council	*Systems*	*Network*
Central Council for Research in *Ayurveda* and *Siddha*	*Ayurveda*, *Siddha*, Tibetian (amchi)	38 institutes, universities, units
Central Council for Research in *Unani* Medicine	*Unani*	22 institutes, including 2 central institutes and 8 regional institutes
Central Council for Research in Homeopathy	Homeopathy	40 institutes
Central Council for Research in *Yoga* and Naturopathy	*Yoga* and naturopathy	coordinating NGOs and hospitals

Source: Websites of respective research councils

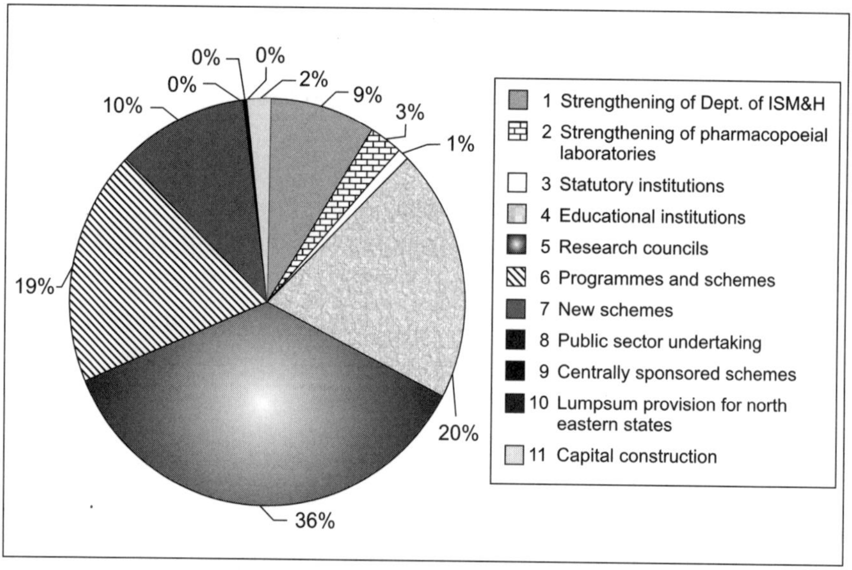

Fig. 9.1: Share of R&D expenditures in total outlay of ISM&H (1997–2002)

Source: Sen (2007)

The five research councils received 36 per cent of the budget in the Tenth Plan (1997–2004). Out of all these research councils, the most important are the ones dedicated to research in *Ayurveda* and *Unani*. The 'Council for Scientific and Industrial Research (CSIR)' under the umbrella of the Department of Science and Technology is another very important institution in research of medical technology. However, in spite of continuous efforts, the citation of Indian TMS in scientific journals is still low, as compared to Chinese TMS (see Table 9.2).

Table 9.2: Citation and patent data on *Ayurveda* and Traditional Chinese Medicine (TCM) as on 10 September 2005

Database	*Keywords*	*Citation*
Pubmed [a]	Indian Traditional Medical Systems (ITMS) excluding *Ayurveda*	361
	Ayurveda	1,045
	Chinese medicine	10,278
Science Direct[a]	ITMS	77
	Ayurveda	87
	Chinese medicine	990
US patent granted	*Ayurveda*/ITMS	3
	Chinese medicine/TCM	195

Note: [a]Pubmed and Science Direct are two standard databases of scientific literature, the former being devoted exclusively to medical journals
Source: Patwardhan et al. (2005)

Investment in building scientific capabilities and manpower in traditional medicine systems

The state played a proactive role by not only investing directly in research in TMS but also fostering an enabling environment for innovation through investment in new educational institutions. Table 9.3 shows the overall education facilities created by the government for the development of upstream scientific capabilities.

Table 9.3: Education facilities in AYUSH in 2008

Facility	*Ayurveda*	*Unani*	*Siddha*	*Naturopathy*	*Homeopathy*	*Total*
Undergraduate colleges	241	39	7	10	182	479
Undergraduate admission capacity	11,375	1,170	350	385	13,385	27,265
Postgraduate colleges	64	7	3	-	36	110
Postgraduate admission capacity	1,052	79	110	-	1,161	2,402

Source: www.indianmedicine.nic.in

Supporting and catering to demand for traditional medicine systems through the public healthcare system

The State has played a pivotal role in creating and maintaining the facilities and hospitals that provide treatments in TMS. As of 2009, India had 718,000 registered practitioners in TMS and 1355 hospitals with a total bed capacity of 53,296. In fact, the urban–rural ratio for medical institutions in AYUSH is 0.44 (actually representing a rural bias), while that for modern medical system is 2.77 (Sen 2007). The 'National Rural Health Mission (NRHM)', launched in 2004–05 mainly for primary health facilities, has stressed on the creation of AYUSH facilities in rural healthcare. Till mid-2009, a total of 5,321 AYUSH doctors have been recruited under NRHM, with the highest numbers coming from Orissa, Gujarat, Karnataka, and Rajasthan. Service of trained nurses or *dais* are being formally used for child delivery in villages under this programme. NGOs are also involved. For instance, Karuna Trust, a renowned NGO, has been successfully integrated under State-provided healthcare system to offer TMS for basic primary healthcare, including mother and child care in 25 primary health centres (PHC) in Karnataka and 9 PHCs in Arunachal Pradesh in tribal-dominated areas (Vivekanada Girijana Kalyan Kendra 2009).

Creation of a library of traditional knowledge

TM forms a part of the 'traditional knowledge (TK)' systems of a country. A noteworthy initiative to protect and disseminate traditional knowledge is the 'Traditional Knowledge Digital Library (TKDL)', formally launched in June 2001, jointly by the Department of AYUSH and National Institute of Science Communication and Information Resources (NISCAIR). The library is seen as a means to prevent organisations from patenting products or processes based on TK, albeit public or common knowledge, and therefore 'wrong' patent-wise from the start. Defensive protection of disclosed TK is the first step in the TK protection policy of the Government of India (Gupta 2005b). Documents are available in local, as well as in five international languages. The TKDL is perceived as having the potential to act as the bridge between modern science, modern medicines, and TK. However, over time, it has taken the shape of a standard information base with detailed classification level for a huge number of medicinal plants. Basically, TKDL serves as a freely accessible non-patent literature database on TMS.

In summary, the Indian government has already recognized the importance of building scientific and industrial capabilities in TMS and is working towards this by focusing on upstream research, protecting TK, and then supporting production through procurement for the public hospital system.

Innovation by private firms in traditional medicine systems with focus on Ayurveda

In India, the TMS industry is extremely fragmented. Nevertheless, the number of production units increased from 8,895 in 2008 to 9,493 in 2009.[3] *Ayurvedic* medicines production units dominate the scene.[4] In 2008, the combined sales figure reached Rs 6,000 crore[5], out of which products worth Rs 1,500 crore were exported. Out of the total export products, 60 per cent were from crude herbs to be converted into medicines elsewhere, 30 per cent from finished formulations, and the remaining 10 per cent comprised products in semi-finished state.

Manufacturing in *Ayurveda* has passed from the stage of small-scale physician outlet to petty/cottage production, and finally to large-scale industrial units. While there are traditional practitioners who make their own medicines as well as small-scale producers, in terms of market shares, they are overshadowed by large firms. Indeed, the leading TMS firms are emerging as a competing alternative to mainstream allopathic medicines. The top 10 companies are Dabur, Baidyanath, Zandu, Himalaya, Charak Pharmaceuticals, Vicco, Emami, Veeswakirthi Ayurvedic Pharmacy, Ayurveda Pharmacy, and Ozone, out of which, the first three account for more than 85 per cent of market share (Sharma 2008). Recently, South India-based companies like Aravindh Herbal Cure, SKM, MSSAssan, etc., are coming forward to produce and export *Unani*-based drugs. There are some leading homoeopathic producers as well.

Among the leading firms, the first phase of mechanised production of TMS medicines till the 1990s, particularly *Ayurvedic* medicines, aimed at making the medicines more palatable, increasing their shelf-life, and providing information on contents and composition of the drug through detailed labels. A second phase of commercialization of the *Ayurvedic* medical sector emerged towards the end of the twentieth century, marked by a move from bulk to mass industrialised production. In this later phase, the production process was completely transformed by the manufacturing firms from the original traditional methods proposed by the *vaidya* or traditional practitioners community (Harilal 2009). Now we have hyper-modern factories producing *Ayurvedic* medicine, where the different phases of TM production are no longer visible. The *Ayurvedic* sector is

[3] www.india.gov.in.

[4] www.indianmedicine.nic.in, accessed in June 2010.

[5] 1 crore = 10 million.

undoubtedly emerging to be medicine-centred as opposed to its basic orientation that was supposed to make it patient-centred, bowing to the influence of the dynamics of conventional pharmaceutical industry (Banerjee 2002).

Ayurvedic product innovations

In TMS, in keeping with the conventional and traditional practices, the 'classically' trained doctors in all systems of TMS prefer to prepare the drugs themselves by procuring the raw materials, and they seldom prescribe buying of drugs from retailers. This is corroborated by a survey of 75 traditional medical practitioners, mainly practising *Ayurvedic* and homeopathy, in and around Kolkata, which revealed that 88 per cent of the doctors dispense the drugs to their patients. Among these, 25.33 per cent manufacture the products themselves. The rest buy essential raw materials from the wholesale markets and mix them in prescribed proportions (Sen 2007). In this context, the objective of the TMS firms is to bypass the TMS practitioner and reach the consumer or patient directly by producing over-the-counter (OTC) drugs and marketing them in pharmacies, retail stores, and online.

The common objective of the corporate TMS producers is to draw the consumer/patient away from the 'classical products,' i.e., products that are produced according to the texts of the respective medical systems, following the exact methods of preparations, towards 'proprietary medicines' under their own brands by mixing the raw materials prescribed in the ancient texts in new ways with other raw materials. For example, the basic *chyawanprash* is a 'classical product,' whereas the newer versions, including chocolate flavoured ones are 'proprietary products'. The majority of sales of leading *Ayurvedic* companies such as Dabur, and Himalaya are from 'proprietary branded' products (the shares are 100 per cent for Himalaya, 97 per cent for Dabur, and 80 per cent for Zandu). The only exception is Arya Vaidya Sala (AVS), which markets only 'classical' products (Bode 2006).

In short, the product innovations from TMS leaders are of two main forms: new formulations and new product variety. In other words, as the bulk materials or basic ingredients cannot be changed, the modern day innovation is focused on formulation stage. These incremental innovations are motivated by three main objectives:

1. *To lower the cost of production*: Firms realize that the preparation of formulations according to the strict adherence to texts from the bulk raw materials is time consuming and costly.

2. *To cater to the tastes of modern consumers*: According to the classical texts, TM are in the form of *taila* (oils), *churna* (powder), *ghrita* (butter), *avleha* (semi-solid mix), and *gulika* (hand-made pills). However, these formulations are not popular with the younger generation and urban consumers. Hence, the TMS firms started producing ready formulations in the form of sweetened coatings, blister packed capsules, and syrups to satisfy the modern tastes of consumers and to facilitate their usage.
3. *To cater to the non-health needs while exploiting demand for natural products*: Lastly, the manufacturing units are moving away from pure healthcare products to beauty enhancing products using TMS and natural products. This type of innovation takes the form of 'increasing TMS product variety'. The companies market 'personal care' range, consisting of shampoos, creams, soaps, etc. The new brands are targeted at the urban middle class with their interest in holistic healing and ready affordability.

Illustration of traditional medicine systems product innovations: Case study of Dabur

Dabur is a leading *Ayurveda* company, which identifies itself as a 'trusted name' in natural remedies for more than 100 years. Its growth in size and profit after tax (PAT) margins took a steady upward path since the new millennium and the latter increased its market from 6.2 per cent in 2003 to 12.7 per cent in 2007. Their products portfolio is presented in Fig. 9.2, which shows that the consumer care section accounts for a massive 68 per cent of total revenue for

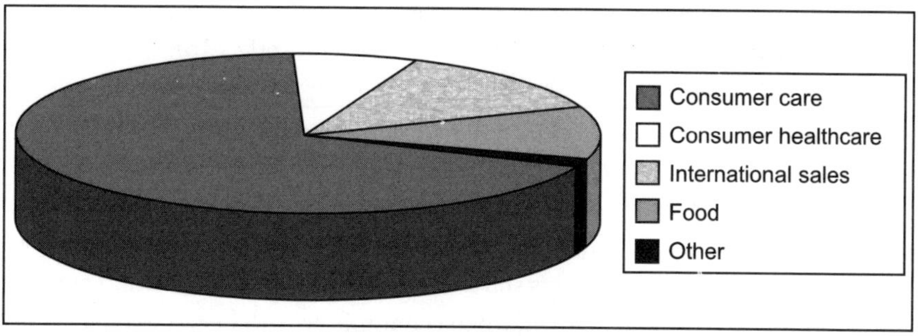

Fig. 9.2: Shares of different sectors in Dabur's total sales in 2006–07
Source: *Annual Report 2006-07*, Dabur (www.dabur.org)

sales, while the consumer healthcare contributes to only 7 per cent. The growth rates of these two niches were 15.2 per cent per annum and 9.2 per cent per annum, respectively, in 2006–07.

The lion's share of Dabur's product portfolio is taken up by healthcare items (see Fig. 9.2). There are products for hair care, health supplements, digestives, home care, and skin care. The company is focused on developing a competitive advantage in 'OTC' comfort medicines such as cough syrup (sold under brand names Honeytus, Shankhapushpi, Shilajit). There are also products for women healthcare, rejuvenators, balms, laxatives, remedy for joint pain, etc. But the jewel in its crown is the daily vitamin supplement *chyawanprash*, which has a dominant market share of 62 per cent in this segment. Dabur has formulated new forms of this hugely popular health tonic, viz., sugar-free version and special variant for cold and cough.

Apart from health-enhancing products, Dabur has also diversified into food products. Within the food section, with a phenomenal growth rate of 27.8 per cent per annum, 'Real' fruit juice and honey are doing very well. Dabur has introduced 'Real Twist', targeting the urban teenagers, and customer-friendly squeeze pack for honey. The evolution of the product diversification of Dabur reveals an interesting trajectory.

Table 9.4: Timeline for Dabur's innovative introductions

Year	*Innovative introduction*
1884	started as an *Ayurvedic* pharmacy
1919	initiated *Ayurvedic* research
1940–50	entry into consumer products: launch of Dabur Amla Hair oil and Chyawanprash
1970–80	entry into oral care and digestives: launch of Dabur LDM and Hajmola
1996–97	entry into health beverages: launch of 'Real' fruit juice
2005–06	acquired Balsara and entered home products market
2006–07	Driving Achievements of Retail Excellence project (DARE)
2007	entry into specialized retailing for health and beauty care

Source: Compiled from *Annual Report 2006–07*, Dabur (www.dabur.org)

Dabur initiated its *Ayurvedic* research in as early as 1919 (Table 9.4) and entered the consumer products market in 1940s. However, the thrust into 'non-classical consumer care' products started only in the post-liberalisation era with the adoption of economic reform, when economic growth began to pick up and the middle class expanded substantially.

Innovative marketing ploy to penetrate 'bottom of income pyramid' markets

The DARE – 'Driving Achievements of Retail Excellence' – initiative launched in 2006, is recognized as a path-breaking innovation in marketing.

- To meet the new challenges in the traditional distribution channels, such as grocery, chemists, *kirana* shops, etc., Dabur moved away from its age-old models of wholesalers and stockists to more trade-focused distribution methods. Urban markets are targeted with chemists, retailers, and speciality shops, while special focus has been given to six rural states.
- Understanding the buoyant growth of consumer care section, Dabur shifted 'pudin hara', a remedy for stomach ache and indigestion, from healthcare category to consumer care department (Harilal 2009). A total of 9000 wholesalers were chosen after a detailed market research exercise to cover 80 per cent sales in this channel, and they were given special incentives under the sub-scheme of 'Dabur Kings' title.
- In rural areas, the campaign was on a very catchy note *Dabur apnao, Lakshmi Lao,* easily reaching out to the rustic consumers. In addition, a resolute institutional sales system has been developed to cater to specific segments like hotels, housekeeping, paramilitary forces, B2B (or business to business) sector, and public sector enterprises.
- The company also aimed to have a face-to-face interaction with its consumers for creating trademark loyalty.

In short, the DARE programme has helped Dabur to build 'social capital' in its distribution chains through innovative incentive schemes, and its penetration of bottom of pyramid (BoP)markets has contributed substantially to its sales. Dabur has initiated collaborations with other kinds of economic actors to promote its products. For example, it has collaboration with the privately run Dhanwantry Educational Society, which owns a college, a hospital, and a pharmacy in Chandigarh, and this is likely to strengthen the demand for its products.

In a nutshell, the thrust of Dabur is not on ethical medicines at all. Instead, product differentiation has been the vital strategy of this company. The same holds for other firms like Baidyanath, Zandu, and Himalaya. Economics of market structure does identify this kind of variety in product market, when the market is characterized by a few big players with substantial competitive power.

Innovation in therapies

Apart from new product varieties and new types of packaging, in TMS, there has been a continuous generation of innovations in therapies and practices.

Combining 'medical equipment' of conventional systems with traditional medicine system treatment

The institutionally trained practitioners, mainly urban-based, are using diagnostic tests for their treatment purposes and the public TMS hospitals are providing these facilities as well (Sen 2007). The practitioners, whether trained or not, whether registered or not, are also mixing their therapy procedures with other systems of medication, such as allopathy. Though this is not allowed legally in India, and the basic paradigms of treatment in TMS in India are totally different from the Western system, this combination, however, is taking advantage of the market for quick and hassle-free remedies.

These new practices are essentially responses to demands from modern consumers for quick remedies. The general nature of 'healthcare demands' of Indians has been changing since the colonial period. Western medical systems have triggered a preference for fast curative healthcare, in contrast to the traditional Indian heritage of slow, preventive medical care, nurtured throughout life by simple but disciplined food habits, meditation, and values that call for limited consumption.

The spa culture

Therapeutic innovations have also taken a relatively new route of provisioning spa services. This is particularly reflected in the mushrooming of *Ayurvedic* and *yoga* centres all over India, with concentration in urban areas. Drug companies, hotel chains, beauty therapists all are opening centres offering these treatments for body detoxification, hair and skin care, arthritis problem, asthma problem, etc. This has given rise to a relatively new 'spa culture' in India, whereby treatments from not only Indian traditional systems, but also from Thailand, Bali, Africa, etc., are being offered and used. In a recent study by the London-based travel and tourism magazine *Condé Nast Traveler*, 'Ananda Spa', Rishikesh was ranked as the top spa destination worldwide. Another Indian brand, Sereno Spa, located at Park Hyatt Goa Resort & Spa, was also among the top 10 overseas hotel spas worldwide chosen for 'Travel Awards' 2006.[6] The luxury hotel chains are tying up

[6] www.indiatravelite.com.

with foreign brands of spas to satisfy their rich clientele (Barman 2006). However, as Table 9.5 shows, India's share in spa markets is only around 5 per cent, far below the shares of Thailand, Malaysia, and Indonesia. However, given the tremendous potential of Indian TMS, this share is very likely to increase over time.

Table 9.5: Potential spa holiday destinations for Asian spa consumers

Name of countries	*Share in %*
Thailand	22
Indonesia	19
Malaysia	13
Australia	7
Singapore	6
Europe	6
India	5
New Zealand	4
USA	4

Source: www. IntelligentSpas.com

The main drivers of innovation in traditional medicine system from private firms

There are many factors that give India a competetive advantage in TMS.

- *Cheap raw materials*: India is acknowledged as one of the mega biodiversity countries in the world, and 3,000 plants with medicinal value are found within its geographical boundary. In India, the codified TMS utilise about 2,000 plant species for medicinal purposes, while the tribal communities utilise 8,000 species of plants, most of which are not known to the outside world (Pushpagandan 2003; Shankar 1992). Three (*Allium sativum*, *Aloe Barbedensis,* and *Panax* sp.) out of 10 top selling herbs worldwide are available in India (Dubey Kumar, and Tripathi 2004). India actually enjoys the varied geographical climate to grow several types of flora and fauna. It has the cold Alpine weather in the Himalayas, tropical wet rainy weather in the east and south, and arid, dry weather in the west – thus having weather conducive to growing all types of herbs. This gives a huge cost advantage to the TM producers.
- *Available qualified manpower*: It is understood that innovation in TMS is less capital intensive and more knowledge intensive than

that in AMS. Hence, qualified scientists are an essential input in innovation creation in TMS. The institutionalisation of education in TMS (discussed earlier) and the increase in the number of medical students learning TMS are producing a large band of scientifically trained practitioners who are replacing the traditional group of TMS practitioners or *vaidyas*. The already existing science and technology manpower in organic chemistry, biotechnology, and allied field in India gives an additional edge in TMS innovation.

- *Very weak and lax regulation*: Owing to almost total absence of regulation and standardisation in the manufacturing sector for TM, the costs of commercializing a new product are very low. The weak regulation in the market also leads to huge cost advantages to the firms, as this sector does not require any clinical testing for marketing. As clinical tests are extremely costly for the pharmaceutical companies for modern medicine, this definitely gives a scope of cost cutting.
- *Lack of R&D success in modern medicines*: There is the reality of new drug-discovery costs and uncertainty in conventional medicine. Pharmaceutical scientists are finding it more and more difficult to find new blockbuster drugs. Most of the drugs are in 'me-too' category, and thus, the cost of R&D is increasing constantly, pushing up drug costs in general. In the five years between 1989 and 2000, only 17 of additional new drugs can be considered as dramatic innovations over existing therapies, while 51 per cent were 'me-too' drugs (Lexchin et al. 2003; Love 2003). Thus, the cost saving issue has been crucial for the global multinational companies to move to innovation using herbs, which have known advantages from classical texts, while the only task remaining is to identify their chemical and biological characteristics, and carry out toxicity and optional clinical trials. The TMS model of innovation 'from mice to man to mice' (referred to earlier) saves the huge cost of fundamental research for identifying the lead molecule.
- *No possibilities for re-engineering*: The Indian pharmaceutical industry emerged following a reform of the Indian patent law in 1972 that permitted commercialization of branded drugs if they were produced using different processes. The 'Trade Related Intellectual Property Rights' (TRIPS) agreement, signed by all member countries of the World Trade Organization (WTO), including India, in 1995, imposes product patents in all sectors, making catching-up via the accumulation

of re-engineering capabilities possible only for drugs whose patents have expired. In India, only the large-scale pharma giants like Lupin, DRL, Wockhardt, Alembic, Torrent, etc., will now be able to afford the high cost of R&D of the modern drugs in India (Dutta 2009). As leading Indian pharma innovators like Ranbaxy and Nicolas Piramal get acquired by foreign multinationals, the number of national firms capable of new drug discovery further reduces. However, even medium and small firms have the option to shift to Indian system of medicines (ISM), as very little of those drugs are yet patented.

- *Fragmented market structure*: In spite of the presence of many small-scale players, the monopoly power in the industry is quite high, expressed by 10 firm concentration ratio of 85 per cent, far higher than the AMS market. Higher monopoly power leads to more innovativeness as the monopolistic firms need the push to maintain their power in the market (Schumpeter 1942). Understandably, the top firms like Dabur, Zandu, Baidyanath, Hamdard, Himalaya, MSS-Assan, and Aravindh Pharma innovate in products and processes. The smaller companies, however, cannot afford to maintain R&D staff and set-up.
- *Nature of demand*: On the demand side, the share spent on healthcare increases proportionately with economic growth and increase in disposable income. Realizing this and in order to reach the consumer directly without going through 'a medical practitioner' or 'a medical establishment,' it is in the interests of firms to commercialize OTC non-prescriptive drugs and boost sales via advertising. Thus, unsurprisingly, the urban areas are flooded with mushrooming TMS clinics, spas, and manufacturers of beauty products. Medicines not only represent the strategies for dealing with illness in a somatic, psychological, and social sense, but also symbolise the cultural ideas about health and therapy (Appadurai 1986; Whyte, van der Geest, and Hardon 2002). TMS represents an optimal strategy for consumers who are drawn to a natural holistic approach and also consumers for whom consumption of TMS enhances social status. In parallel, originating from the West, there is a worldwide trend in cosmetics towards the use of natural products, to exploit the belief that anything natural is safe. Thus, a new segement of beauty conscious rich and upper middle class women from developing and developed countries are brand loyal to TMS-based cosmetic products and foods. The above trends in demand

are triggering bandwagon effects, and thus, incremental innovations in product variety in comfort drugs and beauty enhancers are likely to be the mainstay of innovation of pharmaceutical firms involved in TMS.

Promoting innovation generation from traditional knowledge: Raging debate on IPR

TK will be the basis for all future innovation generation in TMS, and while it is acknowledged that TK needs to be protected, it is not clear exactly what form this should take. Protection can either mean an 'offensive' mechanism to exclude unauthorised use by third party or a 'defensive' method to preserve TK against any sort of erosion (Correa 2002). That TK can be wrongfully exploited to obtain patent applications (and hence future royalties if patent attempt is successful) was clearly illustrated by the patent disputes over the use of traditional Indian products like *haldi* and *neem* by Western firms during the 1990s. Indian scientists and civic associations joined hands to present evidence, which eventually led to the revokation of such patents. For instance, the US Patent Office revoked the patent on *haldi* in 1997 and the European Patent Office revoked the patent on *neem* in 2000. However, current intellectual property regimes (IPR) fail to adequately protect TK, which survives through utilisation of grassroot innovations (Gupta 2005a). Moreover, there is no consensus on the optimal system as 'the modalities for protecting (TK) are still emerging and evolving....Even at the international level, clarity has not yet emerged, and countries are grappling to understand the issue' (GoI 2003). What is clear is that such an IPR system should be sealed through discussion procedures with all stakeholders (Chowdhury 2004).

The first and foremost issue on which there is yet little clarity is the issue of identifying who is to be protected. The domestic corporate giants who are marketing several kinds of TM? The multinational corporations? The public laboratory scientists? TM practitioners and tribal communities? Clearly, the present 'one size fits all' philosophy cannot protect all. The form of protection should depend on a case-by-case context, as well as on the nature of the TMS in question, i.e., whether it is codified or un-codified. In case of codified systems, patent protection is applicable only for novel addition to the existing knowledge in public domain.

Patentability in TM also brings in an ethical issue to be resolved by value judgement. The existing property rights are oriented around the concept of private ownership and individual innovation. They do not consider the indigenous cultures, collective knowledge of the societies, and the attitudes of the grassroot innovators. Particularly the non-codified TMS and rural unregistered practitioners, *vaidyas*, *ojhas*, etc., function in several non-market formats and often are paid not by monetary exchanges, but by pro-bono work and soft loans as in many African communities (Correa 2002)

Going beyond the theoretical debates, at the practical level, it is very difficult to protect TK under the present IPR systems. The novelty of a new medication or therapy based on TK is very difficult to prove because of the very fact that the materials and their usages are known among different communities for ages. Also, the problem of patentability arises because the preparation methods, as described in classical texts, are very simple, consisting of just boiling or grounding, and hence, no complicated 'process' can be identified worth of patenting. Moreover, the practitioners of TK who create the innovations are simple physicians or tribals who neither have the technical or the manpower capabilities to apply for patents nor the deep pockets to pay for the application costs. Most of the TK is held collectively and not individually, and hence, the TK 'holder' is difficult to locate for a patent application. Also, even if patenting were possible, even large Indian companies do not have the technical cum legal cum search capacities to monitor patent infringement and go to court for the same. So simply having a patent would not guarantee automatic protection.

Finally, barring the issue of patenting, it is very difficult to commercialize TK in such a way that it benefits the practitioners. Consider, for instance, the example of the energy-giving extract from a fruit called *Argyapacha*. The story of the Kani tribe and their benefit sharing model began in1987, when some scientists found that the Kanis chew the fruit of the above plant, which gives them tremendous energy and vitality. The case was taken up by Regional Research Laboratory (RRL) in Jammu and then by Tropical Botanic Garden and Research Institute (TBGRI) in Thiruvavnthapuram, Kerala. There the scientists formulated the product from the leaves of the said plant and marketed it under the brand name of '*Jeevani*' in 1994. TBGRI received the royalty and license fee, which was shared with the Kanis in 1:1 ratio. It has been widely accepted that this resulted in partial alleviation of poverty among the Kanis, and their income increased by Euro 200 per month for every household (Mahtur 2003). However, several questions have risen since then. The beneficiaries were limited in number and the membership of the trust, which was disbursing the royalty did not expand to ensure equity in sharing (Chaturvedi 2007). Also the entire community was not ready for commodification of this plant. Also, this type of commercialization process had immense impact on the cultural values of the Kanis, and there have been fights within the community, making them more vulnerable. Some even interpret this to be an illustration of the fact that the government institutions try to push their own agenda in the name of an ethnic community.

Probably what IPR can do best is to involve 'promotive' measures for TMS with many fundamental steps like improvements in procurement and authentication of raw materials, packaging, and distribution (Correa 2002). With increased demands on safety, efficacy, and quality control, greater investments and initiatives are required in R&D, plant capacity, and compliance with 'good manufacturing practices.' In this regard, collaborative research programmes have been initiated by the CSIR with industry organisations in relation with raw material, processing, and formulation (Warrier 1999).

Impact of traditional medicine systems on economic growth or inclusive development?

Contribution to growth and development

The preceding sections indicate that while TMS is an important healthcare segment in India, so far, it has not been marked by the creation of major innovations that have had a health impact. The public sector is focusing on building capacities in basic research and diffusion of knowledge; and so far, the main outputs are in the form of scientific publications and patents, with little industrial impact. Private pharmaceutical firms have also joined the TMS bandwagon, and they are generating innovations in the form of new product varieties to boost brand competition, novel distribution and promotion mechanisms, and innovation in therapies.

This has given rise to market segmentation, as shown in Fig. 9.3. Now the contribution of TMS to economic growth can be better understood. Over the last 50 years, as India's population has grown, the lowest on the quality-price spectrum, segment S1, catering to the poor has expanded. Similarly, as the Indian government supports public hospitals providing TMS, the second segment S2 has also grown and it has also provided an outlet for the TM products of pharmaceutical companies. Finally, new niches catering to middle and higher-income households have emerged in both rural and urban areas and are flourishing as segments S3 and S4. However, the products catering to the last two categories are not therapeutics but OTC products for well-being and

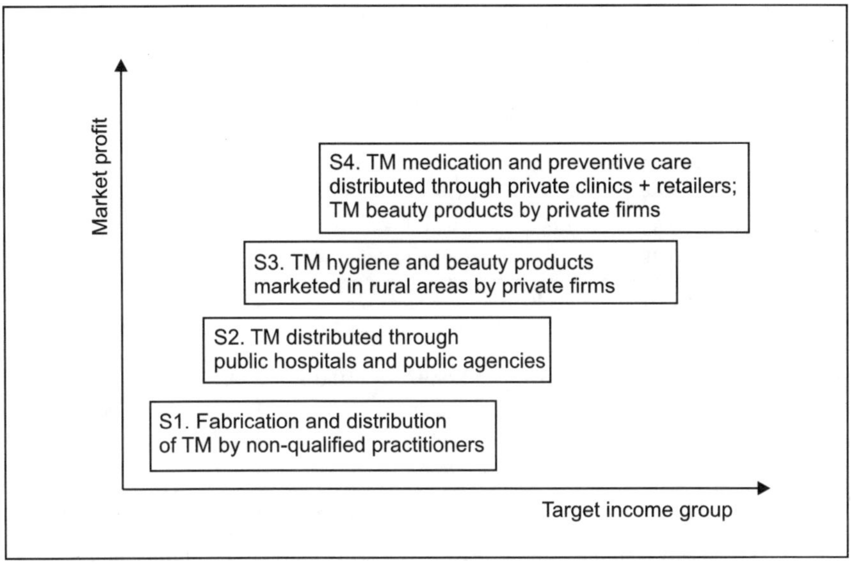

Fig. 9.3: TMS market segments
Note: TM – traditional medicine; TMS – traditional medicine system

beauty. The evidence of innovation is associated mainly with private firms, and it is designed to boost brand competition and market penetration.

On the topic of inclusive development, while it can be argued that market penetration by pharmaceutical firms of the low-income communities contributes to their well-being as it increases consumption, sense of self-worth, and status, the purchase of 'comfort' products does not improve health outcomes. Therefore, the main innovations that have emerged from TMS cannot be unilaterally considered to have inclusive development.

On another note, though the public sector and the informal sector in TMS are not famed for innovation generation, their very presence in the Indian landscape is a great support for the poor and particularly the rural poor. Accessibility is a function of both the availability (quantity and geographical proximity) and the affordability of medicines (and accompanying costs of treatment) vis-à-vis a patient population. Though better availability of TM vis-à-vis AM cannot be confirmed, data on the average expenditure per episode of treatment reveals the relative affordability of TM. This has been an added factor for the rural poor to accept more and more of these systems. Table 9.6 confirms this for TM.

Table 9.6: Average expenditure per illness episode (in Rs)

Nature of illness	*Allopathy*	*Homeopathy*	*Ayurveda & Siddha*	*Unani*	*Combination*	*Rituals*	*Total*
Rural							
• SCD	180.82	114.0	87.84	94.99	148.52	58.56	170.14
• AI	57.05	45.01	23.11	40.0	76.84	34.57	55.71
• CI	210.35	186.8	94.02	-	212.37	276.97	207.71
• Accident	130.35	-	48.62	-	-	-	124.91
Total	91.62	70.40	38.33	78.14	130.23	63.68	89.94
Urban							
• SCD	200.67	142.78	128.97	0.0	240.94	47.33	196.77
• AI	71.83	62.26	60.28	7.33	111.69	10.00	71.69
• CI	204.33	134.63	160.7	-	201.74	113.28	200.03
• Accident	288.63	-	26.44	-	551.78	-	284.45
Total	114.27	82.65	97.05	7.33	165.44	80.00	113.50

Note: SCD – socially communicable diseases; AI – acute illness; CI – chronic illness
Source: Sundar (1995)

The evolution of the wholesale drug price index also indicates that though the prices of *Ayurvedic* drugs increased substantially over time, they are still far below those of allopathic drugs.[7] The trend of the index is shown in Fig. 9.4.

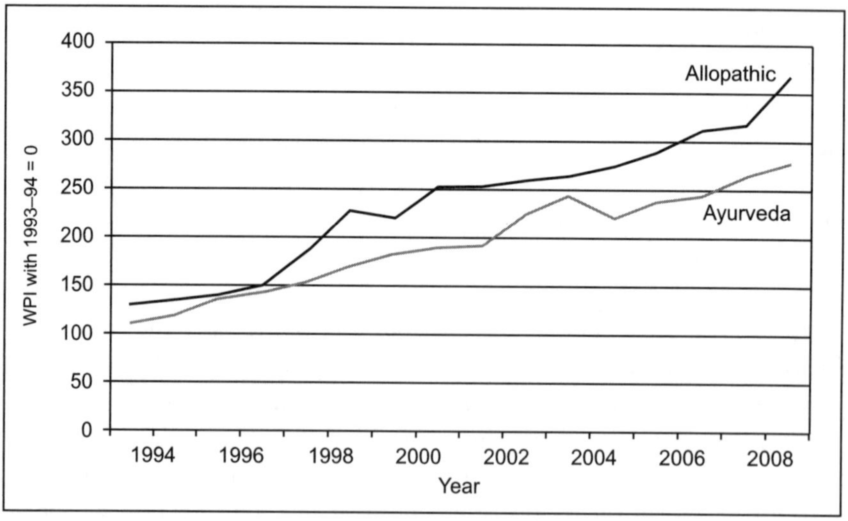

Fig. 9.4: Evolution of wholesale price indices (WPI) in India
Source: www.eaindustry.nic.in

[7] GoI 2010: Office of the Economic Advisor. www.eaindustry.nic.in.

Remaining challenges

Four main challenges can be identified from our case studies, so as to improve the impact of TM innovations on economic growth and inclusive development.

Target patents for required therapeutic areas

In spite of the evolving debate of direct protection of TMS, the government is actively encouraging all research institutes to patent in order to maximise economic returns to public investment. The 'Central Council for Research in *Ayurveda* and *Unani*' has 19 patents on TMS to its credit so far, while another 8 have been filed.[8] The council has also standardised 16 herbal drugs, which are to be used in the nationwide 'Reproductive and Child Health' (RCH) programme of the health department of the government. By far, the major patentee is the CSIR. Out of a total of 419 patents on medicines by CSIR so far, herbal drugs and processes account for 47. However, while some water-borne and vector-borne communicable diseases and infectious diseases cause the greatest damage in terms of 'disability adjusted life years (DALY),' Table 9.7 shows that the major thrust area of patents is not communicable diseases. There is a clear gap between the official rhetoric, which is to generate innovations to tackle these diseases and other orphan maladies, and the reality revealed by patent applications.

Table 9.7: Therapeutic areas of patents on herbal drugs and processes taken by CSIR, as of May 2008

Therapeutic group	*Number of patents*
Nutrients	5
Cancer	8
Diabetes	4
Cerebrovascular	1
Hepato protective	2
Dermatology	4
Arthritis	1
Dental	2
Gastro-enteritis	2
Anti infectives	5
Immuno modulators	2
Others	9
Total	47

Source: www.cdri.org

[8] Council website www.ccras.nic.in, accessed in May 2008.

More effective use of the TKDL

The TKDL is perceived as having the potential to act as the bridge between modern science, modern medicines, and TK. However, over time, it has taken the shape of a standard information base with detailed classification level for a huge number of medicinal plants. Basically, TKDL serves as a freely accessible non-patent literature database on TMS, and its impact is not clear.

Improvement in TM drug safety

Lack of regulation on market sales and distribution of TM prompts the manufacturing firms as well as artisan units to adulterate the drugs, using cheaper herbs in the preparation and/or to change the process in order to reduce the costs of production. Though the government has created provisions of quality control under the 'Rules 160 A-J of the Drugs and Cosmetics Rules, 1945' by opening up inspection machinery in the states where any drug can be tested in public-supported laboratories, there is a parallel '*License Raj*', and corruption and inefficiency are largely present. There is still no provision for accreditation of TMS manufacturers. They also use herbs that are not properly collected and stored (Dutta 2009). Voluntary disclosures can be considered as a possible tool to achieve better quality. For instance, if the producers declare to have followed the procedures according to the classical texts, such as *Granthokta* (mentioned in texts), then the product can be tested just by '*roop-rasa-gandha*' (look, taste, and smell).

Bottom-of-pyramid innovations that promote health security

Worldwide, there are four billion potential consumers at the bottom of the economic pyramid (earning less than US$5 a day), out of which India alone has 171 million low-income households with a combined income of US$378 billion (Anderson and Billou; 2007 Prahalad and Hart 2002). Therefore, if an appropriate product is available at an affordable price and if the low-income households recognize their need for it and can access it from a market, both firm profit and social welfare can grow. As of now, while marketing to the rural consumer, the focus of companies of most firms is on consumer care products such as soaps, shampoos, tooth pastes, etc., rather than on providing therapeutic medicines. Thus, as for mainstream pharmaceutical firms, even for

TMS leaders, there is a need to explore the possibilities for the creation and commercialization of BoP innovations.

Conclusion

To sum up, it is clear that TMS sectors have survived well because of the internal dynamism and market vision of the TMS firms. These, in turn, are supported by a rising demand for natural and safe cures, as well as infrastructural investment by the Indian State in knowledge accumulation in TMS and medical care delivery of TMS products and services. The market share of the organised private sector firms in the TMS sectors and their innovation generation can be expected to increase in the future. But they are not likely to impact the health security of those below and just above the poverty line much. Therefore, the impact of TMS on inclusive development will depend very heavily on the quality and usefulness of innovations from the public sector laboratories for socially needed therapeutic areas. That said, another plaguing problem is drug safety. To promote inclusive development, it will also be necessary to impose more regulation on downstream TMS market and largely expand the provision of TMS offering public health services. Though IPR is important in the long run, at the moment, finding the best IPR system to deal with TMS should not be considered as the most pressing problem.

References

Appadurai, Arjun. 1986. 'Introduction: Commodities and the Politics of Value.' In *The Social Life of Things: Commodities in Cultural Perspective*, edited by Arjun Appadurai, 3–64. Cambridge: Cambridge University Press.

Anderson, J. and N. Billou. 2007. 'Serving the World's Poor: Innovation at the Base of the Economic Pyramid.' Working Paper. Berlin: European School of Management and Technology.

Banerjee, M. 2002. 'Public Policy and Ayurveda: Modernising a Great Tradition.' *Economic and Political Weekly* 37 (12): 1136–46.

Barman, D. 2006. 'Medical Tourism in India.' MPhil Dissertation. Calcutta University: Department of Economics.

Baru, R. 1996. *Studies on Human Development in India*. New Delhi: United Nations Development Programme.

Bode, M. 2006. 'Taking Traditional Knowledge to the Market: The Commoditization of Indian Medicine.' *Anthropology and Medicine* 13 (3): 225–36.

Chaturvedi. 2007. *Kani Case: A Report for GenBenefit.* Preston, UK: University of Central Lancashire. Accessed: June 2012; http://www.uclan.ac.uk/research/environment/projects/assets/cpe_genbeneift_kani_case.pdf.

Chowdhury, Nupur. 2004. 'Patent Bill: Protecting Indigenous Knowledge.' *Economic and Political Weekly* 39 (46, 47): 4984.

Correa, Carlos. 2002. *Protection and Promotion of Traditional Medicine Implications for Public Health in Developing Countries.* Geneva: South Centre. Accessed: June 2012; http://..ps.who.int/medicinedocs/en/d/Js4917e/.

Croizier, Ralph C. 1965. 'Traditional Medicine in Communist China: Science, Communism and Cultural Nationalism.' *The China Quarterly* 23: 1–27.

Dalal, A. K. 2005. 'Integrating Traditional Services in Primary Health Care.' *Journal of Health Management* 7 (2): 249–62.

Denison, E. 1985. *Trends in American Economic Growth, 1929–82.* Washington, DC: The Brookings Institution.

Dubey N. K., Rajesh Kumar, and Pramila Tripathi. 2004. 'Global Promotion of Herbal Medicines: India's Opportunity.' *Current Science* 86 (1): 37–41.

Dutta, A. 2009. 'Traditional Medical Systems in India and China.' In *Economic Reforms in India and China —Emerging Issues and Challenges,* edited by B. Sudhakara Reddy, 375–90. Mumbai: Sage Publishing.

GoI (Government of India). 2003. 'Protection of Biodiversity and Traditional Knowledge: The Indian Experience, Submission by India to the WTO, WT/CTE/W/156, IP/C/W/198, 12.' Annual Report, Department of AYUSH. New Delhi: Ministry of Health and Family Welfare. Accessed: June 2012; www.indianmedicine.nic.in.

Gupta, A. K. 2005a. 'Local Knowledge: Global Institutions.' Paper Presented in International Conference on Biodiversity. 27 January, Paris.

Gupta, V. K. 2005b. 'Traditional Knowledge Digital Library.' NISCAIR Paper presented at the Sub Regional Experts Meeting on Intangible Cultural Heritage, Bangkok.

Harilal, M. S. 2009. 'Commercialising Traditional Medicine: Ayurvedic Manufacturing in Kerala.' *Economic and Political Weekly* 44 (16): 44–51.

Jorgenson, D. 1990. 'Productivity and Economic Growth.' In *Fifty Years of Economic Measurement*, edited by E. Berndt and J. Triplett, 19–118. Cambridge: NBER.

Lexchin, J., L. Bero, B. Djubegovic, and O. Clark. 2003. 'Pharmaceutical Industry Sponsored Research: Evidence for a Systematic Bias.' *British Medical Journal* 326: 1167–170.

Love, J. 2003. *Evidence Regarding R&D Investments in Innovative and Non-Innovative Medicines.* Washington, Geneva, London: Consumer Project on Technology. Accessed: June 2012; http://www.cptech.org/ip/health/rnd/evidenceregardingrnd.pdf.

Koning, M. 1998. 'Biodiversity Prospecting and the Equitable Remuneration of Ethnobiological Knowledge: Reconciling Industry and Indigenous Interests.' *Intellectual Property Journal* 12: 270.

Mahtur, A. 2003. 'Who Owns Traditional Knowledge?' *Economic and Political Weekly* 38 (42): 4471–81.

Omran, Abdel R. 2005. 'The Epidemiologic Transition: A Theory of Epidemiology of Population Change.' *Milbank Quarterly* 83 (4): 731–57.

Patwardhan, B., D. Warude, P. Pushpangadan, and N. Bhatt. 2005. 'Ayurveda and Traditional Chinese Medicine: A Comparative Overview.' *eCAM* 2: 465–73.

Prahalad, C. K. and S. L. Hart. 2002. 'The Fortune of the Bottom of the Pyramid.' *Strategy and Business*. Issue 26.

Pushpangadan, P. 2003. 'Traditional Medicine and Herbal Technology.' Country paper in Expert Group Meeting (EGM) on the Establishment of Asia-Pacific Traditional Network (APTMNET). 16–20 September, Wuhan, China.

Romer, Paul M. 1990. 'Human Capital and Growth: Theory and Evidence–Carnegie-Rochester Conference Series on Public Policy.' *Elsevier* 32 (1): 251–86.

Sen, S. 2007. 'Commercialization of Health Care and Indian System of Medicine: Special Reference to West Bengal.' MPhil Dissertation. Calcutta University: Department of Economics.

Shankar, D. 1992. 'Indigenous Health Services: The State of the Art.' In *State of India's Health,* edited by A. Mukhopadhyay. New Delhi: Voluntary Health Association of India.

Sharma, U. 2008. 'Transforming Ayurveda.' *Express Pharma Pulse,* 16–31 August. Accessed: June 2012; http://pharma.financialexpress.com/20080831/market01.shtml.

Schumpeter, J. 1942. *Capitalism, Socialism and Democracy*. London: Allen & Unwin.

Solow, R. 1957. 'Technical Change and the Aggregate Production Function.' *The Review of Economics and Statistics* 39 (3): 312–20.

Sundar, R. 1995. 'Household Survey of Health Care Utilization and Expenditure.'NCAER Working Paper No. 53, National Council of Applied Economic Research, New Delhi

The World Bank. 2008. *World Development Report* .

Verma, S. and S. P. Singh. 2008. 'Current and Future Status of Herbal Medicines.' *Veterinary World* 1 (11): 347–50.

Vivekananda Girijana Kalyan Kendra. 2009. *Mainstreaming Traditional Medicine in Primary Health Care*. Karnataka: Karuna Trust.

Warrier, A. 1999. 'Branching Out with the Philosophy of Life.' *The Indian Express,* 29 December.

WHO (World Health Organisation). 2006. 'Public Health: Innovation and Intellectual Property Rights'. Report on the Commission on Intellectual Property Rights, Innovation and Public Health.

———. 2003. 'Traditional Medicine: A Report by the Secretariat.' 56th World Health Assembly. 19–28 May, Geneva.

———. 2001. *Traditional Medicine Strategy 2002–2009*. Geneva: WHO.

———. 2000. *Report of the Inter Regional Workshop on Intellectual Property Rights in the Context of Traditional Medicine.* Bangkok: WHO.

Whyte, Susan Reynolds, Sjaak van der Geest, and Anita Hardon. 2002. *Social Lives of Medicines (Cambridge Studies in Medical Anthropology.)* Cambridge: Cambridge University Press.

Working Group on Traditional Health Sciences. 2008. *Summary of Recommendations of Working Group by National Knowledge Commission on Traditional Medicine.* New Delhi: Government of India.

CHAPTER 10

Where is the Toilet Please?

The Sanitation Sectoral Innovation System in Rural India[1]

Shyama V. Ramani and Shuan Sadre Ghazi*

Introduction

A recent survey (UNICEF 2012) reveals that 51 per cent of Indian households do not have access to functioning toilets. Moreover, according to the data collected by the Government of India for the 2011 Census, while 53 per cent of the total population of 1.2 billion have a mobile phone, and 47 per cent have a television set in their homes, about 50 per cent (approximately 600 million people) still defecate in the open. Clearly, this represents a societal failure in addressing a basic need. For those in households without toilets, acquiring and using a toilet is like adopting an innovation. This leads to the query: what went wrong in the Indian national system of innovation (or NSI) with respect to toilets? Who is to blame – the State, households, firms or institutions? The purpose of the present chapter is to explore the answer to the above queries.

[1] Parts of this chapter are based on: Ramani, S. V., S. Sadre Ghazi, and G. Duysters 2012. 'On the Diffusion of Toilets as Bottom of the Pyramid Innovation: Lessons from Sanitation Entrepreneurs.' *Technological Forecasting and Social Change* 79 (4): 676–87 and Ramani, S. V. 2008. 'Playing in Invisible Markets: Toilet Innovations and Empowerment.' UNU-MERIT Working Paper Series no. WP2008-012. Maastricht, the Netherlands: UNU-MERIT..

* Shyama V. Ramani gratefully acknowledges support from WASTE under the FINISH programme and an NWO-ICSSR grant that made this research possible. Shuan Sadre Ghazi gratefully acknowledges a research scholarship from the Ingemar Broman foundation.

The sanitation sector has not received much attention from innovation scholars. This could be because of two possible reasons. First, building a product such as a toilet for the poor is not knowledge intensive except in certain contexts[2] and does not require highly qualified R&D staff in firms or public laboratories for its creation. Second, toilets in some form or another have been known since centuries and have been available to many segments of the population. However, a standard premise in innovation studies is that any artifact that is newly introduced to a community, which never had access to it before, is akin to adoption of an innovation by that specific community – even though it may be a commonly used product by others. Therefore, the framework of the NSI and the sectoral system of innovation can be applied to study the institutional ecosystem surrounding this artifact and its diffusion as an innovation to the poor.

According to the NSI approach, the generation and diffusion of innovations within a country are highly influenced by national institutions and state policy, especially through the provision of country-specific incentives. Clearly, this incentive provision has been missing in the Indian toilet sector. And, how can this problem be remedied? On this question also, the existing NSI literature is relatively silent. Indeed, the current literature on NSI has little to say on the processes of innovation generation *for* or *by* the poor. Thus, by studying which economic actors have addressed the dire need for toilets within the confines of the NSI in India and how they have done it, we hope to provide some insights into the larger subject of the diffusion of essential pro-poor innovations in developing countries.

For the purpose of this chapter, by the poor, we refer to the BoP, i.e., bottom/base of the income pyramid, which includes the largest, but poorest socio-economic groups in the global income pyramid living on a few dollars a day (Prahalad and Hart 2002). The diffusion of pro-poor innovations and adoption of new products and services in the BoP context deserve more attention. Compared to the users in mainstream markets, the underserved communities at the BoP face a higher complexity due to their greater limitations. The current body of literature on BoP innovations offers insights into the features that pro-poor products should have (e.g, Prahalad 2005), the innovative

[2] Designing a toilet for certain underserved communities would require innovation beyond simple modifications to what is available in mainstream market. For example, slum-dwellers or rural populations, who do not have access to running water or waste disposal systems, require specific toilet technologies.

business models that should be considered in BoP context (e.g, Anderson and Billou 2007), and mechanisms to make markets more inclusive for the poor (Mendoza and Thelen 2008). While these studies provide valuable acumen, they have a product-centric view. In contrast, by taking a systemic view, this chapter aims to shed more light on issues in the post-production phase of the BoP innovation that hinder or facilitate adoption among intended beneficiaries.

This study is not only pertinent to India but also to many other regions of the developing world, primarily for two reasons. First, a joint report of the World Health Organization and the United Nations Children's Fund (WHO/UNICEF 2012) reviewed sanitation and health-related challenges of developing countries across the globe and found that about 2.2 million people die every year from health problems caused by drinking contaminated water, often containing traces of faecal matter. The report also pointed out that over 2.4 billion people around the world respond to the call of nature every day in ponds, fields, plastic bags, buckets, rail tracks, and roads due to the lack of access to a toilet.[3,4]

The rest of the chapter is organised as follows. The following section outlines the main findings of the literature on pro-poor innovations. The next section briefly presents the salient features of sanitation drives before and after liberalisation in India. The subsequent section contains the case study of an isolated Indian coastal village, which has been recognized by the Indian government as a success story. This is followed by a discussion on the results of the preceding sections in the light of the evolution of the Indian innovation system. The final section concludes.

Pro-poor bottom of the pyramid innovations

Generally speaking, any new commodity made available or accessible to the BoP can be considered as an innovation with respect to the BoP consumers – even though it may be a commonly used product for higher-income consumers. But not all BoP products have the same developmental impact in terms of improving the lives of the poor. For instance, the long-term impact of the consumption of shampoo in sachets – designed specifically to make shampoo accessible to the poor – will not be the same as that of clean water. Therefore, in line with

[3] Minimum WHO level for improved sanitation for households is a toilet that is connected either to a public sewer, or a septic tank or some pit in such a way that the air, water, and soil in and around the pit are not contaminated.

[4] http://www.unicef.org/media/media_39547.html.

Mendoza and Thelen (2008), we distinguish pro-poor BoP innovations as those that cater to the essential needs of the poor such as healthcare, housing, food, water, and sanitation or enhance productivity and income-generation capacity. Mendoza and Thelen (2008) also reiterate that to be viable, the delivery system of pro-poor innovations must ensure accessibility to the targeted consumers as well as positive financial or reputational returns to the suppliers.

Diffusion of toilets as a pro-poor innovation is a challenge because their successful adoption calls for a change in individual behaviour, daily routines, and perhaps, even social norms. Indeed, such requisites imply that the diffusion of innovations must be distinguished from their mere provision, for the former also involves creating incentives for optimal adoption and effective utilisation by the intended beneficiaries. Indeed, though there is an extensive literature on the processes of innovation diffusion, not much is applicable to the BoP contexts outside of agriculture. Therefore, in order to have a conceptual framework to understand the nature of BoP innovations, we first begin by exploring the demand side, and then we turn to the delivery procedures, highlighting the main recommendations of the literature.

Factors influencing the demand for BoP innovations

The decision to adopt an innovation is influenced by the same factors that usually affect any conventional investment, such as benefits, costs, and risks in a specific institutional/social environment. BoP users are similar to those in higher-income tiers in that they make investment decisions to maximise their expected utility. However, the perceived value of the product and the expected utility obtained are likely to be different given the constraints surrounding them, which can be one or more of the following broad types.

Financial constraints

A high percentage of BoP workers are unemployed, self-employed or employed in the informal sector. Indeed, in developing countries, informal employment comprises one half to three-quarters of non-agricultural employment (ILO 2002). Therefore, unlike mainstream workers employed in formal sectors, the income of BoP households is subject to greater seasonal, temporal, and regional variance. In addition, and because of low/fluctuating income, access to credit or other types of financial services is limited or unavailable for them, which further affects their consumption choices (Banerjee and Duflo 2007).

Lesser knowledge, information, and skills base

Individuals of the BoP communities are likely to have less education, skills, and awareness about the possibilities of innovation (UNDP 2008). They might be unfamiliar with certain goods and services, such as new technologies or financial services. As a result, they might not have accurate information about the advantages (or disadvantages) of new products or might perceive their use as being complicated, which in turn can deter them from using these products. For example, individuals in rural areas are less likely to have knowledge about the benefits of IT services and their demand for them will be low, unless they learn how to utilise them.

Limited access to complementary infrastructure

BoP communities are likely to have less access to complementary infrastructure that creates an enabling environment for effective use of certain products or services. They may have limited access to electricity, water connections, and complementary products needed to install and use certain products and services properly.

The above-mentioned constraints affect the level of demand at BoP. The constraints are likely to translate into a no-demand or lower demand for certain products and services. Nevertheless, it should not be assumed that those at BoP use only low quality commodities and services in keeping with their financial endowments. In addition to being 'consumers', BoP individuals are very careful 'money managers', and planning often goes into saving for the purchase of goods and services that constitute a luxury for them, such as movies and high quality rice for special occasions, which yield high indirect utility or enhance their status and social standing (Banerjee and Duflo 2007).

On optimal characteristics of BoP innovations and their diffusion: Need, technology, and the socio-economic context

Many pro-poor BoP innovations emerge from the identification of a need. The 'need' for any innovation is generated by socio-economic structures and cultural norms. Therefore, the first step is to examine the nature of needs. As Katz (1961) explains, "it is as unthinkable to study diffusion without some knowledge of the social structures in which potential adopters are located as to study blood circulation without adequate knowledge of the veins and arteries."

The next natural step is to confirm the appropriateness of the innovation to serve the need.

There exists an extensive literature on the optimal 'technology design' to cater to the BoP needs. Following the 'small is beautiful' concept a la Schumacher (Schumacher 1973), the 'appropriate technology' works advocate making optimal use of local resources to develop technologies for the benefit of the poor. Early work in the area focused on 'technology' or 'product innovation design' to fit the constraints of the context and the resource base of the final user. By default, it led to the reigning explanation that if an innovation is unsuccessful with the poor, it is because the technology is not designed appropriately.

Other scholars have called for a broadening of the notion of 'appropriateness.' Stewart (1977) pointed out that the new technology must be compatible with the income levels, resource availability, existing modes of production, and existing technologies and costs in the society for which it is designed. It has also been noted that an innovation could be successfully adopted if it has a relative advantage over the solutions already utilised by the users over and above notions of 'appropriate technology.' Such a relative advantage could be lent by factors such as simplicity of use or compatibility with user's norms and existing ways of doing things (Rogers 2003).

Although the above factors are relevant for diffusion of a BoP innovation, it is accepted that in order to ensure a positive development impact, there is a need to take a holistic approach starting from conception of the innovation to its final delivery. This premise has given rise to a vibrant and growing literature, mainly in the form of case studies on different aspects of the diffusion of pro-poor BoP innovations, and its main recommendations can be summarized as follows:

- Confirm that there is an unmet or underserved need for which a technological solution can help (Hart 2005; Christensen et al. 2006)
- Verify that the product characteristics are compatible with the socio-economic context. Favourable characteristics are a proper price/performance ratio, robustness of product for harsh conditions of BoP environment, and de-skilling of product for easy use (Prahalad 2005).
- Build a delivery model that is compatible with the socio-economic context. Beyond fixing of price, the marketing strategy should take into account socio-cultural norms and power (Letelier, Flores, and Spinosa 2003; Kotler, Roberto, and Leisner 2006).
- Identify and use appropriate 'change agents' to adopt and convince others to do the same (Rogers 2003).

- Ensure that the complementary institutions or assets needed for sustained functioning of the innovation are available (Reddy, Aram, and Lynn 1991).
- Confirm the adsorptive capacity of targeted consumers both as individuals and as members of social groups (Jeffrey and Seaton 2004).
- Make sure that the business and delivery model generate positive returns to the suppliers. For this application, private sector management principles could be useful (Ghobadian 2004).
- 'Co-create' value with all stakeholders, including the user community, firms, public agencies, and non-profit organization (NPO) through non-traditional forms of partnership (Franceys and Weitz 2003; London, Rondinelli, and O'Neill 2006; Brugmann and Prahalad 2007).

Though the above literature is very useful, two shortcomings can be pointed out. From a theoretical perspective, the steps to diffuse an innovation are rather clear, nevertheless, the protocols for actual implementation are not evident. It is obvious that every BoP context will be different and possibly require a different solution. Then the question becomes – how can the theory be applied in practice? Moreover, it would seem that by following this set of practically common-sense rules, all practitioners can be successful. Reality reveals that this is not the case, as will be shown in the next sections.

Sanitation drives in India

Pre- liberalisation sanitation entrepreneurs

Since historical times, in rural areas all over the world, answering the call of nature in open fields has been an accepted norm. Open defecation was not perceived as a major social problem or health hazard as vegetal coverage was sufficient to ensure privacy and safe recycling of waste into the soil. In the rural areas, open defecation was collectively practised (in social groups, e.g., groups of men or groups of women going separately) in the fields without any risk. Usually, they went with a pick, dug a hole in the ground (or identified a depression) and after defecation covered it with soil or leaves, so that natural composting occurred without infestation by flies. In most towns, there were only dry toilets (i.e., without running water), where urine was let off into the back street and faeces was taken away by scavengers with long brooms with brittles and the floor was washed with water from the outside. The night soil was transported in wheel burrows and dumped into specific sites for composting. In India, as an offshoot of the caste system, specific communities (called *Bhangis* in Northern and Central India and hailed by other local appellations in the South) performed all scavenging activities, including the cleaning of night soil from these toilets.

Unsurprisingly, the first efforts to address the problem of the lack of sanitation in rural India were neither from the British colonial State nor from the first government of Independent India, but from one of India's greatest social entrepreneurs, M. K. Gandhi, popularly known as Mahatma Gandhi. A champion of oppressed communities, including the members of the manual scavenger castes, he continually and openly insisted, even in packed political meetings[5] that there was no place for scavengers in India and that everyone should clean their own toilets.

In his *ashram* (or retreat), Mahatma Gandhi broke all taboos by experimenting with different models of 'dry closets' and different modes of maintenance of the traditional dry toilet. His biggest achievement was to get his followers to clean the experimental dry toilets, by setting an example himself.[6] While, Gandhiji's experiments did not give rise to any technological innovation, his writings, his speeches, and the routines for toilet maintenance that he initiated created a collective consciousness of the need for a new toilet technology, accessible to all and capable of being maintained autonomously.

Two decades later, where Mahatma Gandhi had failed, Bindeshwar Pathak, a young student of sociology and an ardent follower of Mahatma Gandhi, succeeded. His objective can be summed up as follows: 'To restore human rights and dignity to scavengers by stopping the practice of manually handling human excreta.' Pathak saw his mission of improving sanitation coverage as going much beyond just introducing new technology. For him, it was a veritable movement to fight against casteism and the employment of thousands of members of the scavenger caste in cleaning traditional latrines by sweeping out human refuse. He won his battle by introducing a new model of a toilet that was easy to maintain autonomously without the need for a scavenger. *Sulabh*, the organisation he founded in 1970, has been diffusing this toilet model ever since.

The *Sulabh* toilet looks just like the standard Indian squatting style toilet slab with one hole for flushing, but, instead of the flushed waste going directly into the ground or to a septic tank or to a central sewer canal, it falls into one of the two deep pits that are outside the toilet. When the first pit is full, the family switches to a second pit, while the waste in the first pit gets gradually and naturally transformed into a rich material that can be removed and used as dry, powdery fertiliser. When the second pit is nearly full, the first pit can

[5] http://www.goodnewsindia.com/index.php/Magazine/story/sulabh/P2/.

[6] http://www.lifepositive.com/Spirit/masters/mahatma-gandhi/community.asp.

be emptied[7] and its contents can be used as compost. Therefore, the two pits can be used alternately and continuously. Currently, the *Sulabh* toilet model is used by about 1.2 million poor households and has been declared a 'global best practice' by United Nations HABITAT and Centre for Human Settlements and is being diffused by the UNDP[8] all over the world.

More than a decade after the creation of *Sulabh*, the Indian government also got into action. The first public programme to focus exclusively on sanitation was the Central Rural Sanitation Programme (CRSP) initiated in 1986 by the Ministry of Rural Development. Under this scheme, at the district level, the offices of the District Rural Development Agency (DRDA) financed the construction of toilets to meet set targets, with the beneficiaries being partially or near-totally absolved (depending on their income level) from having to bear the costs of installation. At the district level, officers were given a target number of toilets to be constructed for beneficiaries and these were simply built. Appropriateness was not clear in terms of either technology or the socio-economic context.

The model diffused under the government programme was the 'single-pit latrine', which overflows during the rainy season and which has to be covered or dislodged when full. In the case of the latter, the entire superstructure has to be dismantled and a new pit has to be built. Thus, it is not surprising that most of these single pit latrines were abandoned when they began to dysfunction or when they got full.

A former Secretary of the Planning Commission (Saxena 2005) notes various reasons for the abandoning of the toilets provided by the State. He explains this phenomenon as being due to the lack of demand from people who did not see the need or feel the desire for sanitation; lack of adequate water sources; lack of space; an absence of choice on cost or technology; and a total absence of people's participation. Construction of the latrines was done centrally without hygiene promotion and marketing of the products in the targeted zones. There were no well-planned supply chains, which posed logistics problems wherever materials and skills were not locally available.

A UN report also confirms that the CRSP failed to be effective for several reasons. For instance, only the toilet slab was provided and the households were required to build the superstructure by themselves. However, the cost of

[7] Usually manually and without difficulty as the faecal matter would have dried completely and been transformed into fine dust.

[8] United Nations Development Programme.

the superstructure was out of reach for many rural households. Moreover, the public sector employees were not very motivated, and other actors like NPOs and firms were not involved in the sanitation drive (UN-DESA 2003). There was no reflection on whether needs were supported by effective demand, or whether all the institutions and capabilities necessary for adoption and sustained use were present. The adoption failed because the reality on the ground did not match the requirements for success.

Post-liberalisation sanitation drives

Between 1950 and 1990, a high population growth coupled with steady destruction of vegetal coverage led to open defecation without privacy and in such large quantities that lack of sanitation started posing problems, not only to human dignity, but also to health, even in rural areas. By the start of the new millennium, the lack of sanitation facilities began to draw more and more attention. Alarm bells rang out that open grounds in India receive approximately 100,000 tonnes of human faeces every day, making the lack of sanitation a real crisis (Bhatnagar and Gupta 2008).

With the adoption of economic liberalisation and the initiation of policy reform in 1991, as in other industrial sectors, the Indian State began withdrawing as a direct player on the supply side of the market and became an indirect player, financing sanitation drives. In other words, from being the main supplier on the toilets market, the State became the main financier of the construction of toilets.

In 1999, the Department of Drinking Water Supply, and the Ministry of Rural Development launched the 'Total Sanitation Campaign (TSC)' in order to provide incentives for the development of a private market for sanitation. It involved a demand-driven approach, including education as a major component and actively solicited the participation of *panchayats* (or village-level government bodies), non-profits, women's groups, and youth clubs.[9]

Incentives for the development of a private market for sanitation are provided in a variety of ways. Financing is available for: (i) 'start-up activities' (conducting surveys on the status of sanitation, people's attitudes, and demand before launching a sanitation drive); (ii) 'information–education–communication activities' (enhancing motivation, health, and hygiene awareness, and teaching how to maintain facilities); (iii) 'opening of rural sanitary marts and production centres' (providing materials, services, and guidance required for the construction

[9] http://www.un.org/esa/agenda21/natlinfo/countr/india/CaseStudyIndiaws1.pdf.

of toilets); and (iv) the 'construction of individual household latrines.' Any of these initiatives can be proposed to the district- or state-level authorities by non-profits, self-help groups, women's organisations, and the *panchayat* or the village council. The number of projects launched under the TSC scheme for toilet construction as well as the reimbursement accorded by the government to every toilet built have increased steadily (e.g., reimbursement was Rs 500 in 1999, Rs 1200 in 2005, and Rs 3200 in 2011). This reimbursement is accorded to the local government body to be disbursed among individual households. For toilet complexes, sanitary marts, and production centres, an interest-free loan is given and it is up to the organisation concerned to develop a business model to amortize the costs.

Another policy innovation was the 'Sanitation Tournament'. Since 2003, the Department of Rural Development has instituted the *Nirmal Gram Puraskar*,[10] providing cash incentives to rural *panchayats*, non-profits, and individuals, who help achieve complete toilet coverage in a given district. The cash prizes increase in terms of area or population covered. Even with modest cash prizes, the prestige of being awarded this prize has created tremendous incentives for complete toilet coverage, and the number of applicants for the award has been increasing yearly since its inception.

In 2012, the TSC was revamped as the '*Nirmal Bharat Abhiyan*' (Pure India Campaign) programme and its target is to make India open-defecation free by 2022. Incentives for the construction of toilets have been increased. Now any household that builds a toilet will be reimbursed Rs 10,000 of the cost. In most areas, presently, the cost of a toilet ranges between Rs 8000 and Rs 15,000, depending on the soil and rainfall type as well as distance from a town. Funds from other national programmes such as the Mahatma Gandhi National Rural Employment Guarantee Scheme are being diverted to the *Nirmal Bharat Abhiyan*. Earlier only 'below poverty line households' had been the priority beneficiaries. Now, the provision of incentives for individual household latrine units covers identified 'above poverty line households' (who were previously ineligible for aid), 'Scheduled Castes and Scheduled Tribes households,' small and marginal farmers, landless labourers with homesteads, physically challenged, and women-headed households. Schools and crèches (*Anganwadi*) in rural areas are also covered.

In the sanitation sector, while the Indian government is the biggest financier, it is not the only one. A totally unanticipated evolution of the adoption of

[10] http://ddws.nic.in/tsc-nic/html/nirmal_gram.htm.

liberalisation in India has been the entry of new international agencies in the social sector. Presently, in addition to the well-known financiers of socio-economic development projects such as the UNICEF, UNDP, and the World Bank, international development organisations like Deutsche Gesellschaft für Internationale Zusammenarbei (GIZ), Bremen Overseas Research and Development Association (BORDA), and Water Aid also sponsor sanitation-related activities, while agencies like SuSanA or Sustainable Sanitation Alliance provide consultancy and workshops.

Thus, an enormous market for business opportunities has been created in the sanitation sector for NGOs and/or 'local non-profits' to create awareness on sanitation, to build toilets or to audit the performance of sanitation drives. A variety of business-to-business relationships between the financiers, namely the State and international agencies, and the project implementers, namely the intermediary organisations, have been initiated to make India open-defecation free. These intermediary organisations move from zone to zone as they get new contracts for creating sanitation awareness or building toilets. Sometimes, they get in touch with local social entrepreneurs in a village, who are focused only on a set of villages. At other times, they operate directly in the village, finishing the sanitation project in one or two months. They hold meetings with the village council and the self-help men's and women's groups to decide how to implement the project. This is followed by a short awareness campaign. Then, local masons are trained or a local construction company is hired to build toilets. The supply side chains in operation currently are, therefore, extremely heterogeneous in terms of composition, mode of delivery, and efficiency.

The case study of Kameshwaram: The rise and fall of sanitation

Kameshwaram is a typical coastal village along the Indian Ocean in the Nagapattinam district of the Indian state of Tamil Nadu. On 26 December 2004, huge tidal waves of the devastating Asian tsunami flooded Kameshwaram for a few minutes, but that was enough to wreck total havoc. In the aftermath of the tsunami, people across the world donated funds to charitable organisations to address the immediate and long-term needs of the tsunami victims. According to locals, like in many other villages along the southern Indian coast, this was the first time in living history that anyone could remember the outside world to have taken so much interest in them. Among the do-gooders was a Franco-Indian group, which became formalised after a year as two sister non-profits, Association Un-Ami in France and Friend in Need Trust in India. Both will be referred to as FIN from now on.

The remainder of the case study is based on the data accumulated by the first author[11] over a period of seven years (from 2005 to 2011) in the form of extensive memos, reports to donors, notes by villagers, and e-mails exchanged as participant-observer in sanitation drives both in Kameshwaram and in other villages of India. The notes had been taken during interactive sessions (informal get-togethers, strategic discussions, public meetings, etc.) with different stakeholders in sanitation projects, such as rural households, NPOs, and local government representatives.

In order to develop an inclusive project, FIN held many discussions in the village and especially with the women. This revealed that prior to the tsunami, there had been tree-covered areas, one for women and one for men, which had been used for open defecation. The need for toilets had not been perceived, even though it was well known that during monsoon time, roughly three months in a year, these sites attracted mosquitoes and made mosquitoes-spread diseases rampant. However, with the tsunami and the felling of trees in these areas, vegetal coverage for open defecation was eliminated, putting the women in a quandary to find secluded zones. As they started to retain themselves or go near rubbish heaps at dawn and after dusk, some got bitten by rats, one by a scorpion, and sexual harassment began to raise its ugly head. Hence, suddenly there was a dire need for toilets.

Search for appropriate technology design

Many rural areas do not have toilets because decentralised sanitation requires the technology design to be appropriate to the environment. For instance, Kameshwaram being a high water table area, with heavy rains during the three months of the monsoon, it is not possible to construct either the conventional single pit latrine or the *Sulabh* double pit toilet, for any pit in a high water table area gets waterlogged very frequently and overflows during the rainy season. Indeed, this is the reason why the *Sulabh* toilet model, which is suitable for dry areas, was not adopted in regions with a high water table such as coastal zones.

Identification of an innovative design

Eventually, a new toilet design model in the form of a urine diversion toilet, introduced in India by Paul Calvert, popularly called ecosan, was identified as being appropriate for Kameshwaram. It was brought to the notice of FIN

[11] Also founder-director of Friend In Need.

by other NGOs active in sanitation drives, which were aware of different sanitation models.

Paul Calvert is a British engineer who joined Intermediate Technology Development Group or 'ITDG UK' in 1985 in order to translate his faith in Schumacher's dictum 'small is beautiful' into action and use appropriate technology to uplift disadvantaged populations. In 1989, he was based in Trivandrum in Kerala, as Technical Manager for ITDG's Farmers, Herders and Fisherfolk Programme. He was a boat builder and an appropriate technologist with a strong community, social, and environmental bias.

When Calvert saw the horrific open defecation sites that women had to use, he was moved to do something about it. Through a process of trial and error, he came up with a model that separated urine and faeces after they combined, and while trying to reduce the costs of this model, he realized that there could be substantial efficiency gains if faeces and urine were not allowed to get mixed in the first place, and herein was the heart of his innovation.

In the ecosan, the toilet squatting slab has three holes, one behind the other, with different slopes. The user urinates first and shifts slightly back to defecate, permitting the faeces to fall into a compost pit. A mug of ash or saw dust is then thrown into this hole, facilitating dehydration of the faeces. Then the user moves further back to wash the behind. The urine goes out through a bamboo pipe to irrigate a garden planted around the toilet. The wash water is filtered through layers of gravel so that the water that leeches out into the soil is harmless. Thus, urine, faeces, and wash water are completely separated and recycled.

While there is evidence of experimentation with urine diversion toilets during the 1970s in many parts of the world, and especially in Nordic countries, their virtues were practically unknown outside of the circle of its devout practitioners. Thus, Calvert had to re-invent a version himself and his contribution to the basic model was to have added features that made it user friendly under Indian conditions.

Search for funds

To initiate a sanitation drive, FIN interacted with another regional NGO, which was able to attract sponsorship for construction from UNICEF. UNICEF in turn, was interested, because ecosan represented a 'totally decentralised' and 'sustainable sanitation systems' that closed the loop – completely recycling the waste without any risk of environmental contamination. Therefore, with sponsorship from UNICEF and seed funding from FIN, about 150 urine-diversion toilets, especially conceived for coastal areas, were built with great fanfare in Kameshwaram.

The curse of fame

For the installation of the largest number of ecological toilets in a village (for that time) and on the assumption that the drive would continue till 100 per cent sanitation coverage was achieved, the Kameshwaram *panchayat* was awarded the '*Nirmal Gram Puraskar*' in 2007 by the Indian government. The *Nirmal Gram Puraskar* to Kameshwaram signalled to intermediary organisations active in the sanitation field that this village was one where the residents were ready to make the behavioural switch from open defecation to toilet usage. The non-profits active in sanitation then sought funds from international agencies to build more toilets. For instance, in 2008, about 100 more urine diversion toilets were built by another local non-profit with funding from Water Aid and seed funding from FIN.

This increased the fame of Kameshwaram even more. Then, private actors, namely, Western Christian Church groups and the charity foundations of local industry associations, also got into the act, sponsoring the construction of 350 single-pit latrines and toilets with attached chambers, also referred to as septic tank toilets, between 2007 and 2009. Many septic tanks in rural areas are simply large pits with a few barriers that may or may not be cemented. Without an impermeable septic tank, waste water from toilets can also contaminate the surface water in high table areas with faecal matter. These private actors were not concerned about technology design or contamination. The financiers had simply transferred funds through their networks for the construction of a set of toilets, whose photos could be displayed in the annual reports of all economic actors concerned, and especially those of the charity financiers. The local construction company was happy to get a contract to build 350 toilets and the village government and the households concerned were happy to receive 350 free toilets. There was no dialogue and these toilets could well become 350 more points of contamination in rural India in the years to come. Thus, by 2010, there were 675 toilets in Kameshwaram, of different kinds, namely, single-pit latrines, urine diversion toilets, and toilets with the so called septic tank, covering about 43.93 per cent of Kameshwaram households.

Chance identification of problems

Two years after their construction, during the annual 'Toilet Beauty Contest'[12] in 2009, it was noticed that many of the toilets, especially the urine diversion toilets, were being abandoned. A survey of about 600 houses with and without toilets was conducted and five focus group discussions were undertaken by FIN in 2011 as to why and how this was happening. The main reasons are summarized below.

- *Faulty construction*: The main reason for abandoning of the toilets was faulty construction. For instance, in all toilets requiring repairs, it was seen that the roofs needed to be replaced because of cracks. This meant that during rains, the toilet could not be used and there was always a risk of the roofing slab falling on the user's head when in the toilet. Moreover, it was noticed that all the roofs were of asbestos sheets, a material that is banned in the head office countries of both UNICEF and Water Aid as it is confirmed that asbestos increases the risks of getting cancer. Some of the toilet slabs and pipes needed to be fixed. Most of the doors could not be closed properly because the latches had fallen off or were too rusty. All doors and door handles were made of metal and no anti-rust paint had been used since their construction. In most cases, the steps had cracks too.
- *Shortcomings in construction and technology design*: Either external or internal walls of all toilets had to be repaired as the salt in the air had leached into them. Furthermore, in the compost chambers of all ecological toilets, the ash that the households had been advised to throw in the chamber after usage had leached the inner walls. Many of the conventional toilets with septic tanks had also been causing great problems. Overflowing during the rainy season, they cause a terrible stench, and consequently, lie abandoned.
- *Lack of motivation and generation of negative externalities*: Some households had some problems with clogging of toilets and pipes, but they did not know to whom they could turn to repair the toilets. As a

[12] The 'Toilet Beauty Contest' is a tool initiated by FIN as an instrument to counter the problem of reticence of males to use toilets and generally to dispel the taboo attached to toilets as 'dirty places' which are 'shameful to clean and maintain'. It uses prizes for value enhancement of toilets through growing toilet gardens, decorating the toilet, introducing design innovations, and installing new structures within or outside the family and similar activities.

result, bad smell ensued from the toilet, and they stopped using it, in order to eliminate the smell pervading into the vicinity. This gave rise to the belief that if a toilet were present near a house, there would be a stench. As more households began to abandon their toilets, a band-wagon effect of imitation was triggered. Finally, as toilets came to be used more as storerooms; households where women's empowerment was low, also followed suit, regarding the need to consecrate a room for women's dignity as a waste of space.

- *Socio-cultural features disadvantaging ecosan toilets*: Furthermore, other socio-cultural norms disadvantaged ecosan toilets. For instance, people prefer conventional flushable toilets (also called 'flush and forget' toilets) or open defecation to ecosan toilets because in the former the faeces disappears after it is flushed (or is covered with leaves), which leaves the next user with a feeling of using a clean (or finding a new clean) place to defecate. On the other hand, in ecosan toilets, excrements fall into a pit, which are covered with ash, and the next user has to defecate over the same pit, which many find highly unpleasant. Moreover, ecosan toilets need frequent repairing: the pipes get clogged often with ash, the pipe that disposes of the (ammonia rich) urine into the ground has to be moved periodically to avoid damage to the crops, and the chambers containing the human waste transformed into dry compost have to be emptied periodically. These are all considered extremely repugnant tasks.
- *Association of ecosan with social discrimination*: Though a demand for toilets had been created with the construction of so many toilets, 'social tensions' were also generated between those with a conventional toilet and those with an ecological toilet. The 'ecosan' toilet model is a technological innovation that requires a basic understanding for its proper use, and users also need to make more efforts than those using other types of toilets to maintain them properly. Thus, the groups diffusing the conventional latrines pointed out to the inconveniences of the ecosan in order to promote their own model. Thereby, the households with the ecosan toilet came to feel that they had been dumped with a 'poor man's burden', while other households, which had been granted a conventional toilet had been treated much better, having received a modern toilet, such as the ones that city dwellers had, that would promote the latter's 'upward socio-economic mobility'.

Thus, the village of Kameshwaram, which had won the *Nirmal Gram Puraskar*, with toilets constructed by international agencies and private charities, today stands with incomplete sanitation coverage and abandoned toilets. Interestingly at the same time, the focus group discussions revealed that the majority of households recognize that using a toilet is not difficult in comparison to open defecation, that it is more convenient, and that it has positive consequences on health, social status, and personal dignity. They also acknowledge that ecological or urine diversion toilets are better for the environment than single-pit toilets or toilets with septic tanks, but the majority of households also argue that they are indeed more difficult to use than septic tank toilets. As for maintenance, almost all households agree that maintaining a toilet is not difficult. However, in case of problems with a toilet, a high percentage of households frankly claim that they would prefer waiting for an NGO to come and fix it, rather than call a plumber or a mason themselves.[13]

The results of this survey highlighted that all the sanitation drives in Kameshwaram had clearly created a demand for toilets. But for free. The villagers still prefer open defecation to paying for the repair of the toilets as they feel that toilets are a merit good. They are aware that the government is giving a lot of things for 'free.' They have been especially well served during political campaigns (by all political parties), when they have received alcohol, televisions, food processors, etc., for voting for a specific political party. Thus, their main suggestion is that the government cleans up its act, provides toilets in its political campaigns, or that NGOs repair and maintain the toilets for them for nearly free.

The village of Kameshwaram is not the only village in India in this situation – there are many *Nirmal Gram Puraskar* winners and non-winners all over the country in the same state. India is full of fossils of abandoned and stinking toilets, built by the State, international agencies, NGOs, non-profits, and charities.

Discussion of results

While the literature on BoP innovation offers valuable insights to better understand innovation in the BoP context, most of it tends to propose a product-centric view. In contrast, we have tried to contribute by studying the diffusion of BoP innovation in a more holistic fashion while underpinning important

[13] See http://www.friend-in-need.org/__media/Kameshwaram_survey_june2012.pdf for more details of survey.

factors for diffusion in a systemic view. In the context of the Indian system of innovation, through the case study, we have tried to specifically highlight some issues in the post-production phase of the BoP innovation that hinder or facilitate successful adoption among intended beneficiaries. Now, we share the main findings for the sanitation sector from the preceding sections.

Result 1

On the evolution of the sanitation sectoral system of innovation at the macro-level:

- The sectoral system of innovation in sanitation has got much more complex after liberalisation.
- Isolated, small-scale, and altruism-driven social entrepreneurs were the main engines of sanitation coverage in India before liberalisation.
- Medium and large-sized social enterprises are emerging and active in promoting sanitation coverage after liberalisation. They serve as the business partners of the State, international agencies, and large firms.

We illustrate this proposition with Fig. 10.1, depicting the sanitation sectoral system of innovation before and after liberalisation.

In the pre-liberalisation period, two types of actors diffused sanitation as an innovation: the State and isolated social entrepreneurs in a parallel process without any interconnection. A social entrepreneur is described as a creative,

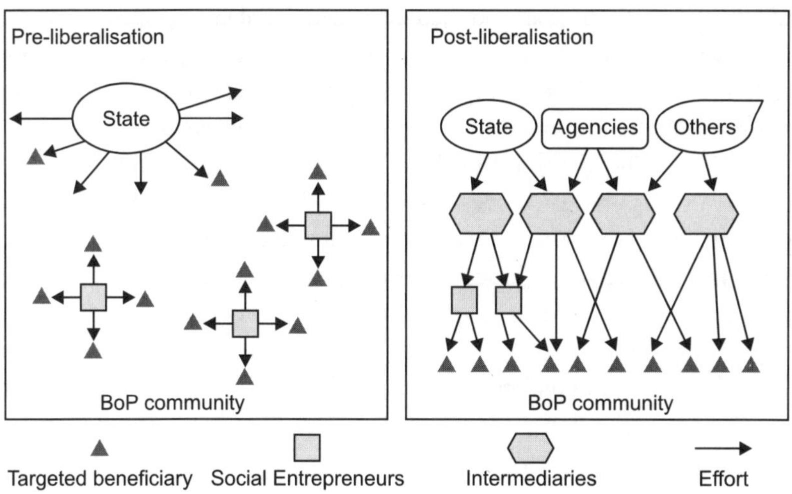

Fig. 10.1: Evolution of the Indian national system of innovation with respect to sanitation

radical, and effective individual who is committed to serve society and achieve specific 'social missions' or 'social projects' to improve the quality of life of a marginalised group or a victim group or a needy group (Shaw 2004). Social entrepreneurship is the intent of an entrepreneur to create social value instead of shareholder value (Zadek and Thake 1997). Initially, social entrepreneurs like Bindeshwar Pathak and Paul Calvert operated in quasi-isolation, mingling mainly with the target beneficiaries. They built and diffused the toilets, and raised funds on their own, without contracts from the government. They managed to survive because they were able to attract resources and personnel with their message, their personal charisma, and their efficiency. They were not only motivated by their social mission but they were also willing to take the financial risk of raising funds to implement their actions either from the target communities or from their circle of family and friends. Therefore, their own incomes were modest and variable, at least during the early stages of their work.

When social entrepreneurs take ownership of activities performed by them in the form of an enterprise, their organisation is referred to as a social enterprise (Shaw 2004; Hines 2005). A social enterprise finances itself through finding opportunities by which 'social missions' and 'social projects' can be implemented. They emerge when both a 'problem' and the 'resources to solve the problem' have been identified (Lévesque 2004). In the post-liberalisation period, the government, international agencies, religious charities, and civil society charities entered as financiers of sanitation. This opened up business opportunities for local NGOs and both small and medium-sized social enterprises.

Like the early social entrepreneurs, these social enterprises of the post-liberalisation period seek to improve the quality of life of the poor through increasing sanitation coverage. But the new wave of social enterprises also strives to maximise the income generated by securing contracts from large financiers. They seem to also have a tendency to switch their social missions as market opportunities evolve. Thus, as the achievement of basic socio-economic targets gains in importance (e.g., through the Millennium Development Goals), the number and variety of social enterprises are likely to increase. The ratio of social enterprises to social entrepreneurs has increased greatly after liberalisation. Indeed, this had been predicted more than two decades ago by Peter Drucker in his seminal work on the '*The Age of Social Transformation*' (1994) that social sector organisations will become partners with government to democratise the creation and diffusion of innovations.

Result 2

Reasons for the failure in diffusion of toilets as a pro-poor innovation at the micro-level:

- From the supply side, failure of sanitation initiatives can stem from problems in a host of issues, such as technology design, product design, construction design, delivery design, and missing institutions, capabilities or linkages.
- From the demand side, failure of sanitation drives can stem from lack of awareness and knowledge to use and/or maintain a toilet, lack of willingness to pay for repair, value perceptions of different toilet technologies, and herd behaviour in imitating what the neighbours do.
- Interconnections and interplay between demand and supply side actions and outcomes.

In terms of technology design, many standard pit latrines in high water areas are abandoned because of waterlogging problems. They are also abandoned in dry areas because of lack of sufficient water to maintain the toilet. A typical example of a failed product design is the initial model under the Total Sanitation Programme. The intended beneficiaries were just given a toilet slab that was fixed on the ground. They were expected to build the rest, which of course many never did.

Construction design defects were rampant in the sanitation initiatives in Kameshwaram. Badly constructed toilets, a familiar sight all over India, seems to stem from commonly held beliefs that every toilet is more or less safe and that anyone can learn to build a toilet in a week. Thus, local construction companies, working in collaboration with NGOs, hire unskilled people who do not know how to pay attention to quality or details. Moreover, such sanitation initiatives tend to face time and budget restrictions, having to follow targeted processes with little attention to quality or aesthetics. Thus, most of the toilets for the poor in a sanitation drive also tend to be ugly clones of one another.

The ecosan toilet is best suited to high water table areas such as Kameshwaram, but it requires additional efforts from the user for proper use and maintenance. Thus, 'education' has to be an important element integrated in delivery designs. At present, this usually takes place in the form of a workshop with the intended beneficiaries. The drawback is that it cannot be assumed that anyone who has attended a one-day workshop has fully mastered the knowledge required. This was clearly illustrated in a survey by FIN. The first NGO partner of FIN had made efforts over a period of time to educate the villagers. The toilets built by that NGO were seldom abandoned despite the need for repairs. On the other hand, most of the toilets built by the second NGO partner, which had held the usual one or two short awareness building workshops were indeed deserted.

An ecosan toilet can be considered as an innovation that is appropriate from the technological and socio-economic points of view, meeting criteria such as safety, acceptability, and affordability. However, diffusion of ecosan toilets was impaired because there were no local organisations for maintenance. This important factor is illustrated in Fig. 10.2.

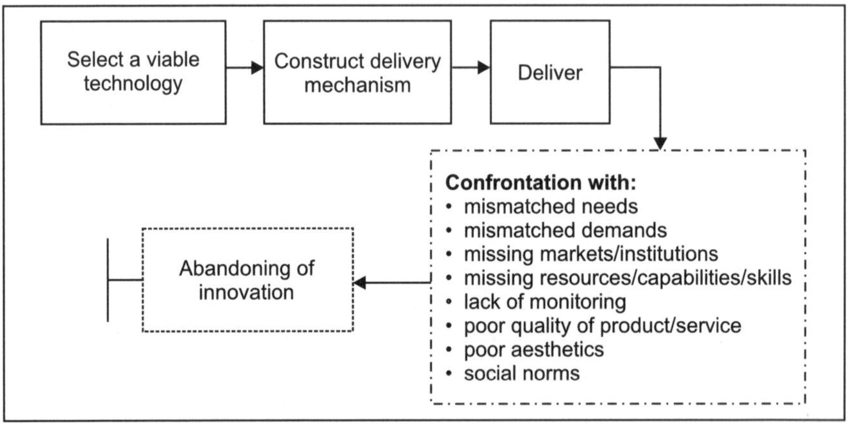

Fig. 10.2: Failure of pro-poor innovation diffusion model from the supply side

On the demand side, despite the fact that the ecosan toilet was best suited to Kameshwaram's high-water table, it was no longer 'desired' by the target population once the toilet with the attached septic tank became available. Then, ecosan was regarded as 'less of a status symbol' as compared to a septic tank model. The provider's value proposition did not match the BoP consumer's perception of added value. Other gaps in the demand included lack of skills to use the innovation, lack of motivation, lack of purchasing capacity, unwillingness to pay, etc. In other words, a host of challenges for adoption that stem from the demand side are due to problems of expressions of demand and their mismatch with perceptions of the value of the innovation. For these problems too, there is a need for solutions such as de-skilling the products and/or enhancing skill/awareness level through sustained education as well as innovative financial models that are compatible with the financial sensitivities of the users.

Finally, there is a constant interplay between the forces of demand and supply that reinforce the emergence of negative outcomes (or positive outcomes), as summarized in Fig. 10.3.

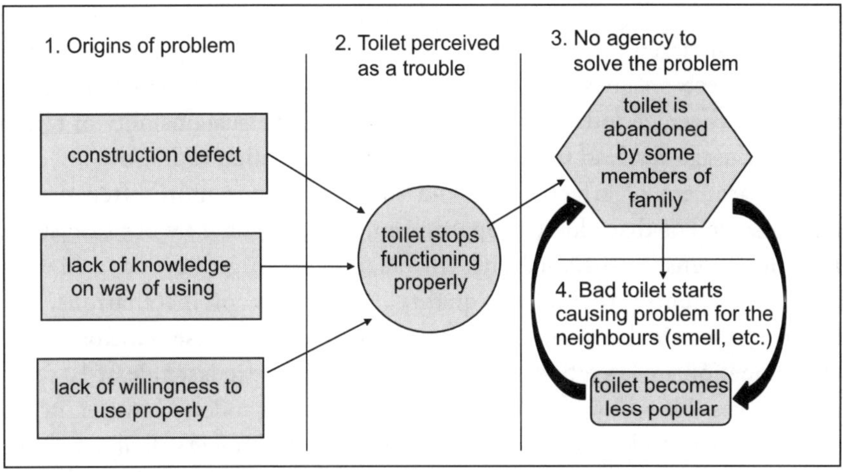

Fig. 10.3: Demand side dynamics in the abandoning of toilets

Result 3

Management of public–private partnerships:

- Application of modern management principles by private partners can be helpful in enhancing efficiency of social enterprises. However, without the application of sustainability audits, monitoring mechanisms, and incentives systems, sanitation initiatives of social enterprises may not yield effective and sustainable outcomes.
- For sustainable sanitation coverage, new micro-social-businesses have to be created at the village level to provide accompaniment and maintenance.

The case study of Kameshwaram shows that behind a sanitation drive, there is a complex network of actors, comprising financiers, facilitators, service providers, and field staff. Only the last category interacts closely with the target community, and that too mainly during the time of the project implementation.

The financiers do not necessarily have the knowledge or skills to penetrate the BoP market, and hence, initiate direct or indirect contractual business agreements with social enterprises to implement sanitation projects. Most of the social enterprises (unlike local social entrepreneurs) do not develop intimate relations with intended beneficiaries, but they usually have an excellent knowledge of the dynamics of project implementation in rural India. Thus, such social enterprises act as intermediaries between large financiers and the

BoP community. Financers push social enterprises to implement projects and achieve their targets for commercial reasons; however, the long-time sustenance of such projects are questionable. Our case study clearly showed that increasing sanitation coverage today without accounting for the sustainability of toilets and their usage can lead to more points of contamination tomorrow.

The problem of effectiveness and sustainability is more often due to mismanagement than due to opportunism, as highlighted by our case study. In order to ensure sustainability, financiers have to go beyond 'achieving targets efficiently' to include 'quality' and 'safety' of installations. For instance, a sustainable sanitation system is one that minimises environmental contamination, makes efficient use of water available, can last at least 10 years, and is accessible to low-income groups in need of a toilet. However, in the village studied, faulty construction, inappropriate technology design, and the use of hazardous materials such as asbestos are leading to abandoning of toilets after 18–24 months. Thus, to ensure sustainability, it is necessary to take into account monitoring, accountability, and standardisation of models.

For sustaining their business, social enterprises move wherever a business opportunity takes them, and in pursuing the business interest, they deliver some social benefits. Like bees, which move from flower to flower to gather pollen, social enterprises move from opportunity to opportunity, trying to make the BoP garden bloom, and also produce honey. In the garden of sanitation, such bee-like actors are useful to maintain toilets, repair them, and accompany users – as there has to be a service-providing unit in each village to which the local residents can turn. At the same time, such organisations should not focus uniquely on toilets, because again the caste stigma will be attached to it and no one outside of the manual scavenging caste will join, and if none of the latter community is present in the village, the organisation will go unmanned. Therefore, toilet maintenance should come under a more general unit, which takes care of maintenance of a number of other pro-poor technologies as well. Again, utmost care must be taken to ensure that within the organisation, there is multi-tasking so that all members take care of toilets in addition to their other duties.

Conclusion

At the 'World Toilet Summit' held in New Delhi in India, the former President of India, Abdul Kalam, noted that complete sanitation coverage could save at least Rs 5 billion every year for India. He based his statement on estimates that the monetary costs of treating health problems related to sanitation is around Rs 2 billion and the loss of working days due to associated illnesses about Rs 3 billion[14] in India. Thus, sanitation is clearly a worthwhile challenge to tackle, not only from the humanitarian perspective, but also for contributing to economic growth.

The objective of this chapter was to understand the root causes of the low adoption of toilets in rural India, considering 'toilets' as an innovation for households without the same. Even in rural India, the low adoption of toilets is giving rise to health problems, because the population density is too high vis-à-vis the vegetal coverage required for ensuring privacy and safe composting. In this sectoral system of innovation, between the 1950s and the 1990s, isolated and committed social entrepreneurs formed and led small teams to make a difference. Then, following economic liberalisation in the 1990s, a variety of financiers for the diffusion of toilets appeared on the supply side. The government is the chief financier. In second rank, come the international agencies. Lastly, there are religious charities and do-gooders whose actions are very difficult to monitor or control.

The new set of financiers is creating opportunities for the extremely competitive market of social enterprises in the social sector market, including sanitation. Thus, boundaries between social entrepreneurs and the directors of social enterprises are becoming fuzzy. From the isolated efforts of social entrepreneurs and the ineffective provisioning of the Indian government, there are now vertically coordinated supply chains manned by NGOs, social enterprises, and social entrepreneurs, forming delivery platforms with financiers. Such collective efforts have given 251 million households access to improved sanitation since 1990.[15]

At the same time, it is not clear, whether these numerous sanitation drives will lead to sustainable toilets or millions of new points of contamination in the future. The case study of Kameshwaram showed that technology and even financial investment are only two components of the solution for successful innovations for the BoP group. Many of the toilets being constructed are not safe or sustainable, because of faulty design, faulty location, and poor construction. Rather than opportunism, it is cost, time, and target-driven pressures in the

[14] http://www.medindia.net/news/Better-Sanitation-can-Help-India-Save-Five-Billion-Rupees-Every-Year-APJ-Abdul-Kalam-28819-1.htm.

[15] WHO/UNICEF. 2012. Joint Monitoring Report: Progress on Drinking Water and Sanitation; http://www.wssinfo.org/fileadmin/user_upload/resources/1278061137-JMP_report_2010_en.pdf.

context of international recession that are driving organisations to quickly build toilets and move on.

To achieve large-scale improvements in sanitation coverage, it is indeed necessary to adopt medium and large-sized social enterprises as business partners. However, the zeal of the small-sized social enterprises, led by social entrepreneurs, has to be maintained through well-formulated contracts that ensure sustainability. Medium and large-sized social enterprises are not 'emotional' entities – their mission is to prove their effective performance in constructing toilets per se, but not in constructing safe toilets or in toilets that will last over a decade or toilets that are completely used. Thus, unless funders insist on 'quality' or 'effective use,' the social enterprises will not have any incentive to ensure them on their own. This creates a challenge for funders also, because they too have a short-term to a very short-term view of impact.

To conclude, despite all the progress made and the rhetoric of all economic actors on sanitation and sustainability on paper, 'sanitation' remains a taboo subject. Despite all the funds, the variety of financiers, and the burgeoning social enterprise sector, there is still a need and a place for social entrepreneurs and civil society in the sanitation sector. Only they have the 'emotional commitment' to adopt a long-term vision of development with dedication to quality, which are unlikely to maximise pay-offs or profit in the short-run.

References

Anderson, J. and N. Billou. 2007. 'Serving the World's Poor: Innovation at the Base of the Economic Pyramid.' *Journal of Business Strategy* 28 (2): 14–21.

Banerjee, A. V. and E. Duflo. 2007. 'The Economic Lives of the Poor.' *Journal of Economic Perspectives* 21 (1): 141–67.

Bhatnagar, V. and S. K. Gupta. 2008. 'India—Secured landfills: The Bucket at The End of The Solid Waste Management Chain.' *Water and Sanitation Program*. Washington DC: The World Bank. Accessed: 12 September 2012; http://documents.worldbank.org/curated/en/2008/04/16708011/india-secured-landfills-bucket-end-solid-waste-management-chain.

Brugmann, J. and C. K. Prahalad. 2007. 'Cocreating Business's New Social Compact.' *Harvard Business Review* 85 (2): 80–90.

Christensen, C. M., H. Baumann, R. Ruggles, and T. M. Sadtler. 2006. 'Disruptive Innovation for Social Change.' *Harvard Business Review* 84 (12): 94.

Drucker, P. 1994. 'The Age of Social Transformation.' *The Atlantic Monthly*, November.

Franceys, R. and A. Weitz. 2003. 'Public–Private Community Partnerships in Infrastructure for the Poor.' *Journal of International Development* 15 (8): 1083–98.

Ghobadian, A. 2004. *Public–Private Partnerships: Policy and Experience*. Basingstoke, Hampshire; New York: Palgrave Macmillan.

Hart, S. 2005. *Capitalism at the Crossroads: The Unlimited Business Opportunities in Solving the World's Most Difficult Problems*. Upper Saddle River, NJ: Wharton School Publishing.

Hines, F. 2005. 'Viable Social Enterprise: an Evaluation of Business Support to Social Enterprises.' *Social Enterprise Journal* 1 (1): 13–28.

ILO. 2002. *Women and Men in the Informal Economy: A Statistical Picture*. Geneva: International Labor Organization.

Jeffrey, P. and R. A. F. Seaton. 2004. 'A Conceptual Model of Receptivity Applied to the Design and Deployment of Water Policy Mechanisms.' *Environmental Sciences* 1 (3): 277–300.

Katz, E. 1961. 'The Social Itinerary of Technical Change: Two Studies on the Diffusion of Innovation.' *Human Organization* 20 (2): 70–82.

Kotler, P., N. Roberto, and T. Leisner. 2006. 'Alleviating Poverty: A Macro/Micro Marketing Perspective.' *Journal of Macromarketing* 26 (2): 233–39.

Letelier, M. F., F. Flores, and C. Spinosa. 2003. 'Developing Productive Customers in Emerging Markets.' *California Management Review* 45 (4): 77–103.

Lévesque, B. 2004. *Entrepreneurship collectif et économie sociale: entreprendre autrement*. Montréal : Alliances de recherche universités-communautés en économie sociale.

London, T., D. A. Rondinelli and H. O'Neill. 2006. 'Strange Bedfellows: Alliances Between Corporations and Nonprofits.' In *Handbook of Strategic Alliances,* edited by O. Shenkar and J. J. Reuer, 353–67. Thousand Oaks: Sage Publications.

Mendoza, R. and N. Thelen. 2008. 'Innovations to Make Markets More Inclusive for the Poor.' *Development Policy Review* 26 (4): 427–58.

Prahalad, C. K. 2005. *The Fortune at the Bottom of the Pyramid: [Eradicating Poverty through Profits: Enabling Dignity and Choice through Markets]*. Upper Saddle River, NJ: Wharton School Publications.

Prahalad, C. K. and S. Hart. 2002. 'The Fortune at the Bottom of the Pyramid.' *Strategy+Business* (26): 54–67.

Ramani, S. V. 2008. 'Playing in Invisible Markets: Toilet Innovations and Empowerment.' UNU-MERIT Working Paper Series no. WP2008-012. Maastricht, the Netherlands: UNU-MERIT.

Ramani, S. V., S. Sadre Ghazi, and G. Duysters 2012. 'On the Diffusion of Toilets as Bottom of the Pyramid Innovation: Lessons from Sanitation Entrepreneurs.' *Technological Forecasting and Social Change* 79 (4): 676–87.

Reddy, N. M., J. D. Aram, and L. H. Lynn. 1991. 'The Institutional Domain of Technology Diffusion.' *Journal of Product Innovation Management* 8 (4): 295–304.

Rogers, E. M. 2003. *Diffusion of Innovations*. New York: Free Press.

Saxena, N. C. 2005. 'Bridging Research and Policy in India.' *Journal of International Development* 17 (6): 737–46.

Schumacher, E. F. 1973. *Small is Beautiful : A Study of Economics as if People Mattered.* London: Blond and Briggs.

Shaw, E. 2004. 'Marketing in the Social Enterprise Context: Is it Entrepreneurial?' *Qualitative Market Research: An International Journal* 7 (3): 194–205.

Stewart, F. 1977. *Technology and Underdevelopment.* London: Macmillan.

UN-DESA. 2003. 'India: A Case Study: Related Best Practice or Lessons Learned in Water & Sanitation.' Accessed: June 2012; http://www.un.org/esa/sustdev/sdissues/water/casestudies_bestpractices.pdf.

UNDP (United Nations Development Programme). 2008. *Creating Value for All: Strategies for Doing Business with the Poor.* New York: UNDP.

UNICEF. 2012. *A Snapshot of Sanitation, Hygiene and Drinking Water Safety in South Asia-2012 Update.* Kathmandu: UNICEF. Accessed: 10 November 2012; http://www.unicef.org/rosa/ROSA_Sanitation_Snapshot_Final_Final_04_09_2012.pdf.

WHO/UNICEF. 2012. 'Progress on Drinking Water and Sanitation: 2012 Update.' Joint Monitoring Programme for Water Supply and Sanitation. New York: WHO/UNICEF.

Zadek, S. and S. Thake. 1997. *Practical People, Noble Causes: How to Support Community-based Social Entrepreneurs.* London: New Economics Foundation.

CHAPTER 11

Pro-poor Innovation Making, Knowledge Production, and Technology Implementation for Rural Areas

Lessons from the Indian Experience

Dinesh Abrol

Introduction

In India, pro-poor innovation making demands efforts, especially for the introduction of solutions that peasants, landless rural workers, and artisans can adopt to protect and improve their own livelihoods and well-being. The population of rural India has increased at the rate of 1.15 per cent per annum, from 743 million in 2001 to 833 million in 2011.[1] Even now, 68.8 per cent of the Indian people live in rural areas. Rural incomes have been growing at a slower rate than urban incomes. At the aggregate level, the rural share of incomes in total incomes (net domestic product) has been declining since the 1970s. The proportion of persons employed in agriculture remains high. About 63 per cent of the workforce in rural India is still engaged in agriculture for its livelihood. In the decade of 1999–2010, rural India showed higher rate of growth of employment in industries. Within industries, the share of rural manufacturing in total manufacturing national domestic product (NDP) rose from 25 per cent to 42 per cent between the period of 1970–71 and 1999–2000 and then remained constant in 2004–05. But the modern and more dynamic sectors of communication, finance, and business services showed a declining share of incomes. Benefits of the dynamic and high earning service sector-led growth are not reaching the rural areas (Unni and Raveendran 2007).

[1] Census of India, http://censusindia.gov.in/2011-commoncensusdataonline.html.

With agricultural growth fluctuating and employment guarantee scheme still having an impact only on a small section of the workforce, in overall terms, employment growth in rural India has not been very encouraging. With the poor growth of agriculture and agrarian distress looming large, even the farm households in rural India are dependent on the non-farm sector (NFS) for their livelihood. Much of these activities in the rural non-farm sector (RNFS) is of low productivity, and hence, earns low income, often termed as informal employment. Though there has been a decline in self-employment over the decade from 55 per cent to 45 per cent in 2009–10 in rural areas, about half of the informal non-agricultural employment is still self-employment. Nearly all of self-employment is own account work, operated with family labour. Such activities are either traditional industry, such as pottery, weaving, agro-processing on a small scale, or small shops and eating places in rural areas. Much of this is unviable and only provides a subsistence living. However, this component is also declining rapidly. Informal employment has increased in rural areas in the form of casual labour in manufacturing and construction.[2]

This raises a serious question about the nature of structural transformation occurring in rural India. In the post-Green Revolution period, especially since the 1990s, there is a technological stalemate in the Indian agricultural sector. Owing to rising costs of inputs, the cost of cultivation has been increasing over time. In the absence of proper remunerative price mechanism, the farmers are increasingly exposed to the price volatility in output markets. The ratio of gross value of output to its cost of cultivation has been declining over time for the Green Rrevolution crops. Because the rural poor cannot depend for the maintenance of their livelihoods on the existing socio-technical regime of high external input system of agriculture, in our view, a change in paradigm of agricultural development is now quite essential for the purpose of technological transformation of rural areas. To list the major problems with regard to the achievement of productivity through the existing socio-technical regime, the major ones are over and disproportionate use of chemical fertilisers, pesticides, and water, depletion of groundwater resources in semi-arid and rain-fed region, insecurity of tenant farmers, increasing cost of cultivation, and inadequate public investment.[3]

Linkages existing at the level of production system between the traditional rural non-farm economy and the emerging farm sector have become weaker.

[2] See the *India Rural Development Report*, 2012 for details.

[3] See the *India Rural Development Report*, 2012 for details.

As big business is rapidly penetrating the economy of rural India and replacing the public sector, it is increasingly transmitting external shocks and becoming a cause of instability in agriculture. Since we expect this new force to actively utilise pro-poor innovation making activities in rural India, it is necessary for the peasants, landless workers, and artisans to challenge this new force socio-technically to maximise their resilience in the farm and non-farm activities in an integrated fashion. For the viability of the increasing number of small and marginal holdings, collectives and cooperatives will have to be promoted to give them resilience against natural variability and external economic shocks by enhancing their bargaining power vis-à-vis big businesses. As producers in the local economies, peasants, landless rural workers, and artisans are fast losing competitiveness and competitive advantage. They are also unable to improve their access to resources, skills, knowledge, capabilities, and markets. For them to be able to competitively use the processes of pro-poor innovation making, there has to be an evolution towards new paradigms of agriculture and agro-industrial production in the local economies, so that already rudimentarily inter-linked farm and RNFS occupations can also develop as systems – enabling peasants, artisans, and landless workers to achieve economies of scale and scope and remain competitive. Practitioners of pro-poor innovation making must create more of such opportunities for the protection and promotion of livelihoods of the poor (Abrol 2009a,b, 2004).

In this chapter, the claim is that the challenge of promotion of pro-poor innovation making should be viewed as the problem of management of transition to a new path of rural development by shifting to the practice of paradigms of sustainable agro-industrial production. Pro-poor innovation making calls for an exploration of context-specific paradigms of agro-industrial production in which peasants, artisans, and rural labour can successfully implement diversity-sensitive agro-ecological approaches for rural development and emerge as social carriers of pro-poor innovations in the emerging Indian context of rural development. Currently, the discipline of innovation studies is rapidly moving forward to focus on the problems of transition management in order to usher in sustainable development in high-income capitalist countries, but the challenge of harnessing technology and innovation for poverty alleviation in developing countries remains under-examined, including by scholars of these countries. In particular, the relevant political aspects in the case of development of rural areas in developing countries require more study. There is an urgent need to learn from the historical experience and explore new alternate trajectories.

In this context, the present chapter examines the experience of implementation of pro-poor innovation making at the level of efforts undertaken, in particular for the development of rural industries to improve rural livelihoods since the beginning of Indian Independence. It traces the evolution of the three main phases of pro-poor experiments in innovation making with respect to the development of rural industries. First, during the 1950s and 1960s, there was a lot of diversity in initiatives, and these were put into place by a variety of actors. After the mainstream Nehruvian leadership decided to give priority to the development of basic and capital goods industry, the same leadership also chose to accommodate in the planning process, the technological upgrading of economic activities being undertaken by the poor in India. Second, a trajectory of gradual niche 'involvements' to uplift the poor followed during the 1970s and 1980s in which the activists of different political traditions took part and tried innovation making. Finally, with liberalisation after the 1990s, the strategy of State, including the knowledge institutions vis-à-vis pro-poor, changed and the scenario of innovation making underwent a process of adjustment and gradual decline. With the introduction of latest science, technology, and innovation (STI) Policy of 2013, the efforts for pro-poor innovation making with the help of the corporate sector are again expected to get a boost (Government of India 2013).

Today, the actors promoting pro-poor innovations in India include not only the State, public sector research agencies and universities, NGOs, and civil society groups, but also new actors and very unlikely champions of the poor – namely large firms, including foreign multinationals. They are sought to be the carriers of development through catering to the effective demands of the poor. But the practice of 'bottom of pyramid (BoP)' or 'frugal' innovations has not been effective in satisfying even the consumption needs of the poor (Kaplinsky 2011). Moreover, findings of the United Nations Development Programmes (UNDP) survey of inclusive business also indicate that large corporations, national or transnational, are mostly unable to include the poor as producers (UNDP 2008). Corporate self-interest dominates. Either through vertically integrated supply chains or as networked systems, corporations are not able to offer to the poor the benefits of competence enhancement and organisation of local production. Reasons can be traced to their failures to build partnerships with the poor as producers, higher transaction costs given the diversity of interactions, difficulties in interpreting local demand, inability to deal with heterogeneity, and underestimation of the investment required in local capacity building for deeper inclusion of the peasants, artisans, and rural workers (Arora and Romjin 2009).

Clearly, the challenge of pro-poor innovations has become more complex. Therefore, with a view to understand the implications of the newly emergent scenario for the future interventions, we will focus on the relevant political aspects of pro-poor innovation making in India. The two crucial political processes, 'the politics of knowledge production' and the 'politics of technology implementation', are basically analysed for their role in the promotion of capabilities and opportunities that became available for the development of pro-poor socio-technical systems because in the Indian context of capitalist development, the State was able to also accommodate in a limited way the mobilisation of knowledge production and technology implementation for the benefit of pro-poor innovation making during the post-Independence period. However, we also show that the efforts put in by these traditions for pro-poor innovation making using State support were subject to the impacts of the closing and reopening of the political economy space.

In what follows, we discuss the role and contribution of the political traditions in the mobilisation of people for pro-poor innovation-making. First of all, the focus is on the socio-technical frames used by the leaders of these political traditions to guide the course of societal interventions for the achievement of a pro-poor socio-technical success in India. Now, the number of social carriers of innovation who can directly contribute in the mobilisation of the people for pro-poor innovation making is greater than ever before. In this connection, we first ask – what are the lessons that can be learnt from the past efforts of the State and the social movements to stimulate pro-poor innovation generation in India? Second, how well or inadequately did the societal interventions and the State's innovation policy perform in the past? What determines success? What causes failure? Finally, what kind of measures does the State need to take to make the transition to a new path of agro-industrial development?

The remainder of this chapter is organised as follows. The first section presents a brief outline of the accommodation of the pro-poor innovation making efforts by the three main political schools of thought in India. The next section then attempts to answer the central question of learning appropriate lessons through the presentation of three detailed case studies of the 'politics of institution building in higher education,' the 'promotion of knowledge production in State sector R&D agencies', and the 'innovation system building by social movements for pro-poor technology implementation.' The subsequent section discusses the main results. The final section concludes.

Evolution of the diversity in the making of social carriers of pro-poor innovation

The three main schools of political thought: 1930s–1950s

During the formative years of politically independent India, the freedom movement was a key influence on the building of institutions for learning, competence building, and innovation making. Indeed, the politics conducted over the vision and strategy of technology development fruitfully utilised the tools of constructive action, unity, and struggle. During the pre-Independence period, the people of India started to develop institutions. There was much space for the making of innovations on the basis of different types of experiments, motivated by different orientations, ideological visions, and understanding of political strategies. As a result, there was a lot of diversity in terms of institution building for knowledge production, knowledge mobilisation, and innovation making. In many ways, the directions of 'pro-poor innovation' activity were constituted in the course of the debates started between the people of Gandhian, Nehruvian, and Left orientation during the formative period of 1930s to 1950s. Even today, among the active social carriers of pro-poor innovations, many of the actors are known to draw explicitly their ideological inspiration from the ideas and initiatives that emerged during the formative period of post-independent India. The politics of attainment of counter-hegemony against colonialism required the leaders of the national freedom movement to build a multiclass alliance on the ground. Therefore, the leaders of post-independent India were very much ready to practise diversity and people-oriented directions in the conduct of their politics of knowledge production.

The Gandhian, Nehruvian, and Left political traditions differed radically from each other in terms of the conception of 'socio-technical imagination', 'vision of path of development', and the 'social carriers of innovations'. The main points of view existing in the immediate aftermath of Independence are summarized subsequently in Table 11.1 in terms of socio-technical imagination and the vision of path of development of Indian economy, strategy of industrialisation, mode of technological and organisational transformation, socio-technical framing of the problem of pro-poor innovation making, etc. Though it is a drastic simplification of the many points of view within each group, it is necessary to highlight their common features and major differences.

Table 11.1: Panorama of political philosophies

Shaping of development process	*Gandhian orientation*	*Nehruvian orientation*	*Left orientation*
Socio-technical imagination, vision, and path of development for Indian economy	A self-sufficient village economy; small is beautiful; redistribution of land through *Bhoodan*; non-party system of democracy; adopting anti-statist stance after facing the crisis of Nehruvian path; neo-Gandhians resorting to post-modernist/neo-traditionalist politics	Priority to establishment of large technical systems as a blueprint of development; planning for capitalist development with land redistribution attempted mainly as rhetoric and its vision of the public sector in commanding heights; small-scale industry; after the emergence of big business as a major player in the economy and with the deepening of the crisis of Nehruvian imagination, shift to market governance	Centrally coordinated workers-managed large technical systems (LTS); in-transition planning as the agency of State capitalism as anti-imperialist step; with radical redistribution of assets, especially land, as transitional demands; later opting for decentralised planning for participatory local area development; State's accountability to the people.
Strategy of industrialisation	Decentralised industrial development; 'textile-first-type' strategy	industrialisation and establishment of basic industries and capital goods sector; import substitution; protection of cottage industries; after the experience of crisis, export promotion; trade as an engine of growth	Extended support to heavy industry strategy development of home market via cottage industry employment and irrigated agriculture for food grains supply
Mode of technological and organisational transformation	Traditional and local knowledge; priority to individual small-scale industry; cooperatives in consumer/ financial/other service areas	Import substitution via replication of duplicative/imitative technological strategy as more of replacement of import effort; S&T self-reliance and technological learning confined to strategic sectors	S&T to meet basic needs of people; self-reliance in large systems (LTS), irrigation, rail, road, telecom, steel, heavy industries, rail, roads, energy, health in public sector; small scale in consumer goods; in agro industry workers cooperatives

Contd.

Contd.

Shaping of development process	*Gandhian orientation*	*Nehruvian orientation*	*Left orientation*
Socio-technical framing of the challenge of pro-poor innovation making	Upsizing of traditional small-scale industry and peasant-based agriculture	Downsizing of modern technology to make it 'appropriate' for small scale / tiny/micro enterprises	Using public sector-based LTS to meet the needs of irrigation and energy, rail, road, telecom; support workers' cooperatives in industries/ services amenable to small scale; trying in Kerala decentralised planning of area development and partial support for People's Science Movement (PSM) experiments in the development of technological models for local economy development
Social carriers of production and innovation for mainstream development	Landowners and big businesses as trustees of wealth; individual small producer	State sector in basic industries and strategic areas; foreign and Indian big business in consumer goods; individual small producer in retail; access to irrigation and power, roads, etc.	Minimise dependence on big businesses; more reliance on public sector; small-scale business in retail and workers cooperatives
Priorities for rural development	Protection of village industries		land reforms; access to irrigation and power; education and health to poor as priority; target poor peasants by adapting technology of Green Revolution
Priorities for the poor	Local self-government; sanitation	Area development; small-scale business; targeting of credit and information for poverty alleviation; provide school education for all	Public distribution of food; availability of transport and health through public sector agencies; provide universal education

Source: Created by the author

During the 1950s, fluidity in the notions of development, national planning, and self-reliance was still visible at the level of the political leadership. Not only experiments for innovation making were possible but their results were also incorporated in the plans under preparation. Ideas of diverse nature with regard to institution building for knowledge production were considered in socially inclusive ways in the committees and commissions carefully set up with appropriate political composition, keeping in mind the aspirations and interests of the people.

Accommodation of the pro-poor innovation making challenge within the political traditions during the 1950s and 1960s

At the time of Independence, a pre-capitalist traditional manufacturing for subsistence living was the dominant mode of production. Cottage industries existed in large numbers as family labour oriented labour units. Moreover, as India entered a period of transition to modern form of large-scale industry, there was a need to promote cottage industries for employment considerations as the economy was in no position to absorb the surplus labour that was being released. Therefore, viability, of cottage industries was important and the mainstream Nehruvian leadership accommodated the Gandhian tradition through the policy of promotion and protection of the cottage industries. Thereafter, governmental protection, economic support, and incremental technological upgrading of traditional manufacturing became the accepted features of the Nehruvian state policy. The cottage industries sector acted as a bargain sector; its existence allowing the Nehruvian leadership to bargain with both the big business as well as the landed gentry.

In terms of its 'socio-technical' frame, the Gandhian imagination followed the heuristics of upsizing by modernising the indigenous technology. The individual producer was sought to be made competitive by upgrading the local/traditional technology. When Schumacher inspired appropriate/intermediate technology movement, the practitioners were using the same frame of Gandhian tradition. During the decade of 1970s, this frame became the heuristics of State-sponsored appropriate technology movement for a wide range of departments and agencies. The governmental agencies encouraged the practitioners to upgrade the technologies embedded in indigenous /local knowledge, local raw materials, skills, and capabilities to make the individual producer competitive. In the Gandhian tradition, the holistic meant an individual producer completing the production process with none or minimum division of labour. In this movement, the 'small was beautiful' but the small could not be made powerful.

Peoples' participation in setting agenda for innovation making activity was also officially accommodated in the planning process by the State formed of mainstream Nehruvian leadership. The State accepted and agreed to give support to the activists of Gandhian orientation for the protection, modernisation, and development of cottage industries in parallel to the mainstream programmes envisaged for industrialisation. In the strategy of industrialisation formulated by Mahalanobis[4] during the Second Five-year Plan, Gandhian experiments got encouragement directly from the State via the Khadi and Village Industries Commission (KVIC). The KVIC had the benefit of policy space and financial support for the practice of 'constructive action' for pro-poor development. Activists of the Gandhian orientation could use the support of KVIC to implement programmes for the development of cottage industries and strive to meet the challenge of technological upgrading of traditional manufacturing.

Scaling up from cottage industries, the micro, small, and medium-scale industry (MSMEs) was sought to be developed through 'downsizing of modern technology'. This technology was to be implemented by the small-scale industry. In order to make the technology appropriate for tiny/micro/small, and medium enterprises, the mainstream Nehruvian thinking actually promoted this frame of 'appropriate'/'economical/'commercial' sizing and designing everywhere in the country. This socio-technical frame was implemented not only in the Council of Scientific and Industrial Research (CSIR) but also in the Indian Council of Agricultural Research (ICAR). It was implemented through the Department of Textiles, Small-scale Industry Development Corporations, and National Small-scale Industry Corporation. In the mainstream science and technology (S&T), it was these heuristics that provided innovators the rules of the thumb on how to adjust with the social carriers of local ingenuity, *jugaad*,[5] and appropriate technology movement. This socio-technical frame allowed the State to accommodate the role of MSMEs in the emerging mainstream economy for the first four decades of post-independent India.

Similarly, the State accommodated the Left through supporting the public sector and small-scale industry from the beginning of the 1960s in plans and policies. The Left tradition supported the promotion of public sector and small-scale industry. By the end of the 1960s, the measures like protection of small producers, promotion of public sector, restrictions on foreign direct investment,

[4] P. C. Mahalanobis. 'Approach to Planning in India', as cited in Myrdal (1968).

[5] Frugal and flexible innovation coming from a Hindi word meaning an improvised solution born from ingenuity and cleverness.

bank nationalisation, and many other such policy developments were an integral component of the policies enunciated by the mainstream Nehruvian tradition in the country. Because of the massive success that the State sector had achieved in the case of the erstwhile Soviet Union in respect to industrialisation, the State was willing to offer commanding heights to the public sector in industry, finance, and many areas of development. Thus, in line with the Left orientation, State ownership of basic industries was enunciated as a principle of policy, and it received a high level of priority in the process of planning up to the mid-1980s. In terms of the socio-technical frame, the Left imagination encouraged the heuristics of workers' participation in the management of the state sector to build the accountability of large technical systems (LTS), strengthen the cooperatives of workers as a mechanism of ownership, management, and control – wherever the organised sector was making an exit – and develop multi-level systems of networked production.

Finally, the State adopted a strategy of import substitution as an important pathway to self-reliance. However, in many sectors, the practice of import substitution was reduced to nothing more than 'import replacement,' which means that the practitioners merely followed a strategy of duplicative and imitative replication of systems obtained from elsewhere. Imports from the developed market economy embedded in the frame of large technical system dependent countries were as such replicated. These products were pushed by the large State corporations and private sector into the local markets, and they were imported intensively at the level of energy, materials, and components.

The impact of accommodation

There were both positive and negative impacts. There were indeed success stories where unorganised petty producers' cooperatives were formed in production and which achieved the economies of scale and scope, promoting the experiments of technological upgrading with a view to develop the competitiveness of peasant–artisan economies with decentralised local area planning. India Coffee House, a chain of eateries set up all over India, was managed as a workers' cooperative, and the Left provided leadership to this initiative of the workers. Many enterprises, including banks, are being run as workers' cooperatives in Kerala. During the 1980s, with the involvement of peoples' science movements (PSM)', experiments succeeded in establishing models that even acquired the status of technology missions. An example is the leather technology mission, which was inspired by the models developed by the PSM members.

On the other hand, for rural industries like milk, food processing, housing and buildings, drinking water, etc., the main problem was the imposition of LTS used in the developed market economies. Indeed, petty producers in many sectors lost economically as LTS were most inappropriate. Adoption of exotic breeds and seeds, nitrogen-cooled vans, refrigeration for the transport and processing of milk, and the neglect of utilisation of local resources and capabilities took place due to the adoption of imported LTS as a blueprint. This blueprint was replicated for the technological upgrading of not only the dairy sector, but also input supply for crop production, fisheries, agro-processing, food processing, and traditional health systems. Sectors where small producers could have been easily upgraded by the State suffered from the adoption of LTS. Indeed, both Gandhian and Left critiques pointed out that the country needed to invest in the creation of appropriate technology and not promote capital-and import-intensive technology systems.

Although the gainful employment provided by the small-scale industry was quite large during the pre-liberalisation period, the fraction of production based on indigenous technologies, which the agencies developed, remained comparatively small. In this socio-technical frame, there was far more success when the social carriers of innovations were capitalists of non-big business origin, who had acquired the social characteristics of large corporations; and also when the innovations involved standard equipment and only needed process know-how. Moreover, when the innovations involved adaptation of the equipment, technologists, and technicians, artisans from families carrying out trade and small business were far more successful than others. Technology implementation was successful when the individuals themselves became entrepreneurs. On the other hand, failures were common whenever the poor individual producers were made to undertake business with inappropriate technologies developed by external public agencies.

There were also objections raised to the practice of import replacement at the socio-cultural level. The Left's preference for public transport and the Gandhian advocacy for non-motorised personal transport vis-à-vis automobile were socio-cultural objections. Indian policy-makers also experienced a resistance to the mainstream thinking within the sectors of buildings, housing, energy, health, etc. But the people who were critical from among these traditions essentially had their focus fixed on the luxury dimension or the socio-cultural dimension of lifestyle challenges. These critiques did not yet have a focus on dimensions such as 'social justice,' 'ecological justice,' 'sustainable production and consumption', and the relations of production under formation vis-à-vis the poor producers through the strategy of import replacement.

Mainstream development path: Closing and opening of the space for pro-poor innovation making during the 1960s

In India, the macro socio-economic landscape faced extreme pressures during the 1960s. These appeared in the form of a food crisis and a crisis of foreign exchange. These pressures also enabled the mainstream Nehruvian leadership to achieve a closure on some of the debates. Although the process of closure remained fragile and vulnerable to be opened up again, the Nehruvian leadership was able to consolidate the socio-technical frames of LTS that are in place even today as dominant frames.

However, winds of change were to blow soon. Since only the big business and landed gentry were the key beneficiaries of the mainstream pathways, the incumbent leadership also could not fully hold on to the closure attained in the beginning of the 1960s for very long. When the Congress party was defeated in several states in the elections of 1967, Indira Gandhi, the political inheritor of the Nehruvian tradition, was compelled to revisit the issues of the poor. Thus, the Congress under Indira Gandhi brought a new development agenda and used the slogan of '*Garibi Hatao*' (get rid of poverty) to win back the people, and the pro-poor agenda of development was given a new meaning. This virtually allowed the State to redefine the notions of planning, technological self-reliance, and national development. Even the Gandhian tradition of pro-poor innovation making was accommodated through the plan drawn up by the National Committee on Science and Technology (NCST) in 1971, for the KVIC. Nationalisation of banking, coal, and many other industrial units was effective wherever implemented. Particularly the step of nationalisation of banks played a key role in the pro-poor innovation-making efforts of the State. To the extent small farmers gained from the support of nationalised banks in agriculture, pro-poor innovation making also received a boost. During the 1980s, the support of nationalised banks to petty producers and small-scale industry was a lifeline for the numerous units spread all over the country.

More opportunities for opening up the closed gates came for the Gandhian and Left traditions. For them, new spaces were opened up in 1977, when the Congress party lost power at the centre. Not only at the centre, changes were there at the state level as well. A non-Congress government of the opposition was in place for the first time. The Left opposition gained power in the state of West Bengal, where it remained in the state government for over three decades. Similarly, the Left has also been frequently in power in the states of Tripura and Kerala. Appropriate technological movements gained from these developments. Experiments that were started by the new government at the centre came to stay in the programmes of the central and state governments. Even when Indira Gandhi came back to power at the centre in 1980, the technology policy statement of 1983 that the government brought out, incorporated the goal of appropriate technology movements. The programme of science and society was launched by the Ministry of Science and Technology in 1983. The scheme of S&T for weaker sections was promoted, which the PSM have also actively used to undertake their own experiments to practise pro-poor innovation making. These steps gave the required space to several voluntary organisations to experiment, which were largely created by the people of Gandhian, Left, and Nehruvian socialist orientations only.

From experiments to system building for the development of pro-poor niches in innovation making: 1970s, 1980s, and 1990s

During the 1970s and 1980s, the niche of pro-poor innovation making was formed and the institutions of State S&T and the followers of social movements were mobilised most actively in the history of post-independent India. Rural technology was enthusiastically taken up by the practitioners of appropriate technology movement. During these two decades, the creativity of social movements and State S&T was unleashed with the help of increased State assistance in the field. During the 1980s, the social movements utilised the niche not only to work on their own notions of pro-poor innovations, but also to develop intermediary organisations with a pro-people orientation. Developmental and educational action related efforts could be conceptualised by them in the field on a massive scale. Pro-poor social carriers of innovation were promoted, and many scientists who had been working on such ideas in the State S&T institutions were brought together. Experimental spaces were created; experiments acted as a tool of ideological debate and political mobilisation. Attempts of the S&T volunteers were supported by State and voluntary S&T institutions. India had not been as yet engulfed by the syndrome of 'there is no alternative (TINA)'!

Indeed for two decades from 1970 to 1990, niches for the ideas that were not socio-technically mainstream were established in some of the finest state sector institutions such as: Application of S&T in Rural Areas (ASTRA) in Indian Institute of Science (IISc), *Gaon Ka Karigar* and Science in the CSIR system of laboratories, PSM, and other S&T-based voluntary organisations. Working in collaboration with the institutions of higher learning (IITs, universities) and CSIR laboratories, these niches have given a meaning to the departments and programmes being set up for the development of rural technologies by the scientists and engineers to encourage the S&T-based voluntary organisations to participate in the making of pro-poor innovations as pro-poor social carriers of innovation.

By the 1990s, the practice of technology blending, which demands that the traditional and the modern shall be appropriately combined by the pro-poor innovation practitioners, eventually became far more acceptable as a socio-technical frame in all the political traditions. With the emergence of new generic technologies, namely, microelectronics, information and communication technologies, biotechnologies, and new materials, a number of new opportunities were opened up for the practice of networked development in which the peasants, artisans, and rural workers could see opportunities for themselves as social carriers of innovation. Technology blending is actually being recognized as a possibility by the social movements. Social space for the development of non-corporate social carriers has been enhanced. Today, there are certainly more possibilities for the social movements to effectively intervene in the sphere of pro-poor innovation making in India. This is a facilitating development that the new social movements can use to build a pro-poor path of development.

It is true that in a way all social movements have been triggered to rethink about the role of technical change in socio-economic development. This idea that at the level of larger landscape, the choice of socio-technical imagination and socio-technical frame of innovation making will have long-term consequences, is beginning to penetrate the mind of all the political traditions. Therefore, among the older social movements, cadres are also getting sensitised to play the politics of development differently now.

In this section, I have described how the Indian State, public S&T organisations, and social movements mobilised great efforts to promote pro-poor innovation making in India. Yet, their combined and sincere efforts have not achieved much. Although all the trends have seemingly undertaken the efforts to organise their followers for the conduct of experiments, perhaps their system building was inadequate because of inappropriate socio-technical frames for the making of pro-poor innovations. It seems likely that socio-technical frames embedded in a primitive conception of competitiveness would not be the best fit for rural India. Furthermore, their technology implementation efforts were poorly developed; pro-poor innovation system building needs a serious reconceptualisation. Therefore, efforts of the State S&T (S&T work being carried out by the KVIC, CSIR, ICAR, IISc, and IITs) failed to make their full contribution.

We illustrate this proposition with some case studies that show how good intentions can fail to have good impact. Three case studies illustrate the struggles and the twists and turns of politics of knowledge and technology generation, which have been an integral part of the processes of conceptualisation and implementation of pro-poor innovation-making activities and the perusal of the idea of a self-reliant India.

Case studies on the politics of pro-poor innovation making

Knowledge production for pro-poor innovation making and the politics of institution building in higher education: The case of Higher Education (Radhakrishnan) Commission

Dreams of a rural university: How it started

The Radhakrishnan Commission (RC) was instituted in 1948 to review the Indian higher education system. Interestingly, RC proposed a plan for the formation of rural universities, an idea for which it was able to evoke the support of all parties, namely, the Nehruvian, Gandhian, and Left leadership. They were proposed as a roadmap to tackle the enormous diversity present on the demand and the supply sides of Indian rural populations through different set of ways of funding, capacity building, organising, and achieving accountability in innovation making.

The RC's proposal for the establishment of 'rural universities' was guided by an understanding that the universities established by the colonial masters – besides some qualitative limitations – had only touched the fringes of the world's newest and most populous democracy. It began by highlighting that as reported by the 1941 Census, about 85 per cent of the population of India lived in villages. It further argued that new, free beginnings were possible, unhampered by the colonial past, by taking advantage of advances in world educational thoughts and practices. Given the tendency of an old and dominant institution to impose itself on any new institution in a similar field, it became a matter of practical necessity that new universities aiming at extending educational opportunities to the great mass of rural India by giving vitality and quality to rural life should have their own independent design and programme.

The way the RC addressed the issue of integration of national purpose into the design of rural universities is quite instructive. India was asked to decide whether to aim at its largest sections, making villages prosperous, interesting, and culturally rich places, with such a range of opportunity and adventure that young people would find more interesting, more culturally advantageous, and more pioneering opportunities there, than in the city; or whether to turn to centralised industries, with labour taking direction either from the State or from private corporations. It noted that there should be no feeling of conflict between the existing and new types of (rural) universities, any more than between engineering and medical education. However, it suggested that because the pattern and spirit of existing universities are so distinctly urban-centric, massive pioneering efforts would be required for evolving new institutions of higher learning to serve the needs and aspirations of our democracy. The commission stressed that rural universities would share many of the qualities and methods of existing universities and further, that there would be cooperation and interaction, whether with the European universities or with institutions in America, and a relation of consultation, friendship, and advice but not of authority.

How it ended

The RC was the first (and last) committee to discuss how India could inculcate national purposes into higher education through the implementation of appropriately chosen university design and joint planning. The RC's model of rural university had aimed at the achievement of functional integration of teaching, research, and extension by linking the process to the construction of a different kind of agriculture. In the recommendations of RC, the model of integration of the functions of teaching, research, and extension was not subordinated to the promotion of high external input agriculture. The model of rural universities was in favour of promoting the development of farm and non-farm production by taking into account the ecological and social diversity and local resource endowments in an integrated way. Extension was not a spillover to be harnessed. Knowledge (re)production would perform by contributing to the goals of rural–urban development in an 'Indian style.'

The establishment of rural universities along the lines of the RC recommendations required the cultivation of a political will and the mobilisation of public support for the construction of India's own development path. However, at that time, Indo-US agricultural cooperation teams were recommending help to be taken from the USA for institution building in agriculture and small-scale industry. The political–bureaucratic apparatus was being advised to balance the help of the erstwhile Soviet Union in heavy industry with the help of USA in agriculture and small-scale industry. Food shortages were another overwhelming pressure factor. War with China had also shaken the government. All these factors made the government rush to take help from the

United States Agency for International Development (USAID). Thus, a drift occurred in educational policy when the Kothari Commission formulated the National Educational Policy of 1966. In the Kothari Commission, the planning and orientation of agricultural universities were handed over to teams led by the Rockefeller and Ford Foundations. These two foundations were actively supported via two agreements signed for Indo-American S&T cooperation in 1955 and 1960. The Kothari Commission almost completely violated the rural university design model proposed by the RC. Though the Kothari Commission insisted that an agriculture university be created in every state, modern input-intensive agriculture was to be taught along the lines of American land-grant agriculture universities (whereby land is granted by the state for a university to build capabilities in agriculture). Thus, the RC committee's vision of a rural university failed to become a reality because it needed the political–bureaucratic apparatus, the scientific community, and the people to jointly stand firm in favour of its implementation. But this did not happen.

Recently, the RC recommendations were revived during the Tenth Five-year Plan. Not only did this revival happen after a lapse of eight or nine plans through the efforts of a very small section of S&T community led by Professor S. K. Sinha, former Director, Indian Agricultural Research Institute (IARI) but because of lack of sufficient mobilisation of the S&T community, the Indo–US Knowledge Initiative for Agriculture came back to raise its head again. While the programmes of agricultural research of the Tenth Five-year Plan envisaged the agenda of setting of agricultural development to shift to the implantation of agro-ecological approaches, the Indo–US knowledge initiative has been geared to the introduction of precision agriculture.

It is clear from the history of unsuccessful implementation of the recommendations of the RC that the challenge faced with regard to the implementation of the visions and strategies of pro-poor innovation is political. Because the politics of knowledge production is no more on the priority of the socio-political forces active among the people, there is no viable political force to take up the political battle of pro-poor innovation making in the Indian agricultural system of innovation.

Other similar experiments that did not succeed

- *City universities*: Similarly, when the RC recommended the formation of 'city universities,' it had clearly recognized that rural and urban universities should supplement each other through a framework of regional development to benefit the development of local economies.

It argued that since universities had so far been established to meet the needs of cities or limited areas, fresh thoughts needed to be given to the formation of regional universities to serve the distinct requirements of diverse natural, linguistic, cultural, or economic regions. The RC was clear that the full genius of a country can develop only with freedom to create variety; that foreign control could best advance itself by regimented uniformity. It also recognized that given a tendency towards bureaucratic unimaginativeness and administrative convenience, regimented authority had become a habit of mind in Indian higher education. The commission, therefore, recommended that when the UGC was to decide the degree of recognition and support to be given to any institution, the criteria should not be the likeness of that institution to others of a standard type, but a judgment about whether that institution is making a substantial contribution to the economic, intellectual, cultural, and spiritual life of India.

- *The Nai-Taleem model*: The Gandhian tradition had put forward the model of *Nai Taleem* to link the education sector with the practice of innovation making for the poor. This model had been conceived as an unconventional mode of learning wherein the students are regularly exposed to both classroom teaching and training in concrete projects. Teachers act as masters who endow student apprentices with both knowledge and skills. Such institutions remained marginal to the plans of education under implementation; only 15 rural institutes exist today. Moreover, a lack of appropriate teachers for the task at hand [i.e., master (*ustaad*)-apprentice (*shagird*) type of learning] remains a big handicap for the model. These institutes have failed to sustain themselves as viable social carriers of innovation. Attempts to overcome this handicap have not had enough support either from the State or from the society in general.

Knowledge production, pro-poor innovations, and the politics of technology implementation: The example of the CSIR

According to its website, the mission of the CSIR[6], created in 1942, is: 'To provide scientific and industrial R&D that maximises the economic, environmental and societal benefits for the people of India.' The CSIR is the largest publicly funded network of laboratories in India. Its motivation in the case of pro-

[6] See website at *www.csir.res.in.*

poor innovation generation came directly from the influence of the freedom movement, but the socio-technical frame it adopted for pro-poor innovation making has become its important limitation in respect to the generation of appropriate rural technologies. Some of its notable pro-poor achievements are, however, presented in Table 11.2 in respect to rural technologies.

Table 11.2: Selected notable pro-poor innovations of the CSIR

Product/Process	*Nature of the innovation*	*Impact on rural/urban households/producers*
Amul – leading brand of dairy products in India	Process developed by the Central Food Technology Research Institute (CFTRI) to make condensed milk powder from buffalo milk – a difficult challenge by the dairy technologists at that time	Amul provided access to milk, butter, and dairy products across the entire country; it also provided livelihoods for hundreds of households
Improved vegetable-tanned leather	Process innovations by the Central Leather Research Institute with the help of which the national leather technology mission has made an impact on the leather industry as a whole	The units working across the country could shift to a low cost solution, which not only solved the problem of pollution facing this important export-oriented Indian industry but also provided for improved incomes, associated value-added products out of fallen carcass utilisation, and social justice
Extracts of citronella, lemon grass, Japanese mint	Agro-processing technologies	This provided livelihoods to hundreds of farmers
Development of improved hand pump, low-cost culverts for rural roads, red clay based sanitary wares, leaf cup production, leaf cup making machines, iron removal plants, technology for the two-pit latrine system.		These technologies contributed to the use of local raw materials and skills and benefited the local economies

Source: Rural technologies (www.csir.res.in/external/heads/achievements/rural_technologies.htm)

The Karimnagar project

During the period of 1972–1979, rural technological development got a big push within the CSIR because the government of the day was inclined to support initiatives for rural technological development. In quite a few CSIR laboratories, rural development projects were identified and taken up. Internally, the leadership came mainly from the then Director General – CSIR (DGCSIR), Dr Y. Nayadumma, who was keen to take up work on rural technologies. With his involvement, the initiative taken for Karimnagar became a major programme for the CSIR system.

The Karimnagar project was initiated in Andhra Pradesh in 1972 with the aim of taking S&T in an integrated manner to rural areas. The approach to technology implementation consisted of the identification of a number of schemes and proven technologies, which could help in socio-economic development of the participating communities irrespective of their economic and social levels. For the CSIR system, extension of the outcomes of the ongoing R&D work with the aim to field-test the pooled results of research was also an important motive. The emphasis was on selecting projects and providing S&T inputs based on them for the following purposes: survey of resources, establishment of industries, and development of public utilities and implementing on priority the technologies for rural water supply, road construction, and rural development.

The DGCSIR appointed an evaluation committee in January 1977 to review the experiment undertaken in Karimnagar in the state of Andhra Pradesh with a view to taking S&T to rural areas in an integrated manner. The detailed evaluation of the process of technology implementation undertaken by the committee, discussed here only in brief, also demonstrates how ineffective the technology implementation approach of a technology push, a modern sector spin-off programme can ultimately turn out to be in the case of rural technologies. Cited below are the evaluative comments of this committee on the performance of the Karimnagar project. For example, the evaluation committee was of the view that 'no project was undertaken or demonstrated which could directly improve the condition of the rural poor and other disadvantaged groups like small and marginal farmers, landless labourers, scheduled castes and scheduled tribes, nomads and rural women as a group'. It also says that 'in the absence of a planned approach and methodology of implementation and with the "scatteredness" of the schemes in whole of the district, the impact of the experiment was getting diluted and diffused'(CSIR 1978).

In order to deal with this criticism, the CSIR decided eventually to confine its work in only seven clusters of the 56 villages of the district. In May 1977, using this methodology, the CSIR even prepared a revised project report. The evaluation committee had the opportunity of also going through this revised document. And it came to conclusion that 'even this approach lacked objectives desired to be achieved.'. The committee felt that 'in the revised document the activities were not clearly defined and mode of implementation was not spelt out.' The committee also commented on the type of partners mobilised by the CSIR system. It said clearly that 'in the new approach, the Karimnagar project envisaged assistance and cooperation of other agencies about which CSIR already had its sad experience.' The committee also commented that

> the CSIR has made its own internal assessment of the project; however, it appeared to be very general in nature. The assessment was not very sharp, well exposed, and objective. No definite programme was worked out to implement the project and translate the idea into action as to which area was having the gap, what technology could bridge this gap, how many demonstrations were required to establish the innovation, who will do extension work and serve the related needs and demands, etc. Whatever projects were implemented again were selected in an ad hoc manner on the basis of whoever was available to try out an experiment as a giver or receiver, whether he had full faith or not in it. (CSIR 1978)

From the comments quoted above, it is certain that often, when an organisation is following the 'technology push' approach, it also generally ignores the need to have even a feedback on the implementation. In fact, it is not possible in this approach to develop the interactive structures needed for the purpose of planning, monitoring, and evaluation.

In the Karimnagar project, the approach taken was that of a technology push, a modern sector spin-off programme for the villages without interaction with targeted beneficiaries or any form of feedback loop. The evaluation committee confirms this by stating clearly that

> one of the objectives of the CSIR laboratories has been extension of the results of the research in the prospective areas in different parts of the country. In fact, it seems to have been the main objective that Karimnagar Project would provide a good ground for field testing of pooled results of research. (CSIR 1978)

This becomes also evident when the committee notes clearly in its final recommendations that

> research and development have little meaning unless these are tuned to the needs of the people and taken to a stage which could help in their successful development and adoption. The next stage comes of making the laboratory techniques and technologies known to and adopted by their prospective users. This is to be followed by 'service after sales' and removing any teething trouble experienced by the users. Further improvement in the technologies could be brought under a continuous process of R&D. The feedback information should be regularly conveyed to concerned scientists or the technologists so that they may be exposed to real field conditions. (CSIR 1978)

From this discussion on the experiences of the Karimnagar experiment, it should not be difficult to conclude that the failure of the CSIR in respect to rural technology development has a direct connection with the methodology of technology push without any form of feedback loop.

Why did the CSIR fail in the Karimnagar project?

The Karimnagar episode raises a number of questions that are fundamental. Is this not surprising that for a country like India with a large rural population whose basic needs are yet not met, a project such as the Karimnagar one utterly failed? Did the later leaders attempt to overturn this understanding? If not, was there any resistance on the part of the CSIR scientists? Is the issue connected in any way with the ethos of the CSIR scientists? Is it connected with the institutionalisation of a particular strategy out of which the CSIR leadership is finding quite difficult to come out? What must the CSIR leadership do to bring about the change? We attempt to answer at least a few of these questions with respect to the Karimnagar project now – starting with the government, then the CSIR system, and ending with the nature of the innovation generated. These explanations aim to provide insights into how 'technology transfer' projects must be designed by public agencies in order to be useful and effective in promoting rural development.

- *The government strategy*: The Indian government faced a number of questions while carving out a developmental path during the 1970s. Should the industrialisation of India continue to mean the development of only those industries that cater to the needs of classes whose lifestyles have become dependent on the imitation of goods being produced in the industrialised countries? Should Indian industrial development strategy continue with a bias against the possibility of rural industrialisation, which has an enormous potential to cater to the requirements of local markets? Can rural industrialisation be undertaken through the strategy of upgrading the existing local skills and local resources? Was not the CSIR system set up to be an organisation devoted to generating technologies for the Indian industrial development? If political will at the top had decided that the answer was 'no' to the above questions, the Karimnagar fiasco would not have happened.

- *Lack of continuity in an investment trajectory as government leadership/strategy changes*: In 1980, there was a change in the central government. The new leadership, on its appointment in the CSIR, was of the view that the CSIR had not been created to generate technical know-how for the rural sector. One can ask, just because the government of the day is changing its course and stressing less on rural industrialisation, should the CSIR drop its efforts for rural industrialisation? The reality is that the CSIR is completely dependent on the funds and executive support being made available by the government under its policies, both for infrastructure development and for establishment of industrial estates. Thus, when the government changed and withdrew the funds, the local district collectors completely lost interest in a programme like Karimnagar. The next government of the day was interested in shifting the focus of rural industrialisation away from the rural industrial estates-based programmes to the support for village industries. In response, the CSIR had to withdraw from the district development programme of Karimnagar. In programmes of industrial development, formulated as a part of the Fourth Plan when the planners formulated a massive area-development exercise, the experiment of Karimnagar was to be implemented through a rural industrialisation programme of creating rural industrial estates. Without the help of district collectors and associated development agencies, it was not possible to continue the programme. Only by establishing a new set of bridging institutions, the CSIR could have kept the experiments alive. But no such efforts were made in the CSIR system.
- *Being clear about the mandate: Is CSIR only for industry or also for rural development?* In a paper published by Dr J. C. Srivastava of CSIR on the 'Role of CSIR in the Development and Application of Technology

for Rural and Backward Areas' (1984), this issue has been discussed with a lot of bitterness. Dr J. C. Srivastava, who used to manage the rural development activities, records that

> with a new Director General taking over the helm of CSIR towards the end of 1978, the CSIR's interest in rural development seems to have almost ceased. Directors of CSIR laboratories (Refer Directors Conference, Bangalore 1978–79) seemed to be of the view that CSIR had little role to play in promotion of rural development which should be normally left to ICAR, Agricultural Universities, State Governments, voluntary agencies and the like. Indeed it appears to the author that CSIR authorities during recent years have even frowned upon its laboratories undertaking R&D programmes related to rural development. Such work was considered to be outside the purview of CSIR, which was expected to confine itself to basic research and development of industrial technology. Work on technology delivery was firmly rejected on the plea that rather than being guided by passion, we should work for what the CSIR has been actually set up. (Srivastava 1999)

- *Problem of leadership or lack of it at the top*: The CSIR leadership dropped rural industrialisation from its basket only because it had a particular role and division of labour in mind. The CSIR leadership had this division of labour in view where by itself the CSIR would take the risk only in basic research. And if the leadership was ever required to be undertaking the risk for the development of technology where the users themselves had to be developed, its view was that it should rather wait for the government to take a lead. Of course, it would be altogether another matter if the CSIR had a strategy of pursuing leadership in the markets. Then, like a product champion, it would have even tried championing the rural technological development programmes to the government. The problem lay in the fact that the leadership of CSIR pursued pro-poor innovation as a defensive strategy, because it gets funding from the government for R&D.
- *A mindset problem of CSIR scientists themselves*: The CSIR scientists tend to view themselves as scientists working to strengthen the technological capabilities of the modern small and medium-scale enterprises – leaving rural development actions to emerge as a spin-off from the mainstream programmes. Under this mindset, rural development is viewed as a burden rather than an opportunity. In the spirit of true spin-off approach, the CSIR system neglects the fact that the technology transfer has to

take into account user capabilities and resources. Since the users are in the unorganised sector, lacking skills to communicate their demands effectively to the CSIR scientists, unless the scientists reach out to understand the rural users, technology transfer will not be effective. Of course, we must look for a consistent explanation for the ethos of the CSIR scientists. A part of the burden of responsibility needs to be taken also by the leadership, which does not value the pro-poor innovation making. The evaluation system that the leadership follows requires a major change in its objectives and methods.

- *Too capital-intensive process technologies*: Many CSIR rural technologies have been based on the processes intended for being used in capital-intensive small-scale industries. These technologies were meant for catering to the urban markets and needed high inputs of non-renewable energy sources to operate machines to be utilised in the course of production. Entrepreneurs who successfully used these technologies tended to come from well-established business families.
- *Lack of 'appropriate design' and 'interaction with targeted beneficiaries' in 'technology push' projects*: Evaluations clearly point out that the R&D personnel of the CSIR do not configure their R&D projects for the appropriate economies of scale and scope of rural contexts. Technologies developed do not meet the real needs of targeted regions and cannot be fitted into the industrial and social infrastructure existing in rural areas. The process of technology transfer is undertaken by *pushing the available solutions* without the technology adaptation effort required for fitting the technology to the conditions of the users. This is because interaction of R&D workers with users is weak. There has been no attempt to understand the users as systems and to manage technology transfer as an interactive process.
- *Weak external networks with non-CSIR experts*: Technology programmes for rural development continue to be weak on the aspect of networking of external resources such as expertise and funding. During the phase of technology development, the interaction with users is very poor. The laboratories continue to depend mainly on the efforts and inputs of their scientists alone and in-house R&D funding for technology development. Involvement of external experts and multidisciplinary background of laboratory scientists significantly improved the technologies. The situation of networking has been no better during the phase of technology implementation.

CSIR today

Today, the CSIR has a programme called 'CSIR for 800 million people' to support the projects for rural development and weaker sections, but the efforts being undertaken in the laboratories are essentially still very much the same technologies and processes, which the CSIR system has been undertaking as a residual activity.

In the case of village industries, the conditions for technology development have not changed in any significant way. From a recent survey of the 500 users of CSIR rural technologies that the author undertook in collaboration with his colleagues, it has been seen quite clearly that even the newly developed technologies are failing to make an impact on the rural scene. Only 18 per cent of CSIR rural technology users were in production and the rest have either not started or have chosen to discontinue the production. This report shows that the programmes of rural development have apparently been going on in the laboratories without any kind of critical evaluation being undertaken by the CSIR headquarters.

The RNFS is a field where the CSIR has had the possibility of contributing a great deal, but even after 25 years of involvement, it remains an area of extremely low priority for the CSIR leadership. In the document titled 'Creating an Enabling Environment for Commercialization of CSIR Knowledge Base: A New Perspective' (1993), it has been revealed that this sector has been completely absent from the attention of the committee. In the latest CSIR Vision document of Dr R. A. Mashelkar, this area does not get any attention at all. Dr Mashelkar can be presumed to have indirectly mentioned it in the document through the reference made to the objective of technology for human welfare. There are no specific targets with respect to this area in the Vision document, though in all the other domains the numbers are clearly indicated in the document. Dr Samir Brahmachari, the new Director General, leads the CSIR to implement *Vigyan Kutirs* in the same way as done by most of the past Directors Generals. Implementing rural development residually, as a part of the CSIR-800 mission programme, through the technology push approach again, has remained the fate of pro-poor innovation making even today.

From the top to the bottom, there has been very little learning in the management of R&D and technology transfer for rural development in the last 20 years. Scientists continue to self-indulgently believe that their technologies are economically fully viable in rural areas. There is very little record being kept by the laboratories on the status of technology utilisation.

Since the top management has been mostly indifferent to the efforts of R&D and technology transfer for rural development, there have been hardly any investments in bringing the users closer through the encouragement to link organisations/agencies that could have acted as the bridge. This lack of investment is reinforcing the habits of CSIR scientists in favour of technology spin-offs without any feedback loops that embed it appropriately in the rural context for which it was designed. Both technology development and implementation continue to be oriented toward the linear model of innovation, where the laboratories produce spin-off technologies without fitting them to the conditions of users or involving them in the adaptive efforts (Pulamte and Abrol 2003).

To conclude this case study, we note again that CSIR leaders take up technology development projects for the rural areas only when the government of the day wants them to pursue this objective. The Karimnagar initiative got formulated when the government of the day was taking keen interest in the implementation of area-development exercises. The government of the day was the driving force for this programme – without any internal momentum within the CSIR. Since the CSIR leadership's lack of commitment to the development of rural technology comes from the fact that the political–bureaucratic apparatus treats pro-poor innovation making as a residual activity for its own design of development, it is important that the CSIR becomes accountable to the Parliament through the introduction of a special research audit system. Of course, this is not the responsibility of CSIR only; but the CSIR will succeed only when its partners in the government and the private sector are also ready to give priority to pro-poor innovation-making activity.

Pro-poor rural non-farm innovations and the politics of technology implementation: An assessment of KVIC, CSIR, and other rural industries development programmes

Much of the RNFS occupations are subsistence employment. They survive because the RNFS occupations are linked to either serving the local rural markets or meeting those needs of the urban poor that the modern industrial sector is yet not able to satisfy. However, millions have lost their livelihood on account of the lack of technological upgrading in the face of increased market competition. In this section, we analyse some of the strategies deployed that failed to stem this problem with examples taken from the KVIC.

In the past 50 years of development in India, the governments of both the states and the Union have sponsored a number of programmes to upgrade the RNFS occupations. The promotional policies and technology support programmes that have been applied directly or indirectly to the RNFS development, can be categorised in the Indian context as policies and programmes for promoting (i) self-employment among the rural poor and (ii) traditional rural industries.

Strategy 1: Subsidising to maximise rural self-employment

During the Sixth Five-year Plan (1980–85), the government launched a new package of incentives for the rural poor to make use of opportunities for self-employment through programmes like the Integrated Rural Development Programme (IRDP), Training of Rural Youth for Self-Employment (TRYSEM), Development of Women Children in Rural Areas (DWCRA), and State-specific schemes for self-employment of scheduled castes, scheduled tribes, and women. These programmes were initiated with the aim of targeting households below the poverty line, and the idea was to provide each of them with a productive asset through a subsidised loan so that they can rise above the poverty line. The assets provided to rural poor under these programmes included milch cattle, goats, sheep and poultry; equipment such as sewing machines, tool kits, camel carts, handcarts, rickshaws or bicycles for hiring out; or working capital for petty trading, tea or pan shops and the like. During the Seventh Five-year Plan (1985–90), this strategy was supported in a bigger way through large allocations in the form of both government funds and bank credit. This emphasis was continued during the Eighth Five-year Plan (1992–97). Throughout this period, cottage-scale units such as those employing individual weaver, the potter, the blacksmith, and the village shoemaker were also subsidised to ensure maximum employment per unit. For example, both Khadi and handloom were protected, the former through subsidised credit and subsidies to consumers on purchases, the latter through non-imposition of excise on hank yarn.

However, this policy had just the opposite effects. The cottage-scale units had no incentives to explore technology upgrading. Low productivity, low wages, and underdevelopment prevailed. Programmes for self-employment failed because of inefficient implementation via wrong selection of beneficiaries, 'leakages,' failures of enterprises, loss of assets, saturation of small markets for produce, etc. Moreover, projects were not on a cost-effective basis; backward and forward linkages were absent; beneficiaries selected among rural poor were not adequately skilled and lacked management competence; the role of the middlemen and contractors was inimical, leading to 'leakages'; and training was done in isolation without integrating them with specific projects, and often without reference to local demand for skills.

Strategy 2: Improving competitiveness of rural industries by modernising traditional equipment

In the 1970s, the efforts for the advancement of the practice of S&T for the industrial upgrading of traditional manufacturing received a big push in India from the government for the first time. But most of the technological solutions that this effort offered were unable to connect well with the local markets and capabilities accessible to the rural poor, and hence, were largely ineffective in improving competitiveness. Unfortunately, this technology push approach persists uniformly in all the agencies even now.

An illustration is readily provided through some examples of technology upgrading attempted by the KVIC. The KVIC's strategy was to improve traditional technologies by scaling them up to intermediate levels and introducing power-driven machines. However, this approach increased the costs, made practical functioning difficult, and adversely affected the formulation of viable projects. For example, although the semi-automatic improved loom was developed in 1972, about 90 per cent of handloom weavers still continue to use the pit loom. The power pottery wheel entered the market about two decades ago (1970), yet the village potter continues to operate the traditional wheel. The traditional *ghani* (oil expeller) is fast disappearing but it has not yet been replaced by a power-driven *ghani*. The large producer using solvent extraction/expellers is more competitive. Improved *gur* (jaggery or molasses) furnaces have been developed but not adopted adequately. These are only a few instances.

The basic problem was that in most cases technologies had been developed keeping the individual enterprise of a small producer in mind while totally neglecting to incorporate local resources – raw materials, engineering materials, energy sources immediately accessible by the people. Local markets were ignored. There were no efforts to develop technologies that could strengthen inter-links in the local economy by developing input–output relations among existing occupations, in terms of specific products and services. There were no attempts to innovate using local engineering capabilities and materials to substitute non-local products, e.g., replacing stainless steel vessels by glazed clayware and, if necessary, metal bottom or internal pipes, etc., for heat transfer. Then, it would have been possible for the agencies to lower investments and scales, and develop inter-links in the local economy. Non-conventional energy sources were not suitably integrated into industrial activities requiring machines. The focus was more on providing non-conventional energy sources for the purpose of cooking and lighting. Technology for co-products and by-product formation was given very little attention (Rao and Yashwant 1994).

Efforts made by the KVIC to overcome the limitations of individual artisan's resources through the use of common facility centres also failed to finally upgrade the production of rural artisans in an economically viable way. These centres have ended up serving only those artisans who are working for large urban markets under the control of large traders or a state sector-owned bureaucratic marketing facility.

Best practices and recommendations

First, analysis based on these discussions, including case studies, indicates that the dominant political mainstream did not accommodate the efforts for pro-poor innovation making automatically. Second, though at regular intervals of almost every 10 years the politics of technology implementation has been revisited by the political–bureaucratic apparatus, failures continue with respect to the process of mainstream development in meeting the expectations of the poor. Third, analysis also clearly shows that till this date, for the expansion of contribution of pro-poor innovation making the nation still largely remains subject to the dynamics of closing and reopening of the political-economy space, as available. As politicians come and go following the changing fortunes of mainstream development process, socio-technical frames of pro-poor innovation making remains an unattended factor. Fourth, counter-hegemonic political traditions need to realize the critical role of the socio-technical frames in the achievement of competitiveness of pro-poor innovations. For the accumulation of counter-hegemonic ability by the relevant socio-political traditions their carriers should undertake their own efforts for pro-poor innovation making as a cultural, political-economy endeavour. Socio-technical frames and the imagination embedded in them must be treated as an instrument of resistance required to be developed by the poor themselves and with collaboration with other economic actors, to achieve the required results with regard to the building of pro-poor innovation systems. In this section, we first examine the obstacles and then outline some success factors for pro-poor innovation making.

Factors that limit the creation of pro-poor innovation

- *Application of a primitive notion of competitiveness to rural producers*: The case studies analysed in the previous section reveal that in India, pro-poor innovation-making activity in rural industries has suffered from the implementation of a primitive concept of competitiveness. The view that innovation in rural industries should be restricted to the transitional objective of poverty alleviation has its roots in the practice of a primitive concept of industrial competitiveness. The approach towards implementation of technological upgrading in the rural industries has been to target the individual small producers and help them to become individually competitive. In this approach, competitiveness of this segment gets judged by comparing the existing relative costs and prices to individual producers because there is no recognition of potential competitive advantages that might accrue through the exploitation of economies of scale and scope, and cluster or network effects. Therefore, these synergies are not generated or captured and in their place, small producers are forced to accept intermediate technologies with higher costs of inputs per unit of production.
- *Lack of cooperation between small producers combined with lack of bargaining power*: Competing large capitalist enterprises are in a position of erecting barriers to competitive access for small producers in the input and product markets and often use such barriers. Mutual competition amongst small producers also leads to narrowed access to inputs and technologies and affects the effective demand for their products. In those cases where small producers are being organised only for access to inputs and credit, the absence of cooperation in production results in mutual competition at the marketing stage that in turn breaks their existing alliances. Often both, the traders and the large-scale capitalist enterprises (competing with them in the market for the sale of products), have an interest in aiding the processes that can disrupt the cooperation of small producers. As in turn this failure of cooperation among small producers leads to the establishment of a vicious circle for the utilisation of intermediate technologies in these sectors, it creates an impact of the slowing of innovation diffusion as a whole in the economy of a country like India.
- *Viewing the rural producers as passive recipients rather than co-producers of innovations*: A common feature of failed ventures evoked in the case

studies was that the process of technology implementation was largely guided by the notion of the end-users of innovation, namely, poor peasants, artisans, and agricultural labourers, to continue as passive recipients of the technologies selected for promotion by intermediaries. Predominantly, machine training and process demonstration are the practices in technology implementation. Technology support has been conceived in the form of one-time injection of improved hardware. This means that in the case of rural poor, user support is not oriented to the task of improving the 'participation' of these end-users in the process of development of intermediate or appropriate technologies.

- *Applying a pure market cost–benefit analysis without taking into account externalities*: In the case studies presented here, technology evaluation was built around market cost–benefit analyses. Assessment of the negative externalities was rejected as a manageable problem through regulation and compensatory solutions. Technology assessment at the level of impacts of the systems on ecological and social justice is another recent development that has to be considered. Even today in these political traditions, technology assessment is not a major feature of their pro-poor innovation interventions. The Green Revolution technology was critiqued for the landed gentry preferring technical change to land reforms. However, it is important to note that the selected technical solution of high external input agriculture threatened the poor peasants, artisans, and landless labourers more. But it is only now evident that poor peasants, artisans, and rural labourers have lost a lot on account of the adoption of inappropriate technological choices even when implemented through the State or cooperative sectors.
- *Vested interests*: For a vibrant pro-poor innovation making, diversity and experimentation must be encouraged and it must also be recognized that this may go against the interests of some actors in the innovation system such as elites (who may change according to context and issue considered), and some big businesses, politicians, bureaucrats, foreign firms, power groups of all kinds, etc. And the constraints posed by the vested interests should never be underestimated. Ultimately poverty elimination needs the poor to organise themselves for the cooperation in production to get redistributive justice. The poor will have to compete and bargain with all the other actors in the innovation system, including the corporate sector and private business. When in business, it is not possible to escape the iron law of competition in the market-driven economies.

Factors that contribute to success

Technology and market are interlinked issues. Often, for the markets where the rural poor can compete with ease, even the ready-to-transfer technologies are not available. For the rural poor, the problem of access to markets is a systemic problem. They cannot compete with everyone anywhere and everywhere. The markets for which the rural poor should compete must be within their reach. Further, for the enterprises of the rural poor to succeed, the programmes of industrial modernisation must have a strategy of developing the local economy as a production system in itself. Despite these challenges, there have been success stories and they yield the following insights.

- *Useful technologies are created through the assessment of conditions of competition facing the targeted beneficiaries*: For many traditional industries, the users come mostly from among the rural poor. Therefore, there must be an assessment of the needs of potential users, especially in terms of the type of competition they face and their market and non-market opportunities as well as their current level of access to markets, resources, and capabilities. There must be a close collaboration between the scientists and the prospective users, otherwise it will not be possible for S&T agencies to help them become competitive users of the technologies developed. Analysis of the pattern of success indicates that the formulation of strategy for system transition starts from the standpoint of how to improve the 'participation' of the end-users in the system of innovation.
- *New technology is effective if it is compatible with a higher returns yielding business model appropriate for a local context*: There are efforts of public agencies to improve the competitiveness of rural firms because their efforts tend to conceptualise the problem of technology support as only of a one-time injection of improved hardware. Moreover, the concept of improved hardware was limited to creating the machinery and equipment through mainly downsizing of large-scale modern technologies, upgrading of traditional technologies, or blending modern with traditional technologies with the aim of making an individual producer efficient. But, since the improved technologies were poorly connected with their existing local production as well as technological systems, the targeted users could not be convinced to adopt the new technology. Finally, it was assumed that for every improved technology, there was an obvious business model that would increase the profitability of the user. But this assumption was never explored or constructed.

- *Scale, network, and cluster effects should be explored, initiated, and used*: Calculations of the economic viability of technologies cannot be made with the assumption that small producers in village industries are incapable of organising themselves for a better access to the higher scales of production. Selection of technology development objectives and choice of partners for user development need to be designed with the perspective of organising small producers in village industries to interface for achieving economies of scale and scope through their appropriate organisation and assuring network and cluster effects, so far ignored in competitiveness evaluation.
- *Technology implementation has to be guided by a systemic conception of competitiveness*: The production and technology systems of the targeted users must be identified such that any new technology package reduces their costs by either resorting to alternative input-using technologies or using the present inputs more efficiently. It is necessary for the social carriers of pro-poor innovation to actively cooperate among themselves in production with the aim of breaking the monopoly of large capitalist enterprises. In order to be systemic in the approach to innovation, small producers not only have to utilise those technological opportunities that connect well with the local markets, capabilities, and resources, but also have to organise, develop, and obtain technology for cooperation in production among themselves. This also implies the maximum utilisation of local advantage while linking them internally and developing their capacities to plan for better market access and egalitarian organisation of production. They need to undertake technology choices that enhance cooperation in production via the development of production linkages, value addition, improved production, and greater diversification.
- *Change in the paradigm of rural development*: History tells us that the contribution to rural industrialisation process was seen by the dominant leadership of scientific community working in the CSIR system of laboratories as being outside their domain. In their understanding, the development of modern industry did not include the technological transformation of rural areas. Rural industrialisation was seen through the lens of a developmental concept that was embedded in the dichotomy of traditional and modern industries. Since the framing of upgrading of traditional industries was embedded in the individual, small producers becoming competitive, which did not fit well with the

> imagination of large technology system, the scientific community did not consider this task to be working on the frontiers. Today when the world is seeking eco-restructuring of the development process, a vision and strategy of multi-level industrial development is considered to be high technology and a frontier to be captured.

As an alternative socio-technical regime agro-ecology-based approaches to agricultural and rural industrial development have the potential to resist agri-business and agro-food companies in the retail market. Agro-ecological approaches have the wherewithal to make the Indian agriculture economically, ecologically, and socially viable. Agro-ecological approaches have the capacity to reduce the farming risk on the input side by moving the peasantry away from the non-local inputs and on the output side in the market on account of the diversification of outputs. Agro-ecology-based approaches allow peasants to move away from the existing system of monocropping with increasing trend to an integrated system of primary production, integrating all the relevant subsystems of primary production, namely, crop production, livestock management, aquaculture, and silvi-pastoral cover areas. The core principles of agro-ecology include recycling nutrients and energy on the farm from within the local systems of ecology and economy, rather than introducing external inputs from the outside; integrating crops and livestock; diversifying species and genetic resources in agro-ecosystems over time and space; and focusing on interactions and productivity across the agricultural system, rather than focusing on individual species.

Agro-industrial production systems have the possibility of contributing materials of different kinds of use or values besides just food. Biomass-based dispersed industrialisation is conceptually and technologically considered feasible by the engineers (Datye 1997). Agro-ecology is highly knowledge-intensive, based on techniques that are not delivered top-down but generated within the local economy on the basis of knowledge and experimentation of peasants and landless workers and through the development of group enterprises of rural labour and artisans. Since the approach of agro-ecology as a mode of agricultural development is capable of strongly contributing to the broader goals of regional economic development, public health, and wider ecological health transformation, now the pro-poor innovation making can be geared to provide solutions for the evolution of new kind of agriculture and rural industries as a part of socio-ecological and socio-technical restructuring to be undertaken in the case of the Indian economy.

Concluding remarks and reflections on new trends

The politics of Indian economic development has been shaped by debates and bargainings between the cadres of Gandhian, Nehruvian, and the Left right from pre-Independence times. Debates have not ceased even today and are continuing in a vigorous way between the cadres of these three orientations in the new context of struggles emerging for the achievement of ecologically and socially just development among the people of the country. But there are challenges for the present State policy that now favours an open, liberalised economy in contrast to the original Nehruvian, Gandhian, and Left traditions that are not in any way close to reaching the stage of becoming a viable political alternative. Even today, the force of their challenge is a function of the crisis and of the ability to overcome the political and economic dimensions of the crisis facing the mainstream. Since everyone is required to work within the framework of parliamentary democracy, the State has been accommodating and adjusting with the niches under formation in the country through the efforts of the people of varying political persuasion. Nevertheless, the capacity of the different factions to bargain with each other and the mainstream and vice-versa has always been important. Within all these bargaining games, pro-poor innovation making has been treated right through as a residual activity by the State and even today it is no different and practices of pro-poor innovation making are yet to gain the required success and ascent. This article has tried to offer insight into why and how it is so.

Starting from the assumption that there should be an improvement in the technological resilience of local economies to prevent their continuing destruction through the expansion of the 'modern' capitalist sector, this chapter examined the Indian experience in pro-poor innovation making. It first demonstrated that the practice of innovation making has been influenced by the choice of pathways of development, strategy of technological modernisation, and the heuristics of innovation making adopted by the Gandhian, Nehruvian, and the Left leadership. Then, it examined the activities of economic actors such as S&T institutions, development agencies, and public sector corporations in the making of pro-poor innovations with a view to derive insight into the interaction of policy, practice, and theory of innovation making.

Our analysis demonstrated that pro-poor innovation generation and diffusion in India have not been successful, because the mainstream pathway of development and the socio-technical models of technological transformation adopted by the social carriers of innovation in the case of pro-poor innovation making have continued to be mostly guided by a non-systemic concept of competitiveness. Technology platforms, ecologies, and pro-poor innovation systems were not constructed with the pragmatic aim of achieving ecological and social justice and economic empowerment of the poor – but as a residual socio-technical system without ensuring any systemic competitiveness. Thus, the past experience of technology implementation provides valuable insight into how not to address the challenge of creation of models of governance of pro-poor innovation making. It also confirms that the policy-makers are perhaps expecting too much to be accomplished in respect of poverty alleviation through innovation, especially given the fact that poverty is embedded in structures of exploitation.

In India, the politics of knowledge production has been driven by the strategic objectives set up by the State grounded in the assumption that only large-scale technical systems can foster economic development and other socio-technical frames can play only a transitional role. The impact of this understanding is clearly reflected in the pattern of investments undertaken in R&D in atomic energy, space, defence, industry, agriculture, health, etc., and the maintenance of broad-based capabilities in basic research and generic technologies. As far as the politics of technology generation and implementation for the benefit of Indian economy is concerned, the innovation policy agenda has been driven far more by the response necessitated from the State to overcome immediate crises of the day rather than by the political, economic, and ecological challenges facing the Indian society as a whole. In India, paradoxically, at regular intervals of almost every 10 years, the politics of technology implementation has been revisited by the political–bureaucratic apparatus due to the crisis facing the governments at the centre, with the yardstick being the current expectations of the poor. In particular, the agenda of pro-poor innovations has been explicitly revisited and used by the political and bureaucratic apparatus to reframe the innovation policy. However, indigenous efforts in S&T have been better translated into opportunities for the incumbent large corporations, which continue to depend otherwise on foreign technology, than into the strengthening of pro-poor innovation systems.

Public–private partnerships and the corporate sector are being roped in increasingly as partners of the State in bringing about economic development. However, given its own record of pro-poor innovation generation and delivery, policy-makers need to be wary of the possibility of adverse inclusion of the poor by the corporate sector. It is self-evident that policy-makers should not expect the corporate sector also to perform miracles through pro-poor innovations. Even in the past the state sector agencies did end up treating the poor adversely by encroaching on the local economies by making the petty producers either the consumers of their goods and services or the sources of cheap labour and raw material providers. Policy-makers are seemingly putting too much confidence into the ability of the corporate sector. In the so far practised demand-led models of innovation, corporate governance means homogenisation of the systems of production and consumption rather than tackling the issues of diversity in demand, flexibility in production, resource regeneration, skill enhancement, and organisational and technological capability building. Economic inducements alone do not determine the bias of technological change. Public values matter considerably to the social design of innovation.

A vision and strategy of pro-poor innovation making are required to aim at the development of a new production paradigm and a radically new mode of innovation creation. This vision must be founded on the critical lesson from the Indian historical experience, which shows that the poor must be allowed to become effective social carriers of innovation. This, in turn, can happen only if the rural poor are aided to organise themselves into co-operatives and collectives and given opportunities to improve their bargaining power so that they can establish a degree of control over the pace and direction of technological and structural change. Of course, this means that the model of 'pure market governance' is unlikely to do any better than the 'technology push' approach in facilitating rural development. Finally, policy-makers have to remember history – and remember it well. Innovation making in the social sectors such as food, fuel and fodder, electricity and transport, water and sanitation, health, and education involves the development of physical infrastructure with lumpy supporting investments having a common pool character. In the past, system failures occurred in such sectors to the detriment of access to the poor, due to the initial selection of ill-suited technological paradigms and their continuation even after becoming clear that the selected systems of technologies are not sustainable. Such mistakes of the past should be avoided.

Select Bibliography

Abrol, Dinesh. 2009a. 'Framing of the Progressive Interventions and Agrarian Transition: Lessons from the Indian Experience.' Proceedings of Seminar on Agriculture sponsored by VigyanPrasar, Jan VigyanVedika (JVV), Ananatpur, Andhra Pradesh.

———. 2009b. 'Indian Agrarian Situation and the Socio-Technical Challenge: Towards the Formulation of a Peoples' Democratic Perspective.' Proceedings of Seminar on Agriculture sponsored by VigyanPrasar, Jan VigyanVedika (JVV), Ananatpur, Andhra Pradesh.

———. 2005. 'Embedding Technology in Community-Based Production Systems through People's Technology Initiatives: Lessons from the Indian Experience.' *International Journal of Technology Management & Sustainable Development* 4 (1): 3–20.

——— 2004. 'Lessons from the Design of Innovation Systems for Rural Industrial Clusters in India.' *Asian Journal of Technology Innovation* 2 (2): 67–94.

———. 1983. 'American Involvement in Indian Agricultural Research.' *Social Scientist* 11 (10): 8–26.

Abrol, Dinesh and L. Pulamte. 2008. 'The Challenge before Indian Agricultural Universities: Impact and Relevance of the Kothari Commission's Vision and Strategy.' In *Perspectives on Education and Development Revisiting Education Commission and After*, edited by Ved Prakash and K. Biswal, 653–700. New Delhi: National University of Education, Planning & Administration (NUEPA).

Arora, Saurabh and Henny Romijn. 2009. 'Innovation for the Base of the Pyramid: Critical Perspectives from Development Studies on Heterogeneity and Participation.' UNU-Merit Working Paper series, Maastricht Economic and Social Research and Training Centre on Innovation and Technology, United Nations University, Maastricht, The Netherlands. Accessed: June 2012; http://www.merit.unu.edu/publications/wppdf/2009/wp2009-036.pdf.

Ayres, R. U. ed. 1998. *Eco-restructuring: Implications for Sustainable Development.* Tokyo: United Nations University Press.

Bhalla, A. S. and A. K. N. Reddy. 1994. *The Technological Transformation of Rural India*, 113–30. London: Intermediate Technology Publications; Practical Action Publishing Group

Bhalla, A. S. 1996. *Facing the Technological Challenge*. London: Macmillan.

Bhalla, A. S. and James Dilmus. eds. 1988. *New Technologies and Development: Experiences in 'Technology Blending.'* Boulder, Colorado: Lynne Rienner Publishers.

Bhattacharya, S. N. 1980. *Rural Industrialization in India*. Delhi: B. R. Publishing Corporation.

CAPART (Council for Advancement of People's Action and Rural Technology). 2004. *Rural Technology*. New Delhi: CAPART. Accesed: June 2012; www.***capart***.*nic.in /*.

CSIR (Council of Scientific and Industrial Research). 1978. *Karimnagar Project: Evaluation of Rural Development Experiment*. New Delhi: CSIR.

———. *Technologies for the Rural Sector.* New Delhi: CSIR. Accessed: June 2012; www.csir.res.in.

CSSTD. 1981. Proceedings of the Workshop on 'Gaon ke Karigar aur Science' (Village Artisans and Science), 28 October–2 November 1979, Bardoli, Gujarat, Centre for Management Development in R&D, Planning Division, CSIR.

Datye, K. R. 1997. *Banking on Biomass for Dispersed Industrialization*. Ahmedabad: Centre for Environment Education.

Fisher, T. and Vijay Mahajan. 1997. *The Forgotten Sector: Non-Farm Employment and Enterprises in Rural India*. Southampton: Intermediate Technology Publications.

Gandhi, M. K. 1959. *Rebuilding Our Villages*. Ahmedabad: Navjivan Publishing House.

Gill, Kanwaljit Kaur and Sucha Singh Gill. 1990. 'Agricultural Development and Industrialization in Punjab.' *Economic and Political Weekly* 25 (45): 2507–09.

Government of India. 2013. *Science, Technology, and Innovation Policy 2013*. New Delhi: Ministry of Science and Technology, Government of India. Accessed: June 2012; http://www.dst.gov.in/sti-policy-eng.pdf.

———. 1988. *Village & Small Industries Sector Framework: Policy Issues & Perspective*. New Delhi: Planning Commission, Government of India.

———. 1979. *Evaluation Study of Rural Industries Projects*. New Delhi: Programme Evaluation Organisation, Planning Commission, Government of India.

———. 1955. *Report of the Village and Small Scale Industries*. Planning Commission, Second Five Year Plan Committee. Delhi: Government of India. Accessed: June 2012; *planningcommission.nic.in/plans/planrel/fiveyr/2nd/2planch20.htmlý.*

IDFC (Infrastructure Development Finance Company Limited) Rural Development Network. 2012. *The India Rural Development Report-2012*. New Delhi: IDFC.

Jequier, Nicolas. 1976. *Appropriate Technology: Problems and Promises*. Paris: OECD.

Kaplinsky, R. 2011. 'Bottom of the Pyramid Innovation for Pro-poor Growth.' Paper prepared for Innovation and Growth, PRMED Division of the World Bank, The Open University, Milton, Keynes, UK. Accessed: June 2012; http://www.bndes.gov.br/SiteBNDES/export/sites/default/bndes_pt/Galerias/Arquivos/empresa/download/inovacao_paper_02.pdf.

Khuntia, S. A. Parida, and V. N. Misra. 2003. *Downsizing Technology for Rural Development*, Vols. 1 & 2. New Delhi: Allied Publishers.

Kothare, N. R. 1978. 'Agro-Industries Corporations.' *Reserve Bank of India Bulletin* XXXII (2): 80–88.

KVIC (Khadi and Village Industries Commission). 2002. *The Rural Employment Generation General Programme (Rural Industrialisation Programme)*. Mumbai: KVIC.

Myrdal, G. 1968. *Asian Drama: An Inquiry into the Poverty of Nations, Vol II*. New York: Pantheon.

NRDC (National Research Development Corporation). 2001. *Rural Technology Development Programme*. New Delhi: NRDC. Accessed: June 2012; *www.nrdcindia.com/*.

Rao, Pandit and A. Yashwant. 1994. 'Experiences of KVIC.' In *Technological Transformation of Rural India*, edited by A. S. Bhalla and A. K. N. Reddy, 113–30. London: Intermediate Technology Development Group, Practical Action Publishing Group

Parthasarathi, Ashok. 2005. 'Rural Industrialisation Programme: Looking at Khadi and Village Industries Commission.' *Economic and Political Weekly* 40 (44/45):4763–67

Planning Commission. 1962. *Project for Intensive Development of Small Industries in Rural Areas*. New Delhi: Rural Industries Planning Committee.

Pulamte, Lalsiemlien and Dinesh Abrol. 2003. 'Technology Transfer for Rural Development: Managing R&D at CSIR.' *Economic and Political Weekly* 38 (31): 3315–18.

Rao, V. K. R. V. 1979. 'Industrialization and Integrated Rural Development.' *Man and Development* 1 (2): 18.

Sharma, K. D. and M. A. Qureshi. eds. 1975. 'Alternative Technology.' Proceedings of the Seminar held under the joint auspices of the IAS, Simla and CSIR, New Delhi, September. Shimla: Indian Institute of Advanced Study.

Singh, Baldev. 1988. 'Development of a Self-Reliant Agro-Food Industry.' *Economic and Political Weekly* 23 (14/15): 683–85+87.

Singh, Tarlok. 'Joint Planning for Agriculture and Industry—Some Implications.' *Man and Development* 1 (2): 48–49.

Srivastava J. C. 1999. 'Application of Science and Technology for Rural Industrialization.' In *Indian Economy after 50 Years of Independence: Experiences and Challenges*, Vol. 5, edited by D. K. Das, 190–204. New Delhi: Deep and Deep Publications.

UNDP. 2008. 'Creating Value for All–Strategies for Doing Business with the Poor.' *GIM Global Reports-First Global Report*. Accessed: June 2012; http://www.growinginclusivemarkets.org/publications/global/.

Unni, Jeemol and G. Raveendran. 2007. 'Growth of Employment (1993–94 to 2004–05): Illusion of Inclusiveness?' *Economic and Political Weekly* 43 (3): 196–99.

Willoughby, K. W. 1990. *Technology Choice: A Critique of the Appropriate Technology Movement*. London: Intermediate Technology Publications.

CHAPTER 12

Concluding Remarks*

Shyama V. Ramani

The investment of the Indian state has created a strong infrastructure of schools, universities, and public laboratories that is bearing innovation-fruit, as the various chapters of this book clearly show. Indian firms and public laboratories, and even non-profit organisations are innovating under extreme resource constraints and developing cost-effective new products, new processes, and new business routines. Innovations have contributed to income and employment generation as well as economic development. At the same time, despite the progress made, there is still quite some way to go to catch-up with the leading countries in each sector. For example, in agriculture, neither public laboratories nor local firms have successfully created and commercialized a genetically modified trait on their own.[1] In pharmaceuticals, there is a lack of technical skills in clinical trials and new drug discovery involving biotechnology. Public sector firms have low R&D capabilities in the power equipment sector. There is a manpower crisis in the patent bureaucracy. And the list can go on. However, what is absolutely clear from the different chapters of the book is that innovations have engendered systemic transformations in India – changing the goals, beliefs, perceptions, and cognitive schemas by which economic actors are formulating and implementing their strategies.

[1] A trait is a plant characteristic such as being resistant to drought. Though Indian private companies such as JK Agro, Metahelix, and Nathseeds have commercially approved GM events in the market, these are not selling well. The market leader is Monsanto-MAHYCO Biotech Ltd (GEAC Report 2012).

* Acknowledgement: I thank the contributors of the various chapters of this book, and especially Ajay Thutupalli for useful comments on this note.

In this book, the evolution of innovation capabilities in any sector is considered to be the result of games played between players whose strategies are interdependent and whose choices jointly determine the final outcome. Innovation generation is not viewed as being due to the efforts of just one actor, say a firm, but the result of the profile of actions chosen by the state, firm clusters (national and international), laboratories, public agencies, private intermediaries, and other entities, which put together constitute the innovation system. By studying different sectors operating under the same national system of innovation (NSI), but being shaped by different sectoral dynamics, the present work aims to further our understanding of the evolution of innovation in the Indian economy. By examining the inter-linkages between the national and sectoral systems, an attempt is made to provide insight on the factors and forces that determine their co-evolution.

The different chapters of this book confirm that the variety of innovations generated has enormously increased over time. They now range from complex original innovations to re-engineered innovations, incremental re-engineered innovations, and innovations in business and delivery platforms.

Catching-up began with recombination of existing technologies in all sectors in the aftermath of Indian Independence. Firms learned to replicate and recombine available elements of 'state-of-the-art technologies' before going on to create new ones. The different chapters showed how cost-cutting processes, innovations or novel delivery platforms preceded new product innovations because of the need to develop cheaper production processes with scarcer resources and to offer products on the market at affordable prices. Then, as firms ascended the learning curve, higher value-added innovations were introduced in the market. Each chapter presents its own story of how this was achieved.

The pre-liberalisation success stories were in agriculture, pharmaceuticals, and power generation. While agriculture (Chapter 4) and pharmaceuticals (Chapter 6) were confronted with dramatic advances in biotechnology originating from outside India, especially in the post-liberalisation era, hi-tech sectors like software (Chapter 5) and energy (Chapter 8) took off fairly well in terms of innovation generation. However, bigger surprises were in unexpected niches like traditional medicines (Chapter 9), with corporations like Dabur, drawing doctors and patients away from products made by practitioners to 'proprietary branded products' through their new product formulations.

Two basic premises of evolutionary economics were confirmed. First, the sectoral studies made it evident that contrary to mainstream economic theory, which assumes that economic actors make well-informed, rational choices, innovation-related strategies are often based on routinised behaviour. More often than not, they are made under imperfect and incomplete information, being guided by set goals, beliefs, perceptions, and cognitive schemas.

Sanitation (Chapter 9) provides a perfect example of this conjecture. In deciding whether or not to adopt sanitation as an innovation, the poor are not at all conscious of the parameters of the game. They are also only partially aware of the pay-off structure in terms of the trade-offs between the present investment in a toilet and future benefits in terms of better health and income generation capacity. Therefore, education and awareness creation can make the information base of targeted beneficiaries more complete and perfect. However, this observation does not hold only with respect to the decision-making of the poor and the illiterate. The same dilemma is confronted by top scientists and policy-makers when dealing with a totally new technology paradigm like nanotechnology (Chapter 7). How should the funds be allocated among the different possible niches? Investment decisions have to be made under conditions of imperfect information on the technological and market potential of the different options.

In short, from illiterate villagers, to top scientists and managers in public laboratories, firms, and government bodies, each economic actor has to decide upon an innovation strategy so as to move closer to his own (or his group's) objectives, given his own (or own group's) beliefs, cognitive structure, resources, information base, and common knowledge about the game environment.

Second, the Indian NSI is a complex, non-linear system (like that of all other countries) rather than being a well-ordered and predictable one.[2] It is also marked by path dependencies and 'butterfly effects'.[3] All the sectoral studies reveal that while the state is the major player and controller, the Indian innovation system is self-organising through co-evolution of the different players as well as through a transformation of the variety of agents and artifacts.

The state (or international conventions) may impose rules of the game in the form of regulations, but the interaction between economic actors can further generate endogenous rules that emerge as norms of the game. At the sectoral

[2] See Surie (2011) for details of these concepts as applied to innovation systems.

[3] The butterfly effect in chaos theory has shown that in non-linear systems with sensitive dependence on initial conditions, small changes at one point can result in large differences in later states.

level, this was starkly illustrated in the case of pharmaceuticals (Chapter 6). The intellectual property regime was changed in 1970 to permit the public sector firms to acquire technological capabilities, but it was the private firm cluster which took off. An example of a butterfly effect was also provided by the pharmaceutical sector (Chapter 6). The Hatch-Waxman Act was passed in the USA (outside of the Indian NSI but part of the Indian pharmaceuticals sectoral system of innovation), but in India, it was recognized as a 'window of opportunity' by visionary firms, which then developed world-class capabilities in producing generics.

Though it is extremely difficult, if not impossible to summarize all the results of the chapters, five main results on the evolution of the Indian system of innovation can be identified, despite the heterogeneity of the sectors studied. We now turn to these.

In all sectors, the games between the economic actors in the different sectors have become much more complex, especially after liberalisation

The Indian NSI has become more complex, housing a greater variety of actors, bound together in networks that are denser, and through a greater variety of relationships. Even the rules of the game, as well as the number and variety of rule makers, have exploded.

For instance, agriculture (Chapter 4), a sector that one would normally not expect to be a candidate for the explosion of complexity, has indeed become much more complex. In the immediate aftermath of Independence, from 1947 to the mid-1960s, the main actors cooperating to create innovations were the government laboratories of the Indian Council of Agriculture Research (ICAR), each working in tandem with a network of state agricultural universities (SAU) and exchanging knowledge with and getting support from US agricultural universities. Then came the Green Revolution, with technology transfer from foreign institutions like the CIMMYT[4] and IRRI[5] in public–public international collaborations. This marked an important technology paradigm shift in agriculture, and most agree that the Green Revolution saved India from famine and anchored its national autonomy. From the 1970s till the 1990s, the agriculture system of innovation continued to be transformed substantially with the entry of new actors such as the state seed corporations supplying modern

[4] Centro Internacional de Mejoramiento de Maiz y Trigo in Mexico.

[5] International Rice Research Institute at Manila.

varieties of seeds (or Green Revolution seeds) and new Indian small-scale seed firms producing and selling hybrid millet, maize, cotton, and vegetable seeds. New kinds of links were also forged with the CGIAR[6] network of agricultural research. Then during the 1990s, with liberalisation came the foreign multinationals like Monsanto with their genetically modified plant varieties, which marked the second major technology paradigm shift. While the state seed corporations and agriculture universities worked to create and sell cereal varieties in rice, wheat, pulses, and oilseeds (namely the essentials for food security), innovations in maize, millets, sunflower, cotton, vegetables, and flower seeds increasingly emerged from private corporations. New relationships were forged between multinational ag-biotech firms, Indian firms, and the ICAR network of laboratories, through licensing and contract research. Similar trends of increasing complexity has marked the other sectors of the Indian innovation system as well.

The strategies and role of the key actors within the national system of innovation has evolved, and in some sectors has changed radically over time

The leading firms in the pharmaceutical sector (Chapter 6) make an interesting study. Initially, during the 1960s, they were not investing in R&D but lobbying politicians to obtain licenses for imports and expansion of their manufacturing base. Then during the 1970s and 1980s, they began to invest greatly in developing re-engineering innovation capabilities. Thereafter, they combined this with the enhancement of their manufacturing capabilities and regulation handling capacities, which fed into their innovation capabilities. In the 1990s, they began to try to cross-over from synthetic chemistry to biotechnology. In the new millennium, they began to initiate a variety of collaborations with multinationals as well as dedicated Indian and foreign biotechnology firms. Thus, the variety of strategies put in place to acquire and enhance innovation capabilities increased greatly. Similar stories can be told of other sectors, of experimentation with innovation strategies not only by firms, but also public laboratories, informal economy entrepreneurs (Chapter 9 on traditional medicine), social entrepreneurs (Chapter 10), and civil society movements (Chapter 11).

In the process of such evolution of player strategies, the roles of different economic actors in the innovation system also changes. For instance, let us consider the relationship between the public sector laboratories and firms in

[6] Consultative Group on International Agricultural Research.

the innovation creation process. Under the traditional mode, also termed in innovation studies as the Mode 1 pattern of knowledge creation and utilisation, knowledge is generated within types of institutions different from those in which it is later applied. For instance, universities are engaged in pure research; public research identifies those that can be transformed into technology and develops the same at a pilot scale, which are then transferred to firms for scaling-up and manufacturing. There is thus both a chronological and an institutional separation between the locus and time of the processes of knowledge generation and utilisation. Under Mode 2 form of interaction, as opposed to the classical mode 1, new knowledge issuing from public laboratories is rapidly applied either through start-ups or through technology transfer and this gives rise to feedback loops in which industrial applications call forth new types of basic research (Etzkowitz and Leydesdorff 2000). Moreover, knowledge intensive firms do some basic research, while public research institutions engage in creating industrial applications. Thus, both the chronological and the institutional separation of the traditional model become fuzzy.

In some sectors, like equipment for power generation (Chapter 8), Mode 1 form of interaction still dominates the sectoral system of innovation. The impact of universities is limited to producing qualified personnel with a general background in the subject, and the role of public laboratories is limited to redesigning foreign technology for the Indian context. In agriculture (Chapter 4) and pharmaceuticals (Chapter 6), it is clear that the Mode 1 operated in the 1970s, and by the 1990s, with liberalisation, knowledge circulation had switched to Mode 2. This pattern is also evident in the sunrise industry of nanotechnology (Chapter 7) where Mode 2 model of public–private cooperation is already taking off in the form of spin-offs.

The increasing prevalence of Mode 2 has led to the establishment of the Triple Helix Model (i.e., trinity of government, university, and industry) in the developed world, whereby the government, universities, public research organisations (last two being directly sponsored by government), and firms are the dominant players, constantly interacting to create new knowledge for innovations. So far, the Triple Helix Model is not very evident in Indian agriculture or pharmaceuticals. In the former, cooperation between public sector entities and private sector seed firms is rare; and in the latter, with the exception of biotechnology-related projects, the government has not financed large research programmes. However, in the latest sunrise technology, namely nanotechnology, the government is playing a very active role in promoting innovation, thereby carving the Indian variant of the Triple Helix Model.

The success of innovation generation and commercialization by Indian firms will depend on how they manage relationships with three other key (non-state or non-public) actors, namely foreign firms, NGOs, and social enterprises

In some sectors, there is a real risk of domination by foreign firms in terms of technology dependence. This is already the case in certain agricultural biotechnology niches (Chapter 4). In pharmaceuticals (Chapter 6), there has been a spate of acquisitions of Indian firms starting with the leader, Ranbaxy. Once acquired, research-intensive local firms like Ranbaxy and Shantha Biotech have been put out to pasture, to specialize in manufacturing low-cost generics for local and international markets. The challenge, therefore, for Indian pharmaceutical firms is to develop an R&D vision and an innovation strategy given their limited resources (vis-à-vis their developed country counterparts) to remain competitive, and at the same time, not be bought-out. Many integrated drug companies are trying the contract research route to learn and acquire capabilities, but it is not yet clear to what extent they will be successful. In the power equipment sector (Chapter 8), the local public sector giant, the Bharat Heavy Electricals Limited (BHEL) is facing stiff competition from Chinese multinational firms like the Shanghai Electric Group and domestic companies that have joint ventures with foreign partners such as Mitsubishi, Alstom, and Toshiba. If, and how, the technology dependence can be overturned, and technology spillover can be integrated through collaboration between local and foreign firms is anyone's guess.

In India, NGOs have actively played the role of citizen's watchdogs. In agriculture they have raised a voice against genetically modified plant varieties (Chapter 4). In pharmaceuticals, they regularly protest against the ever-greening of patents (Chapter 6). In materials, they are fighting for the banning of asbestos, which will require a major regulatory change and which cannot be done without a consideration of regulation for all nanomaterials as well (Chapter 7). At the same time, NGOs helped greatly in the diffusion of cook-stoves (Chapter 8). Thus, NGOs sometimes can make or break the success of innovation adoption and their stance with respect to the social value of innovations. The management of relationship between the NGOs and innovation generators, therefore, can be crucial for market success.

The new agents of the government and the large firms (both Indian and foreign) are the small and medium-sized social enterprises (Chapters 9–11). It is clear that these social enterprises are not cast in the same altruistic mould as the civil society associations (Chapter 11) or social entrepreneurs, but they are familiar with private management principles and are more amenable to undertaking projects that require satisfying targets within a given time period. However, whether they will be better economic carriers of a positive socio-economic transformation through innovation adoption remains to be seen.

Initial conditions, triggers, and domino effects are driving the Indian system of innovation towards 'divergence' in some niches rather than 'convergence' vis-à-vis the dominant paradigm established by the leader countries. Such 'divergent' trajectories could become models for other countries as well

Behind the term 'catch-up', there is a notion of a 'dominant paradigm' established by a leader or winner that must be 'emulated' by a follower country. Usually, developing countries take developed country policies (as well as technologies) as those to emulate to catch-up and cooperate and compete with them. Indeed, the 'catch-up' literature takes catching-up as the primary driver of knowledge and technological capacity accumulation, but two points tend to undermine this hypothesis considerably.

In the process of catching-up, there might be learning that causes a new route to be taken, which becomes a better way of innovating than the 'catch-up trajectory' that was sought to be emulated in the first place. That is, even while some economic actors, including the state, try to 'emulate' the leader countries in catch-up efforts, the system can move towards divergence rather than convergence with respect to a dominant paradigm. This is happening in India, because innovations are being developed here under conditions of scarcity and institutional inefficiency, to cater to both effective demands and needs that cannot be supported as effective demand due to lack of purchasing power. Emulation of a catch-up trajectory in these cases may be neither possible nor desirable. TeNet (Chapter 2), the Jaipur foot (Chapter 3), traditional healthcare firms (Chapter 9), and ecosan toilets (Chapter 10) are examples that serve to illustrate this inference.

Such feats of 'frugal' or 'Gandhian' or '*jugaad*' innovations are making headline news both within and outside of India. According to the Radhjou et al. authors of *Jugaad Innovation*, an international bestseller,

> Western corporations can no longer just rely on the old formula that sustained Jugaad innovation and growth for decades: a mix of top-down strategies, expensive R&D projects and rigid, highly structured innovation processes. Jugaad Innovation argues that the West must look to places like India, China, and Africa for a new, bottom-up approach to frugal and flexible innovation.[7]

[7] http://jugaadinnovation.com/

Moreover, for the first time, there are scholarly discussions on 'reverse innovations' by which innovations originating in emerging economies are making their way to developed countries – just the opposite of the traditional trend. Most of this (though not all) is still either *by* or *for* Western or Japanese multinationals. General Electric, for example, has developed a US$1000 hand-held electro-cardiogram device conceived for the rural Indian market (and in India) which is much cheaper than its existing counterpart (Immelt, Govindrajan, and Trimble 2009). Though designed in India, it is now in use in many ambulances and hospitals in the developed world. India is also responsible for a significant chunk of below-the-radar 'invisible innovations' created via the outsourcing of R&D, which serves to increase the product variety and innovations offered by multinationals all over the world (Kumar and Puranam 2011). This was confirmed also in the chapters on software, pharmaceuticals, and even traditional medicine (Chapters 5, 6, and 9). One might argue that these paeans of praise to the skills of innovators in India are basically about innovations being churned out by underpaid scientists, for the purposes of generating further wealth for multinationals, without having much impact on the local poor. Though this is very true for the large part, it is still nurturing the build-up of techno-entrepreneurship, technological learning, business skills learning, confidence, and coordination skills in addition to income generation.

Butterfly (trigger) effect has already been mentioned in the pharmaceuticals sector (Chapter 6). Another illustration of this effect, at a micro-level, is discussed in the sanitation sector (Chapter 10). The residents of a village were happy to use an ecological toilet (ecosan) till others in the village got a conventional toilet. As sanitation coverage of the village increased, a totally unexpected outcome was the abandoning of ecosan toilets by their users.

Domino effects have been noted in practically every sector. In pharmaceuticals, (Chapter 6) firms watch one another closely and imitate; in agriculture (Chapter 4), farmers observe one another and imitate; and finally in the social sector, because of informational constraints, households imitate the behaviour of other successful households (Chapter 9, 10, 11). Such herd behaviour leads to domino effects.

Commercializing innovations for the poor is no longer the only objective of the state. The Triple Helix Model reigns strong in the social sector–marking another 'divergent' trajectory to which other countries may aspire to catch-up to

The chapter on pro-poor innovations (Chapter 11) is a very comprehensive study of the efforts of the state, public research, and alternative associations and agencies to serve the needs and demands of the poor, and at the same time, build capabilities in the rural economy. It chronicles the successes as well as the failures of top-down initiatives while pointing out the reasons for these, which boil down to a lack of will, consistency, focus, and understanding at the top levels of the bureaucracy; and a lack of a will, bargaining power, collectivisation, and cooperation at the grassroots level.

In contrast, the case study of TeNet (Chapter 2) proves that isolated academic socio-techno entrepreneurship is active and very much alive. When academics become change leaders to mobilise active and intimate networks between a set of innovation stakeholders, sharing common expectations of high rewards from cooperation and coordination, then a veritable technology cluster can be born. Inspiring leaders and teams within facilitating organisations can sometimes compensate for the absence of funds and have much better impact than large scale government programs.

Many scholars consider the lack of an active venture capital system to be the greatest failure of emerging country innovation systems and a drawback that is holding back the development of knowledge intensive sectors like biotechnology and nanotechnology (Niosi 2010). As Dossani and Kenney (2002) summarize, 'In the last decade, one of the most admired institutions among industrialists and economic policymakers around the world has been the US venture capital industry.' They go on then to suggest how in countries like India, a venture capital industry can be activated – heedless of its history. The truth of the matter is that an active venture capital market seems to be an American entity, which cannot be replicated in the same fashion elsewhere, including in India. In India, as in most other countries, the less exciting substitute that serves the same function is government funding.

However, what is extraordinary today is that the social sector in India very much resembles a Silicon Valley phenomenon. The government, international agencies, and large firms (see Chapters 1, 9, 10, 11) are generating business opportunities for NGOs, social enterprises, and even small social entrepreneurs. These financiers are like the venture capitalists in the USA in that they are willing to invest some funds in entrepreneurial organisations wanting to achieve a social mission; they are promoting experimentation; they are trying to identify best practises and they are tolerating failures. Social enterprises are gaining entrepreneurial capabilities as well as sector specific knowledge. The innovations from the social sector are being showcased even to developed country firms to be emulated.

Missing resources and capabilities and need for effective institutions to forge ahead

In most of the sectors studied, clusters of innovating firms have emerged with minimal funds from the central government and VCs. The outcomes described in the studies confirm the availability of techno-entrepreneurial talent. However, there is a dearth of human resources in important agencies like patent offices. Technological capabilities are also missing in crucial areas, for instance, in the construction of many instruments used to apply biotechnology and nanotechnology (which are still mostly imported).

There is a need for effective institutions; for example, regulatory bodies that perform better in terms of ensuring drug safety. There are missing protocols, for example, in regulation of nanomaterials for environmental safety. The established biosafety regulatory system for agricultural biotechnology is ignored by politicians and by farmers, and thus needs special attention Finally, it is worthwhile to note that it is impossible to develop and work upon a vision for sustainable development if there is a high turnover of public servants (due to contracts whose terms are unpredictable and often for a maximum of only three years in a particular post) and over-interference of politicians with the functioning of scientific establishments and public agency missions – as is the case in India.

Thus, the main message of this book for policy-makers is that given scarce resources, any 'resource investment' has to be constantly accompanied by 'institutional reform' to fit the evolving local context in order to maximize innovation generation and its socio-economic impact. These reforms have to be based on monitoring and performance audits rather than being top-down imposed changes or designed to satisfy political whims and fancies. It is useful to keep in mind that according to the sectoral tales of this book, whatever the policy, there will always be some individuals and organisations who experiment to find creative ways to pursue opportunities, and if successful, they will become role models, provoking emulation and imitation, and more than policy they will be the agents of transformative change.

References

Dossani, R. and M. Kenney. 2002. 'Creating an Environment for Venture Capital in India.' *World Development* 30 (2): 227–53.

Etzkowitz, H. and L. Leydesdorff. 2000. 'The Dynamics of Innovation: From National Systems and "Mode 2" to a Triple Helix of University–Industry–Government Relations.' *Research Policy* 29 (2): 109–23 .

Immelt, J. R., V. Govindarajan, and C. Trimble. 2009. 'How G.E. is Disrupting Itself.' *Harvard Business Review October.* Accessed: June 2012; http://hbr.org/2009/10/how-ge-is-disrupting-itself/.

Kumar, N. and P. Puranam. 2011. *India Inside: The Emerging Innovation Challenge to the West.* Boston: Harvard Business Review Press.

Niosi, J. 2010. *Building National and Regional Innovation Systems.* Cheltenham: Edward Elgar Publishing Limited.

Radjou N., J. Prabhu, S. Ahuja, and K. Roberts. 2012. *Jugaad Innovation: Think Frugal, Be Flexible, Generate Breakthrough Growth (1st edition).* San Francisco: Jossey-Bass.

Surie, G. 2011. 'Emergence of New Markets, Distributed Entrepreneurship and the University: Fostering Development in India.' *International Journal of Technoentrepreneurship* 2 (3/4): 362(n dash)81.

Contributors

Shyama V. Ramani is Professorial Fellow at United Nations University at Maastricht and Professor at Maastricht University, The Netherlands. Her fields of specialization include the economics of innovation, development economics, and applied game theory. She has published in leading journals like *Research Policy, Technology Forecasting and Social Change, Industrial and Corporate Change, Theory and Decision, Rationality and Society,* and *Nature Biotechnology.* A firm called Tecknowmetrix has been established in France on the basis of her publications on technology indicators and she is one of its co-founders. After the Tsunami of December 2004, she founded two non-profit organizations, Association Un Ami in France and Friend in Need in India, to help an isolated fishing village named Kameshwaram along the Indian coastline attain the millennium development goals.

Dinesh Abrol is Chief Scientist at Council of Scientific and Industrial Research – National Institute of Science, Technology and Development Studies (CSIR-NISTADS), New Delhi. He has served as a member in the policy-making committees set up by the Planning Commission, Department of Science and Technology, CSIR, Ministry of Health and Family Welfare, Ministry of Rural Development, Department of Electronics, Ministry of Environment and Forests, and State level S&T Councils in India. He has been consulted by WHO India on the status of policy-making on public health, innovation, and intellectual property in India. He has researched extensively on the dynamics of the systems of innovation in India as a part of the study programme on BRICS group of countries for IDRC.

Manish Anand is Associate Fellow at The Energy and Resources Institute (TERI), New Delhi, and currently leads the Science, Technology, and Innovation Area at TERI. His areas of research include innovation policy for emerging technologies, harnessing science and technology for sustainable development, innovation systems, technological capacity building, and the interface between science–technology–society. His current research focuses on capabilities and governance issues related to nanotechnology and innovation for green growth and development in the context of developing countries.

Ankur Chaudhary obtained his Bachelor in Technology in engineering physics from IIT Delhi in 2009 with the Director's Silver Medal and the Mudit Sharma Memorial Gold Medal for academic performance. Following a brief stint in technology-focused consulting, he joined the Indian Institute of Technology as a researcher focusing on innovation and development issues, with a particular emphasis on energy, environment, and technology. He has worked on the *Low Carbon Strategies for Inclusive Growth* report of the Planning Commission of India and is engaged in a project dealing with electromobility and wind sectors in India, China, and Europe.

Arijita Dutta received her PhD from the University of Calcutta. Her research interest is in development economics in general and health economics in particular. Apart from teaching postgraduate students, she actively participates in field-based empirical research in health economics.

Samira Guennif is Associate Professor in University Paris 13 and a researcher in the Centre d'Economie de l'Université Paris Nord (CEPN). Her work is related to the access to essential drugs in developing countries and particularly to HIV/AIDS treatments in southern countries hit by the epidemic. She examines legal and industrial impediments to essential drugs accessibility: implementation of TRIPS provision and free-trade agreements provision in domestic patent law, on one side, and regulation regarding quality and price control impacting local production of affordable essential drugs, on the other.

Nagesh Kumar is Chief Economist of the United Nations Economic and Social Commission for Asia and the Pacific (ESCAP) and Director of ESCAP Subregional Office for South and South-West Asia based in New Delhi. Before assuming the latter position, he was Director of the Macroeconomic Policy and Development Division of ESCAP in Bangkok. Prior to joining ESCAP in May 2009, Dr Kumar served for seven years as Director-General of the Research and Information System for Developing Countries (RIS), a New Delhi-based public-funded development policy think-tank. Dr Kumar was a recipient of the Exim Bank of India's first International Trade Research Award in 1989 and a Global Development Network (World Bank)'s Research Medal in 2000.

Latha Nagarajan is Economist at International Fertilizer Development Center (IFDC) and Department of Food, Agricultural, and Resource Economics at

Rutgers, the State University of New Jersey. She is involved in the studies on the economic impact of developing and spreading drought-tolerant rice in Asia, seed sector assessment in West Africa, and the globalization of agricultural research and innovation by agricultural input industry.

Carl Pray is Distinguished Professor II in the Department of Agricultural, Food and Resource Economics, Rutgers University, New Brunswick, New Jersey and Adjunct Professor and Advisor of the PhD Programme of the Center for Chinese Agricultural Policy, Chinese Academy of Science. During the academic year 2008–09, he was the visiting scholar at the Agricultural and Resource Economics Department, University of California, Berkeley. He did BA in history at Carleton College in Minnesota and then worked as an agricultural extension agent in the Peace Corps in India. His PhD is in economic history from the University of Pennsylvania. The focus of his current research is agricultural science and technology policy in China, India, South Africa, and other developing countries. The results of this research have been published in *Science, Nature, The American Journal of Agricultural Economics, Economic Development and Cultural Change, Research Policy,* and elsewhere.

Shuan Sadre Ghazi is based at the Maastricht office of the United Nations University in the Netherlands. His research interests have been primarily in innovation management and business strategy, with a specific focus on their implications for low-income markets. In the UNDP initiative Growing Inclusive Markets (GIM), he was responsible for case studies of Inclusive Business in the Middle East and North Africa (MENA).

Ambuj Sagar is the Vipula and Mahesh Chaturvedi Professor of Policy Studies and Dean of Alumni Affairs and International Programs at the Indian Institute of Technology (IIT) Delhi. Ambuj's interests lie in science and technology policy, environmental policy, and development policy, with a particular focus on the interactions between technology and society. While his current research focuses mainly on energy innovation and climate policy, he also studies, more broadly, various facets of technology innovation, environmental policy and politics, and engineering education and research.

Smita Srinivas is Director of the Technological Change Lab (TCLab) and Assistant Professor in the Urban Planning Program at Columbia University. She is a faculty associate of the South Asian Institute (SAI), Columbia University. The focus of her work is the current comparative practice of and historical relationships between technological advances and employment and welfare regimes. She is especially interested in re-evaluations of theory and economic history based on the current practice in so-called developing countries, and the strong influence of European industrial history and rhetoric on these economic plans.

Gita Surie is Associate Professor at Adelphi University. She researches globalisation, cross-border innovation, distributed entrepreneurship, the emergence of novel organisational forms and markets, and national and technological innovation systems, particularly in the context of sustainable energy technologies. Her book, *Knowledge, Organizational Evolution and Market Creation: The Globalization of Indian Firms from Steel to Software* was published in 2008. She is a senior fellow at the Wharton School's Mack Center for Technological Innovation.

Adam Szirmai is Professorial Fellow at UNU-MERIT and Professor of Development Economics at the Maastricht Graduate School of Governance of Maastricht University. His research focuses on international comparisons of growth and productivity in manufacturing in developing countries. He has been involved in research projects in manufacturing in Indonesia, China, South Korea, Tanzania, Zambia, South Africa, and Japan.

Index